ELECTROMAGNETIC
DISTANCE MEASUREMENT

Electromagnetic distance measurement, by using light and microwaves for direct linear measurements and thus circumventing the need for traditional methods of triangulation, may well introduce a new era in surveying. This book brings together the work of forty-eight geodesists from twenty-five countries. They discuss various new EDM instruments—among them the Tellurometer, Geodimeter, and air- and satellite-borne systems—and investigate the complex sources of error. The book is therefore a unique and comprehensive source on the subject. UNESCO and R.I.C.S. have assisted financially in its production.

INTERNATIONAL ASSOCIATION OF GEODESY

ELECTROMAGNETIC DISTANCE MEASUREMENT

A SYMPOSIUM
HELD IN OXFORD
UNDER THE AUSPICES OF
SPECIAL STUDY GROUP NO. 19
6–11 SEPTEMBER 1965

PUBLISHED WITH
FINANCIAL ASSISTANCE
FROM UNESCO AND R.I.C.S.

UNIVERSITY OF TORONTO PRESS

TORONTO

Published in Canada by
UNIVERSITY OF TORONTO PRESS
1967
Published in Great Britain by
HILGER & WATTS LTD.
1967

Reprinted 2017
ISBN 978-1-4875-9832-7 (paper)

ORGANIZING COMMITTEE

Major-General R. C. A. Edge (Ordnance Survey), *Chairman*
Mr C. R. Argent (Royal Society)
Commander R. Bill (Tellurometer (UK) Ltd)
Mr W. Collins (Ordnance Survey)
Dr K. D. Froome (National Physical Laboratory)
Mr J. R. Rawlence (Directorate of Military Survey)
Dr A. R. Robbins (University of Oxford)
Mr J. W. Wright (Directorate of Overseas Survey)
Major M. R. Richards (Ordnance Survey), *Secretary*
Captain J. Niblock (Ordnance Survey), *Assistant Secretary*

SYMPOSIUM EXHIBITION

Throughout the Symposium an exhibition of EDM instruments was held in the Department of Engineering Science with the following exhibitors:

Exhibit	*Exhibitor*
The Geodimeter model 6	AGA Signals Ltd, Beacon Works, Brentford, Middlesex
Laser Rangefinder	G. &. E. Bradley Ltd, Electral House, Neasden Lane, London NW 10
The Mekometer	Hilger & Watts Ltd, 98 St Pancras Way, Camden Road, London NW 1
Tellurometer models MRA 4, 101 and 3	Tellurometer (UK) Ltd, Survey House, Windmill Road, Sunbury-on-Thames
Distomat DI 50	Wild Heerbrugg Ltd, Switzerland
E.O.S. Telemeter	C.Z. Scientific Instruments Ltd, 93/97 New Cavendish St, London W 1
The Ordnance Survey Thermistor	Ordnance Survey, Leatherhead Road, Chessington, Surrey
Gallium Arsenide Modulated Light Source	Tellurometer (UK) Ltd, Survey House, Windmill Road, Sunbury-on-Thames
NPL Mekometer II	National Physical Laboratory, Teddington, Middlesex

FOREWORD

Special Study Group No. 19 of the International Association of Geodesy was set up as the result of a decision of the Eleventh General Assembly of the Association at Toronto in 1957. The original purpose of the Study Group was to investigate the new instruments for electromagnetic measurement of distances on the Earth's surface, which had recently made their appearance, and to make recommendations regarding their use for various geodetic purposes. The instruments at that time were the Geodimeter, invented by Dr Bergstrand of Sweden and first used in 1947, and the Tellurometer, invented by Dr Wadley of South Africa and first used in 1956. The Geodimeter making use of a modulated light beam has revolutionized base measurement. The Tellurometer, using microwaves in place of light waves, has enabled electromagnetic distance measurements to be performed in daylight, and has also extended the range over which it is possible to measure.

The impact of these instruments upon geodesy has been great. It is now possible to apply a rigorous control of scale to triangulation systems. The precise geodetic traverse has become a task rapid in execution. In all phases of survey work, valuable applications have been discovered and the task which faced the Study Group has not been one that could be quickly completed.

It has been necessary to study the principles, new to most geodesists, upon which these new instruments operate and to understand the nature and assess the magnitude of the errors that affect them. It was necessary to devise means for minimizing these errors both by better design of instruments and by improved operating techniques. The Study Group's function has been essentially to record and disseminate information about research and development carried out by many organizations and individuals.

There has been a continuous process of improvement. The Geodimeter has been re-designed, rendered more portable and more accurate, and modified to work in daylight. The Tellurometer now operates on the three-centimetre as well as on the original ten-centimetre wavelength. Electromagnetic distance measuring instruments are now being made by other manufacturers in a number of countries on both sides of the Atlantic. The study envisaged in 1958 has become more complex, both in the number of instruments to be studied and in the number of problems that have emerged. In particular, the importance of refraction and the accurate measurement of refractive index have been realized and much work has been done in this field and on the connected problems of micro-meteorology. Studies have been made of the phenomenon of ground reflection of microwaves, and light has been thrown on this problem, which at first seemed beyond detailed analysis.

Foreword

The application of electromagnetic techniques has been extended. At the Twelfth General Assembly of the IUGG in Helsinki (1960), the study of airborne systems such as HIRAN and AERODIST was added to the scope of the Study Group. The technique has now been extended to satellite geodesy, as for example in the SECOR and Doppler systems. Finally the laser has entered the field, with potential applications as a highly concentrated and penetrative light source and also, with its coherent beam, as a method for applying the technique of interferometry over much greater distances than has hitherto been possible.

The work of the Study Group has thus continued to expand, and the purpose of the Symposium was to throw light upon some of the many problems that have been revealed.

CONTENTS

Contents

OPENING CEREMONY

Professor A. Marussi opened the proceedings and said how much he regretted the absence of Brigadier G. Bomford, who was prevented by illness from attending. He explained that the Symposium was arranged by Special Study Group No. 19 of the International Association of Geodesy, the IAG being a section of the International Union of Geodesy and Geophysics. The IUGG was in turn part of the International Council of Scientific Unions. He welcomed to the Symposium Professor H. W. Thompson, President of ICSU. He then called upon Sir William Hayter, Warden of New College, and representing the Vice-Chancellor of the University, to open the Symposium.

Sir William Hayter welcomed the delegates to Oxford University and declared the Symposium open. Professor L. Asplund replied on behalf of the IAG.

The closing speech was made by Major-General R. C. A. Edge, President of Special Study Group No. 19, and a visit was made to the Instrument Exhibition which was also formally opened by Sir William Hayter.

General Subjects and Reports

EDM Research in Austria

K. RINNER

Technical University of Graz

This report deals with research work at the Technical University of Graz. It investigates the accuracy obtained with EDM instruments in the Lower Alps and discusses a device for reducing the influence of reflections. The geometric strength required for continental or world networks is considered and electromagnetic tacheometry and the geometry of radar pictures are discussed.

1. *Introduction*

The use of EDM techniques has proved a valuable help to Austrian land surveyors in performing their tasks, which are so important to the economic development of the country. In this way, it is quite easy to simplify and accelerate the tasks which are in most cases difficult to solve in a mountainous terrain, as far as the determination of the site of basic points is concerned. On the other hand, in view of the complicated measuring conditions which are always arising in the mountains, a thorough consideration of the influence of meteorological anomalies, of the reflections, and of the geometric and physical reductions is necessary. This is the reason why we observe a direct, practical interest in the results of EDM research in Austria, in addition to the general scientific interest arising from the well-known tradition of the country in the sphere of geodesy.

The research aspirations are being encouraged by the Austrian Surveying Commission and the Austrian Research Council. The work is being executed at the technical universities and by surveying authorities (Federal Office for Weights and Measures and Office for Geodesy). Unfortunately, the report on EDM research in Austria is still not coordinated and it is available only in two different parts. This report refers exclusively to the activities at the Technical University of Graz. The work at other institutes is being reserved for a special report.

2. *Research Programme*

The Technical University of Graz has instigated a series of investigations which are still under development:

(*a*) Determination of the relative obtainable accuracies for different measuring instruments and methods when using a test-net with sides up to 20 km, in mountains of medium height.

(*b*) Development of equipment for reducing the influence of reflections.

(*c*) Examination of the geometric strength of different net forms for the tasks of continental and world surveying, having the aim of evaluating optimal results with the minimum effort.

(*d*) Consideration of the possible developments for electromagnetic tacheometry and of the geometry of electromagnetic images of the terrain.

The results which we have now obtained are summarized in the following report. In addition to this report, special reference is made to a volume to be issued in the near future under the title *Die Entfernungsmessung mit elektromagnetischen Wellen und ihre geodätische Anwenung*, which in the present report appears as reference [1] (i.e. *Handbuch der Vermessungskunde*, by Jordan, Eggert and Kneissl, of which it is the sixth volume); further reference should be made to the reports, which are still in preparation, of the appropriate authorities. These reports will enumerate the results in detail.

3. *Examination of the Test-net of Graz*

The accuracy of distances fixed by way of electromagnetic waves is dependent on the instrument used, on the method of measuring, on the meteorological circumstances and, finally, on the land profile over which the survey is executed. Therefore, if we want to compare the accuracy of different instruments and processes, such a comparison can be executed—if we want to be exact—only over the same profiles and under similar meteorological circumstances. For this reason, it will be suitable to take into consideration test-nets with profiles of different length and topography, and possibly different meteorological conditions. Supposing we have to measure, within the net, both the directions and distances, then it is possible to study the propagation of errors with EDM equipment and the relations of weight for different quantities, under ascertained geometric conditions.

The test-net of Graz is located in mountains of medium size, at heights from 360 up to 1440 m. The net includes twenty-four sides, the lengths of which fluctuate between 0·7 and 18·2 km (Fig. 1). The total length of the net sides is about 200 km. An exact description of the points and profiles is available in the report submitted to the EDM course held in 1964 at Zürich[2], and, therefore, is omitted here. We want, however, to repeat that the net contains profiles having characteristic meteoro-logical conditions and reflecting planes.

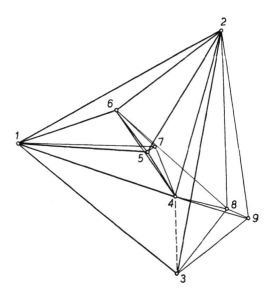

FIG. 1. The Graz test-net, 1:200 000.

1 Pleschkogel KT	6 Eggenberg Nagel TH
2 Schöckl KT West	7 Harthopfer Rohr 1
3 Buchkogel Pyr.	8 Schlossberg A
4 Plabutsch KT Bolzen	9 TH Observ. Pf. SW.
5 Hartbauer Rohr N.	

Within the area of part of the net, which consisted of seven points, all directions were measured and, in this way, a joint adjustment of directions and distances could be executed. The sides of this net were measured during the period beginning in 1961 with different types of Geodimeter (G), Tellurometer (T), Electrotape (E) and a Distomat (D). In addition to this, some new types of instruments were tested while under development. The measurements were taken by working groups of the institutes which placed the instruments at our disposal and which were quite familiar with the measuring techniques, as follows:

1961: Bundesamt für Eich- und Vermessungswesen (Federal Office for Weights and Measures, and Office for Geodesy), Vienna (direction measurements with Geodimeter NASM-2A and 4B).
Technical University of Graz (Geodimeter NASM-4B).
Zentralanstalt für Meteorologie (Central Office for Meteorology), Vienna (meteorological measurements).

1962: Deutsches Geodätisches Forschungsinstitut (German Research Institute for Geodesy), Munich (Tellurometer MRA-1).
Technical University of Hanover (Electrotape DM-20).
Messrs Wild, Heerbrugg (Distomat).
Messrs Aga, Stockholm (Geodimeter NASM-4D).
Bundesamt für Eich- und Vermessungswesen (Federal Office for Weights and Measures, and Office for Geodesy), Vienna (Geodimeter NASM-2A and 4D).
Technical University of Graz (Geodimeter NASM-4B, 4D).
1963: Deutsches Geodätisches Forschungsinstitut (German Research Institute for Geodesy), Munich (Tellurometer MRA-1 and Electrotape DM-20).
Technical University of Graz (Geodimeter NASM-4D).
1964: Technical University of Munich (Tellurometer MRA-3).
Technical University of Graz (Tellurometer MRA-2).

During 1965, some measurements were expected to be executed with a Geodimeter 6 and Distomat (Ertel, Munich).

All of the measurements were reduced in the same manner and to the same points. The distances measured in this way are compared, as in Fig. 2, with the distances which were fixed by way of a Geodimeter

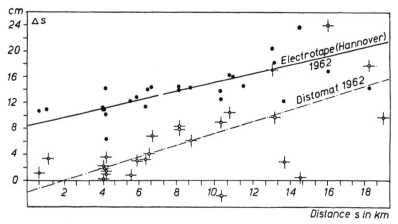

FIG. 2. Comparison of measurements by the Electrotape and the Distomat.

4D (Hg light) in 1963. For each of the instruments, we evaluated the coefficient of a function

$$\Delta s = a + bs$$

and the mean square error, in which a is the zero point constant, and b is the scale error.

The results which we obtained in this way point to the possibility of errors in an Electrotape DM-20. However, the observer who used this instrument denies this possibility, saying that extreme meteorological circumstances were believed to be the cause.[4] A report [2] describes the time needed for the measurements. From the comparison, which is in some way rather problematic, we obtained some good results with the Distomat within a very short time.

The parallel measurements executed in 1964 by a Tellurometer MRA-2, with a carrier wave of 10 cm (by the Technical University of Graz), and a Tellurometer MRA-3 (of the Technical University of Munich) are particularly interesting. A report on this work has been published.[3] Simultaneous measurement was possible with no disturbances while the measuring time was generally lower for the non-digitalized MRA-2 instrument.

A detailed report enumerating the results of parallel measurements[2] has been prepared, and also describes the experiment of using a helicopter for evaluating the average values of temperature and pressure by flying along the path of the waves during the measurements. The result of continuous measurements by use of light waves shows a correlation between the reduced lengths and the time of day, pointing to an incomplete perception or consideration of the meteorological influences. The measurements taken by helicopter show the possibility of measuring the average values of the refractive indices by flying along the ray path, and it is also a practical example of the inadmissibility of averaging meteorological data measured at the end points. The summary report[2] on the test-net of Graz will also include a detailed discussion.

The adjustment of the measured data was made by variation of co-ordinates with different weight assumptions for distances and directions.

For the directions, the following matrix was used:

$$R + v_R = (R) + o + \mathrm{d}R$$

where

R = measured direction
(R) = preliminary orientation
o = orientation constant
v_R = direction residual

As two points of the net were kept as fixed points, i.e., the co-ordinates were taken from the Federal surveying net, we had to introduce a scale factor for distances.

Therefore, the matrix for the distances had the following form:

$$s + \lambda s + v_s = (s) + \mathrm{d}s$$

where

s = measured distances
(s) = distances from preliminary co-ordinates
λ = scale factor
v_s = residuals

In order to make the examination apply generally, real zero point constants were introduced, as we have to assume them when comparing the measurements with different instruments. A constant λ_o has, therefore, been added to the matrix in a second adjustment:

$$s + \lambda_o + \lambda s + v_s = (s) + \mathrm{d}s$$

The weights of the measured data were enumerated from ascertained figures by a roster taken from different assumptions, and evaluated in such a way that all of the possibilities determined in practice were included. Even the most extreme events were taken into consideration.

For a joint adjustment of directions and distances, we have the following law:

$$\sum (pvv)_R + (pvv)_s = \min$$

in which we have to introduce the weights:

$$p_R = C/m_R{}^2, \qquad p_s = C/m_s{}^2$$

with the same constant C and the mean square errors in the same dimension as the residuals v. The adjustment was executed for two groups, each of four weight assumptions, which were taken from the following set-up:

$$p_R = 1, \qquad p_s{}' = C'/s^2, \qquad p_s{}'' = C'' \text{ (const)}$$

The determination of the dimensions was made in such a way that successive weights have a geometric difference of ten. That means, between the first and fourth weights we get the proportion of $1 : 10^{-3}$.

The pure direction adjustment was introduced by the formula of $p_s = o$. The nine weight assumptions used for the adjustment are enumerated in Table 1.

TABLE 1

No.	mm/km	C'	No.	m_s (cm)	p_s
o	o	o			
1	1·0	200	5	±1·0	2
2	3·2	20	6	3·2	0·2
3	10·0	2	7	10·0	0·02
4	32·0	0·2	8	32·0	0·002

For the adjustment of the distances only, we use $p_R = o$, and introduce the weights shown in Table 2.

TABLE 2

No.	p_s
9	$10 : s$
10	$100 : (56 + 0.2\ s^2)$
11	$100 : (56 + 2\ s^2)$
12	$10 : (4 + 0.06\ s)^2$
13	$100 : (4 + 0.6\ s)^2$
14	$10 : (4 + 0.001\ s^2)^2$
15	$10 : (4 + 0.01\ s^2)^2$
16	1
17	$100 : s^2$

With these assumptions, the figures which are already fixed vary in geometric differences of ten. Assumption No. 9 corresponds to an error proportional to the square root of s, No. 17 corresponds to an error increasing linearly with s, and assumption No. 16 corresponds to an error independent of the distance. All of the nine complete measurements of the net with the Geodimeter, Tellurometer, Electrotape and Distomat were adjusted for these seventeen weight assumptions with a zero point constant being introduced for each distance net. Summarizing all this, the net adjustment was executed with twenty-six different weight assumptions for each instrument. At every adjustment, we determined the unknown quantities: the co-ordinates dx, dy, the orientation constant o, the scale factor λ and the zero point constant λ_0 and, in addition to these, the residuals v_R and v_s of the measured data and the mean square errors of the unknowns, as well as of the distances and directions after the adjustment.

In order to compare the results of the different adjustments, we introduced two average error factors for the net:

1. *An average point error:*

$$m_p = m_0 \sqrt{(Q_{pp})} \qquad Q_{pp} = 1/n_p \sum (Q_{xx} + Q_{yy})$$

where

n_p = number of new points, and

Q_{tt} = weight coefficient.

2. *An average relative side error:*

$$\bar{\mu}s = m_0/\bar{s} \sqrt{(Q_{ss})}; \qquad \bar{s} = 1/n_s \sum s; \qquad Q_{ss} = 1/n_s \sum Q_{ss}$$

where n_s = number of sides.

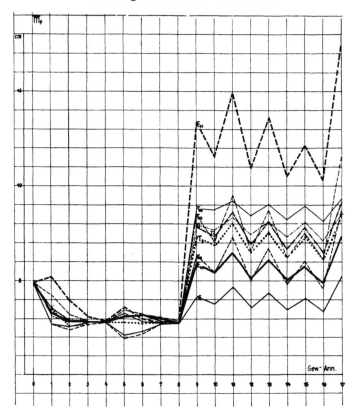

Fig. 3*a*.

For the parallel measurement with the instruments MRA-3 and MRA-2, which were affected similarly by reflection effects as a result of equal profiles and equal meteorological conditions, a special comparison was executed. The result is as follows:

1. In the combined adjustment, the pure direction net for all of the weight assumptions is inferior to the combined net with directions and distances. If we also use approximate distance measurements, the accuracy of the point determination can be increased. The combined net with directions and distances is also superior to the pure distance net (Fig. 3*a*).

2. Distance measurements by light waves led to the best results.

3. The weight assumptions, which are different by three factors of ten for the distances, cause a fluctuation of the average point errors

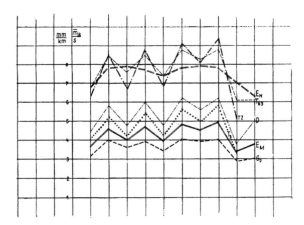

FIG. 3*b*.

only of approximately ±2 cm, and of the relative distance error of ±1·5 mm (Figs. 3*a*, *b*, *c*, *d*, 4*a*, *b*, *c*).

4. For the pure distance adjustment, without a zero point constant, there is a well-defined order of rank, according to the instruments. After the Geodimeter, there are two 3-cm carrier-wave instruments, then the 10-cm instruments and finally another 3-cm instrument (Figs. 3*b*, *c*, *d*).

5. The introduction of a zero point constant changes the result. Now, all of the 3-cm instruments rank after those using light waves and before the 10-cm instruments. We recognize a well-defined order of rank in accordance with the length of the carrier wave (Figs. 4*a*, *b*, *c*). In view of the fact that a determination of the constants of the Electro-tape was executed before and after the measurement in Graz, a change of the zero point constant is less admissible. The supposition, however, that during the measurement quite extraordinary turbulent meteoro-logical conditions existed is also rather unacceptable, since the anomalies which would be required for such an explanation were not evident. Therefore, the observation that the introduction of a zero point constant into the adjustment can considerably improve the result can only be accounted for by the combination of several unexplained circumstances. The elimination of a systematic or pseudo-systematic constant part is justified, even though no physical explanation can be given for it. It is a similar case to that of photogrammetric triangulation, in which the practically determined pseudo-systematic figure can also be proved by the theory of errors.

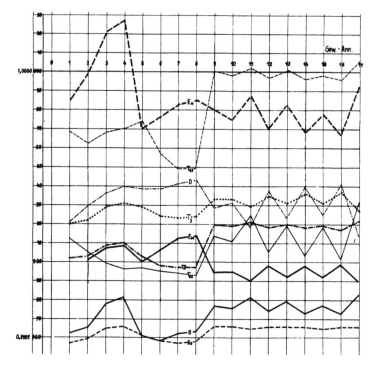

Fig. 3c.

6. Simultaneous measurements with 3-cm and 10-cm instruments show the superior accuracy of the former. These also statistically proved that the 3-cm instrument measurements are less influenced by ground reflections.

7. For the determination of points, distance nets supported by directional data are recommended. Weights of directions and distances can be introduced as constants, their ratio being in the range 1·0–0·2 depending on the instruments used (weight assumption No. 6).

8. It is intended to enlarge the net in the near future by adding sides of up to 50 km. Meteorological investigations as well as systematic analysis of the influence of ground reflections will be carried out.

9. The sides of the test-net at Graz have been measured with all available types of instrument so that standard values are therefore available. Moreover, care has been taken to create homogeneous conditions for the measurements and to obtain reliable meteorological

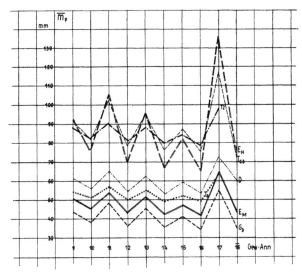

FIG. 3*d*.

data. A programme for an electronic computer (Univac 490) permits instantaneous evaluation of the measured results. Thus, the net is suitable for testing new instruments and for carrying out scientific research.

N.B. Firms and institutes are invited to make use of the test-net of the Technical University at Graz. At the same time we propose that a recommendation by the Symposium be issued to make this possibility known.

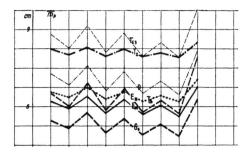

FIG. 4*a*.

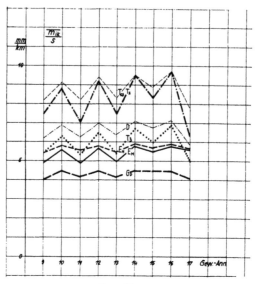

FIG. 4*b*.

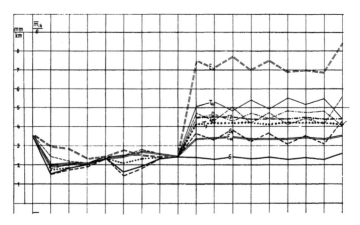

FIG. 4*c*.

4. *Device for Diminishing the Influence of Ground Reflections*

The well-known effect of reflections in EDM can theoretically be eliminated to a great extent if pulse-type signals are used. Because of the inferior accuracy of pulse distance measurements, this has been of no practical importance. However, it is now possible to combine the advantages of pulse and phase distance measuring procedures by using continuous waves with a fixed frequency relation. As such waves have the same transit time, but different phase differences, the influence of ground reflections can be reduced by a compensation method. A device which can be installed in every EDM instrument has been constructed by Prof. Benz (Graz) and an Austrian patent No. 225.753 (61) has been applied for.

5. *Analysis of the Geometric Strength of Networks with Distances*

At the present time, long distances can be measured accurately enough for geodetic purposes by microwave techniques only. Between terrestrial points, the range of such measurements is limited to about 100 km by the need for inter-visibility. Longer distances must either be subdivided or measured by the well-known method of line crossing. The measured spatial distances can be used to connect points on the Earth's surface and elevated points above them, thus constructing a polyhedron inscribed in or circumscribed about the Earth.

In addition to long distances, astronomically orientated directions along the edges of the polyhedron can be determined. This greatly increases the stability of the polyhedron and the positional accuracy of the points of intersection. Such combined networks are independent of the Earth's potential field and are therefore suitable as a basis for the geometric survey, not subject to hypotheses about continents and the Earth as a whole. By suitably combining the available measured values and a proper selection of network shapes, nets with optimum performance and minimum number of measured quantities can be determined. Similarly, studies for the detail nets for ordnance survey can be carried out.

At the Technical University of Graz, this problem is being studied within the frame of a research programme. Starting from the geometry of the method of line crossing for measuring long distances, the geometric strengths of various shapes of networks are systematically investigated, and proposals for their practical application worked out. This work is in progress, and some results are presented below.

5.1 *The Geometry of Line Crossing*

The method of line crossing, as used in SHORAN and similar systems, is well known. From a measuring instrument moving at constant height (aeroplane, rocket or satellite) crossing in a straight line and at approximately right angles to the line to be measured, the distances s_1, s_2 to the end points of the line can be measured. Then the minimum value of the sum of the distances $(s_1 + s_2)$ is determined as a function of the flight time; this value is reduced to an osculating sphere and thus can be derived the length of the geodetic line on the reference surface. Prerequisites for this method are knowledge of the altitudes of the end points of the line and of the measuring instrument, and the assumption that the minimum value of the sum of the distances as well as the line to be measured is located in a plane perpendicular to the osculating sphere. In this case, not only the geodetic distance L in the reference surface, but also the spatial distance D, can be determined from the rectangle defined by the centre O of the sphere, end points P_1, P_2, and the point M from which the minimum value of the sum of distances has been measured. Closer examination, however, discloses discrepancies of the selected model from reality, which impair the accuracy. First, the position of the minimum-value point M depends on the direction of the flight and on the altitudes of the points; only when the direction is perpendicular to the line, and with equal altitudes of both points, will it be located in the normal plane. In the general case, the point will be outside this plane, and the reduction will not yield the geodetic line on the surface or in space. The error formulae have been investigated, and part of the results will be contained in the handbook[1] to be published soon, and will not be treated here.

Accurate determination of the distance from the multitude of values measured during the crossing is possible in two ways: either by projecting the individual measurements on a reference surface, utilizing the altitudes of the measuring points and end points, or by additionally measuring directions.

As the length of the crossing line is only 1/10 to 1/20 of the distance to be measured, the reference surface along the line can be approximated by an osculating sphere. When the altitudes are known, the measured distance can be transferred to this sphere by radial lines. This leads to the two-dimensional spherical problem to determine, from a number of spherical distance sums, the geodetic line which can be solved with strict accuracy, independently of the course of the measuring instrument (as a minimum problem). From the geodetic line, through the chord of the sphere and introducing the altitudes, the spatial distance can be deduced. The only difference between the usual approximation method and the strict method is that the individual measurements are projected, instead of their minimum value. As this projection can be carried out by a computer, the additional work involved is unnecessary. Similarly, the spatial distance can be determined as a minimum value.

For this method, heights must be known, but they cannot always be measured with sufficient accuracy. Their measurement can be avoided if, from the end points of the line, the directions to the measuring instrument are astronomically determined simultaneously with the distances. This can be done, as is known, by photographing light flashes on the background of the fixed stars in the sky. In this case, the angle between the sides s_1, s_2 is determined for each measurement and, from each measurement, the spatial distance can be calculated.

By thus improving the line crossing method, it is possible to determine the chords of a polyhedron which is inscribed in the Earth. This method is an indirect one, whereas the direct method yields spatial distances between elevated points.

5.2 Spatial Networks

Spatial networks can be constructed from direct or indirect measurements. In the first case, some of the points of intersection (elevated targets) are moving, and only the terrestrial ones are fixed. The networks enclose the Earth. A network determined by indirect distances contains only fixed points of intersection, and is inscribed in the Earth. The former type of network is more simple to adjust, but contains more points of intersection, some of which are variable for each group of measurements. For indirectly measured distances, a separate determination of weights is necessary, but the networks formed by them are fixed, and there are fewer points of intersection.

In both cases, there are different types of network, with different performances. For determining these, the overdetermined shape is adjusted by variation of co-ordinates, noting that the inverse matrix contains the weight coefficients:

$$Nx - l = 0, \qquad x = N^{-1}l = Ql$$

The spatial error in the position of a point, independent of the selection of the system of co-ordinates, is calculated from the formula:

$$m_p = m_0 \sqrt{(Q_{xx} + Q_{yy} + Q_{zz})}$$

in which m_0 is the average error of unit weight. With observations of equal weight, m_0 is the average error of an adjusted measurement, whereas the root depends only on the geometrical shape of the network. We define, therefore, the geometrical strength of the network at a point by the reciprocal:

$$L = 1/\sqrt{(Q_{xx} + Q_{yy} + Q_{zz})}$$

Thus a high strength of network corresponds to a small average point error, and vice versa.

The simplest shape of a spatial network is an arc section. The geometrical strength of sections determined from both four and twenty-six distance units, with symmetrically located basic points, has been numerically calculated for various altitudes and positions of the points. From the curves connecting equal point errors, it follows that, with the increase in the number of basic points and distances from four to twenty-six, the point error decreases with the square root. Selecting a unit distance of 500 km, the side length of the basic square is 1000 km, the altitude of the new point is assumed to be 2500 km; the dimensions of the rectangular base are 3250 × 1000 km. For an elevated point with x and y co-ordinates of 1500 km and with a distance error of $m = \pm 1$ m, a point error in the first case would be $\pm 5{\cdot}2$ m and, in the second case, $\pm 1{\cdot}6$ m. Analogous diagrams have been prepared for a point in a cubic spatial network of side length = 5 × length of the basic area.

As a second task, the transferring of four points at ground level has been investigated, first with four and then with twenty-six elevated points. This gives a reply to the question, whether it is more advantageous to use limited overdetermination coupled with high geometrical strength or a high degree of overdetermination with low strength of the network. The results show that it is an advantage to use a greater number of points, and confirm the use of SECOR for the adjustment of all satellite distance records.

Following the simple transferring of points, it is intended to investigate the repeated transferring in a strip circling the globe. For this case too, a mathematical model is first prepared and calculated. The introduction of suitable error distributions for the measured values makes it then possible to simulate the conditions prevalent in nature.

For investigating networks covering the whole Earth, various pyramidal icosahedrons are used as models. By lengths only, this polyhedron, well known in statics, is many times overdetermined. Adding directions for some of the edges further increases the stability or network strength. Finally, a combination of both types of measurements, advantageous from the viewpoint of error theory, will show a valuable way to achieve optimum performance with reduced measuring work.

Part of the study of basic types of spatial nets is Dr Killian's question as to networks solely determined by distance measurements to elevated points, while distances between terrestrial points are not measured at all. If n_t and n_h are the number of terrestrial and elevated points, respectively, the relation

$$n_t n_h - 3(n_t + n_h) + 6 = 0$$

applies to such systems. The only integral solutions are $n_t = 4$, $n_h = 6$ and $n_t = 6$, $n_h = 4$ and define the only possible network shape with ten points. If, in addition, s distances between terrestrial targets are measured, the relation is:

$$n_t n_h + s - 3(n_t + n_h) + 6 = 0$$

In this case, for every value of s, two solutions are obtained. These interesting basic shapes become indeterminate if the points of the network are located on surfaces of the second order. This fact must be kept in mind when planning networks for the transfer of points by distances using this very simple method.

For extensive networks, it is desirable to introduce a criterion for the quality of the network with regard to its geometrical performance. It is proposed to use, as a criterion, the average error m_p of the position of the point, calculated as the root mean square of the position errors for all new points:

$$m_p = \sqrt{[\sum(m_p{}^2)/n]} = m_o \sqrt{[\sum(Q_{xx} + Q_{yy} + Q_{zz})/n]}$$

where

n = number of new points

m_0 = mean square error of a measurement

As another error criterion of a network, the mean distance error after adjustment can be introduced:

$$m_s = \sqrt{[\sum(m_s{}^2)/n]} = m_o \sqrt{[\sum(Q_{ss})/n]}$$

More suitable is a mean square relative distance error determined as follows:

$$\bar{\mu}s = (\bar{m}_s/s) = \sqrt{[\sum(m_s/s)^2/n]} = m_o \sqrt{[\sum(Q_{ss})/s^2]}$$

The same can also be calculated (as has been done for the verification net at Graz) from a different relation:

$$\bar{\mu} = \bar{m}_s/\bar{s}_s = m_o \sqrt{[n\sum(Q_{ss})/\sum s]}$$

but, for systematic reasons, the first relation is more appropriate.

6. *Electromagnetic Tacheometry and Image-forming*

Geodetic tacheometry measures spatial or planar co-ordinates of points of the terrain with relation to local polar systems. As the use of electromagnetic waves permits the measurement not only of distances but also differences, sums, or quotients of distances, it is possible to measure, in addition to polar co-ordinates, elliptic and hyperbolic co-ordinates as well as co-ordinates based on quotients, of the points on the terrain in local systems. These can then be either directly used for geodetic purposes or transformed into a uniform system. Geodetic purposes make it necessary to adopt co-ordinate systems other than those used at present, the co-ordinate lines being hyperbolae, ellipses or apollonic curves. These possibilities make it desirable to investigate systematically the geometric properties of these systems, in order to be able to use them for geodetic purposes in the case of possible developments of electromagnetic measuring systems.

This problem has been studied for plane, spherical and spheroidal curves, and the results reported[1] where conditions prevailing in space are also discussed. Further investigations are intended.

Worthy of special interest are the methods using quotients of distances, as—if propagation conditions are homogeneous—no absolute values are required, but only ratios determined by quotients of transit times.

It is well known that pictures can be obtained through the use of electromagnetic waves (radar). Owing to qualitative improvements, these pictures have gained importance for geodetic and cartographic purposes. In order to use them for geodetic purposes, it is necessary to know the laws of image-formation and of geometric distortion. Investigations in this field have been carried out, and their principles reported[1]. Further investigations, especially on the geometric effects of electromagnetic transit time distortion, are in progress.

Summary

A report is given on research programmes completed or in progress at the Technical University of Graz. An invitation to use the verification net in Graz is made. Some of the research results are reported in Vol. VI of the *Handbuch der Vermessungskunde*.[1]

References

[1] KNEISSL, M., JORDAN, W. and EGGERT, O. *Handbuch der Vermessungskunde* (Vol. VI), Die Entfernungsmessung mit elektromagnetischen Wellen und ihre geodätische Anwendung.

[2] RINNER, K. Entfernungsmessungen mit lichtelektrischen und elektrischen Geräten im Testnetz Graz, DGK-München. In the press.

[3] REINHART, E., 1965. Erfahrungen mit dem Tellurometer MRA-3, Mk II, Beobachtungen im Testnetz Graz. *Alg. Vermess. Nachr.*, **8**, 289–95.

[4] SEEGER, H., 1965. Die Ergebnisse einer Erprobung des Electrotape DM-20. *Z. Vermess. Wes., Stuttg.*, **7**, 222–31.

DISCUSSION

G. Jelstrup: What was the difference in temperature along the line compared with the mean of the two ends?

K. Rinner: The maximum difference was 6°, it being higher as measured by the helicopter.

R. C. A. Edge: Was humidity as well as temperature measured by the helicopter?

K. Rinner: No, only the air temperature.

R. C. A. Edge: Where on the helicopter was the thermometer carried?

K. Rinner: On the tail.

Electromagnetic Distance Measurements in Finland

S. HÄRMÄLÄ

National Board of Survey

and

T. J. KUKKAMÄKI

Finnish Geodetic Institute

Presented by

PROFESSOR L. ASPLUND

The paper briefly describes the various orders of control nets in Finland, electronic distance measuring equipment having been used since 1963 for scale checks on the work done by traditional methods. First-order sides of average length 29·6 km gave a mean square difference of 1 : 475 000. Second order gave 1 : 280 000 and third order 1 : 130 000.

The first-order triangulation of the Finnish Geodetic Institute consists of triangle chains, which form loops of 500 to 800 km perimeter. The National Board of Survey has now started to fill the loops with nets, called main-order nets, with an average side length of 30 km, in which the angles are measured as well as the distances. After preparing experiments, the National Board of Survey started the electronic distance measurements in 1963 and now uses Tellurometer MRA-3 in main-order measurements.

Every distance is measured twice so that at both ends the instrument is used as the master instrument. This method is used in the first place in order to control the frequencies currently. In addition, the frequencies are calibrated at a Laboratory when there seems to be any reason to suspect a change, usually three times every field season. Most of the distances have been measured between triangulation towers rather high above the ground over wooded areas; perhaps this is the reason why only slight traces of reflections can be found in the results. Special attention has been paid to the determination of the refractive index. The air pressure has been measured with Thommen barometers which have been calibrated frequently. The temperature and the humidity have been determined with carefully calibrated Assmann-type psychrometers. The psychrometers whirled by hand have been rejected for their inaccuracies.

The accuracy of the measured distances has been investigated in different ways. In 1964, thirty-six first-order triangulation sides were measured in connection with the main-order network. Nineteen of these sides had already been adjusted and were then used for comparison. These sides were situated in different parts of the first-order net, but no remarkable systematic differences between different areas could be discerned. Instead, a large systematic difference between first-order triangulation and the corresponding tellurometer measurement was apparent, amounting to 1 : 164 000, the tellurometer distances being shorter. After the elimination of this systematic difference, the remaining mean square difference was ± 62 mm on sides with an average length of 29·6 km, or 1 : 475 000.

As another attempt to investigate the accuracy, an adjustment of a net may be mentioned. The main-order net of nineteen new points was adjusted independently of the first-order net. Then the mean square error of a distance of 30 km was ± 50 mm or 1 : 600 000.

The triangulations of the second order have now been carried out as a combination of triangulation and trilateration. The Tellurometers MRA-3 and, especially, MRA-2 have been applied. The adjustments of five 15-point blocks gave a relative accuracy of 1 : 280 000 on average. In the third-order nets, no towers were erected and so the observations were made on the ground. Mostly the angles were measured only at the point to be determined and, in addition, a couple of distances with Tellurometer MRA-2. The relative mean square error of a third-order point was about 1 : 130 000 on average. The density of second-order points was about 1 point per 100 km^2 and that of the third order 4–5 points per 100 km^2.

All the different steps of the computations have been made with electronic computers. The readings have been recorded in field books especially planned for the use of computers. The data are punched and then reduced and, finally, adjusted with an electronic computer.

The Finnish Geodetic Institute started its investigation into trilateration with experiments held in May–June 1965 on the Vihti enlargement net. The longest side of the net had the length of 29 114·1372 ± 0·0573 m according to the triangulation. Two Tellurometer units MRA-3–Mk II were used as master and as remote instrument alternatively. All eleven sides of the enlargement net were measured and the results deviated from the side lengths of triangulation by ± 24 mm on average according to provisional computations. For the extrapolation of the humidity and of the temperature for the actual Tellurometer beam along the whole distance, these atmospheric factors were measured at two altitudes of 2 m and of 10–24 m above the ground at both ends and in the middle of the measured side. The periodic factor of the index error in the instruments was determined at a 432-m distance on the Nummela Standard Base by varying the measured distance during the whole 20 min. period of the A-reading.

DISCUSSION

There were no questions.

The Caithness Base Investigation

Presented by

M. R. RICHARDS

The Ordnance Survey of Great Britain

In 1964, the Ordnance Survey conducted a large-scale practical investigation into the determination of refractive index for microwaves. Stations for measuring meteorological conditions were established along the length of the Caithness Base (24·8 km), previously measured by invar in catenary. Precise measurements of barometric pressure, and of wet-bulb and dry-bulb temperatures at various altitudes, were made simultaneously with distance measurement of the base by Tellurometers MRA-3 and MRA-2. A Geodimeter NASM-4 was also employed. General weather conditions (rain, cloud cover, wind speed and direction and visibility) were also recorded. Observations were spread over two periods of ten days each, while over 900 distance measurements were made. The resulting data are being subjected to analysis. Anyone who wishes to carry out research into any aspect of the investigation may obtain copies of the field data from the Ordnance Survey. The paper sets out all the necessary background information, and includes samples of the available field data.

1. *General*

The Report of Special Study Group No. 19 of the IAG, for 1960–3, contained the recommendation that research into the measurement of refractive index should be energetically continued. The particular aspect of research with which this paper deals is the repeated measurement of a line under varying meteorological conditions. Many previous investigations have been carried out on this subject by numerous organizations. However, these have generally suffered from one or other of two serious shortcomings. Firstly, the line being measured might be of very accurately known length, but relatively short, over flat ground, and thus not typical of normal observing conditions. Alternatively, the line might be long, with elevated terminals and more typical of lines normally measured, but the length would not be known to better than about 10 parts per million (ppm); this is the order of accuracy that might be expected from geodetic triangulation.

It was felt that there would be considerable advantage in carrying out an investigation on a line that was relatively long, of accurately known length, and having elevated terminals that would give a more typical profile. In Caithness, Northern Scotland, the Ordnance Survey have such a line. It was measured in 1952 by invar tapes in catenary. It is 24·8 km long, and is thought to be the longest in the world. The estimated standard error of the measurement is ±2 cm, or approximately 1 ppm. The terminals are 125 and 176 m respectively above mean sea level, and the intervening ground lies between 30 and 75 m above sea level. The whole length of the line is readily accessible.

The Ordnance Survey of Great Britain, jointly with the British Military Survey Service, considered that they were in a position to make a positive contribution to the research into refractive index measurement urged by Study Group No. 19. The opportunity was perhaps unique in the survey sphere, as few other organizations could provide the large manpower force required to carry out a large-scale investigation. It was therefore decided to mount a practical investigation, the main purpose of which would be to study, and provide data on, the behaviour of the lower layers of the atmosphere under different conditions. The effect of the meteorological variations on the derived refractive index would also be studied, in conjunction with a series of electronic distance measurements of the line. Provided that the necessary equipment could be obtained, it would not be difficult to measure meteorological conditions along the line, and at some distance above the ground. From this, some conclusions on the optimum conditions for observations might be obtained.

During the investigation, different types of EDM instrument would be used, and a comparison made of their performance. The 'ground swing' patterns obtained from microwave instruments could also be studied. Previous indications had been that the line might prove difficult to measure owing to abnormal and erratic swing characteristics, but it was hoped that the 3-cm instruments would overcome this. Finally, as a by-product of the very extensive observations that it was intended to make, a great deal of data would be produced which could be made available for others to use for research.

The investigation was staffed by a contingent of Military Survey personnel from 19 Topographic Squadron R.E. together with a party from the Ordnance Survey. The observations were split into two periods of ten days each in order to vary the conditions as widely as possible. The first was in July 1964 when it was expected that there would be fine summer weather. The second was in September–October 1964 when cooler more wintry conditions were expected. In the event the weather differed only slightly between each period; July was cold and wet, October surprisingly mild.

The observations were taken throughout the whole of each day. The twenty days of field work produced over 900 distance measurements of the line, recorded on a corresponding number of booking sheets. The meteorological data, having been abstracted and corrected for calibration, fill over 300 sheets approximately three times the size of this page. These data are available to any person (or any organization) wishing to carry out research into any aspect of the investigation. The only proviso is that the Ordnance Survey shall be provided with a copy of any results which may ensue.

The object of this paper is to set out in detail what was actually done. All relevant information will be given, together with samples of the available data. No conclusions have been drawn, but one aspect of the investigation forms the subject of a separate paper.

2. *Base Line*

The Caithness Base is the primary triangulation side between the stations Spital Hill and Warth Hill. The accepted spheroidal distance between terminals, as derived from the catenary measurements, is 24 828·000 m. Auxiliary stations, to accommodate the numerous EDM instruments at each terminal, were established for the July period as in Fig. 1 (see also Appendix A on p. 35).

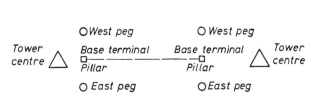

FIG. 1. Layout of the Caithness Base, showing auxiliary stations in July, 1964.

Stations described as East Peg and West Peg were established precisely at right angles to the line of sight, and 3 m from the base terminal. Stations, each consisting of a 30-ft (9-m) survey (Bilby) tower, were established immediately behind each base terminal and on the line of the base.

Auxiliary stations for the October visit were established in slightly different positions as in Fig. 2. The auxiliary stations were established in order to vary the distance measured slightly, so that any possible index error which was a function of the A reading could be recognized as a systematic error, and eliminated. In addition, on each Bilby tower a second position was established on the handrail so as to prevent reflections from the metal work.

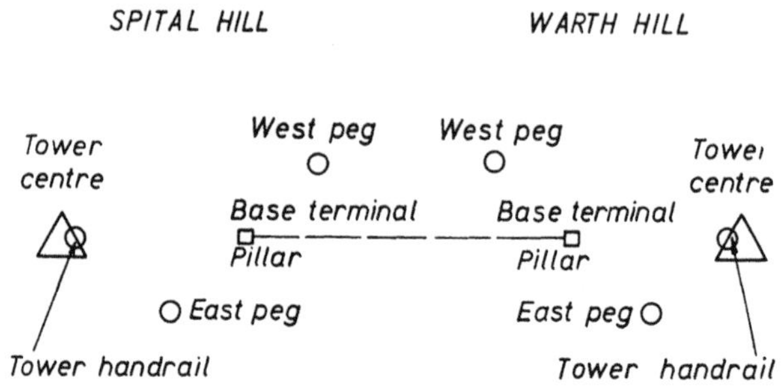

FIG. 2. Layout of the Caithness Base, showing auxiliary stations in October, 1964.

To reduce the measured distance to the spheroidal distance the following corrections must be made:

$$\text{Slope correction} = -\frac{\Delta h^2}{2S}$$

$$\text{Reduction to sea level} = -\frac{S \cdot h_m}{R}$$

$$\text{Chord to arc correction} = +\frac{S^3}{43R^2}$$

where h_m is the mean height of the instruments above sea level, Δh is the difference in height of the two instruments, S the spheroidal distance, and R the radius of the Earth.

The corrections that should be applied to the measured air distances are set out in Table 1 and include corrections for the eccentricity of the auxiliary stations. Individual corrections may vary slightly from the tabulated values according to the height of instrument above datum, but variations will be insignificant.

3. *Instruments*

Two pairs of Tellurometer MRA-3 instruments and one pair of Tellurometer MRA-2 were available for each period.

MRA-3 instruments Nos. 308 and 311 were designated the standard instruments for the July period, and were used exclusively between

TABLE 1a. *Corrections to be applied to air distances, July 1964*

All corrections are in metres.

Type of measure	Correction for			Total correction	Correction for eccentricity at		Overall correction to air dist.
	Slope	MSL	Chord to arc		Warth	Spital	
Pillar to pillar	−0·0545	−0·5874	+0·0087	−0·633	Nil	Nil	−0·633
East Peg to East Peg	−0·0544	−0·5869	+0·0087	−0·633	Nil	Nil	−0·633
West Peg to West Peg	−0·0542	−0·5864	+0·0087	−0·632	Nil	Nil	−0·632
Tower centre to tower centre	−0·0550	−0·6191	+0·0087	−0·665	−3·895	−4·007	−8·567

TABLE 1b. *Corrections to be applied to air distances, October 1964*

All corrections are in metres.

Type of measure	Correction for			Total correction	Correction for eccentricity at		Overall correction to air dist.
	Slope	MSL	Chord to arc		Warth	Spital	
Pillar to pillar	−0·0544	−0·5874	+0·0087	−0·633	Nil	Nil	−0·633
East Peg to East Peg	−0·0545	−0·5873	+0·0087	−0·633	−1·477	−1·486	−3·596
West Peg to West Peg	−0·0534	−0·5862	+0·0087	−0·631	+1·516	+1·482	+2·367
Tower centre to tower centre	−0·0551	−0·6192	+0·0087	−0·666	−3·880	−4·004	−8·550
Tower handrail to tower handrail	−0·0554	−0·6200	+0·0087	−0·667	−2·923	−3·112	−6·702

the terminal pillars, making a measurement each hour (timed to start on the hour) throughout the period. Instruments Nos. 270 and 274 were used for a similar purpose during the October period.

MRA-3 instruments Nos. 270 and 274 alternated with MRA-2 instruments Nos. 773 and 774 making measurements from the offset positions during the first period. Nos. 308 and 311 changed with Nos. 270 and 274 in the second period so that the latter became the standard instruments. Measurements were made hourly with these instruments, timed to start at each half hour. Measurements were not permitted to overlap, i.e. the measurement with the standard instruments had to be completed (or abandoned), before the secondary pair took over, and vice versa.

The Tellurometers MRA-2 Nos. 773 and 774 were overhauled and had the crystal frequencies calibrated and adjusted by Tellurometer (UK) Ltd, before and after both periods of fieldwork, and crystal frequencies were certified as correct. The MRA-3 instruments were unfortunately available to be checked only after the second period. The actual deviations from standard of the A crystal frequencies found on the final check, and prior to any adjustment, are indicated in Table 2.

TABLE 2. *Modulation frequency errors (c/s)*
(*as determined at final check*)

Crystal	Instrument					
	MRA-2 773	MRA-2 774	MRA-3 270	MRA-3 274	MRA-3 308	MRA-3 311
Master A	−14	−7	−4	−8	0	+6
Remote A+	− 4	−8	+5	−6	+11	0
A−	−10	−2	−2	−6	− 2	0

It would appear probable that the maximum error introduced into any of the measurements would be from MRA-2 No. 773, of the order of 1·4 ppm, and hardly significant. Errors introduced by any of the other instruments should be still less significant, and negligible in the context of the investigation.

It was not possible, owing to shortage of time, to test the instruments for zero error before the experiment, although this is now being done. From available evidence, it seems unlikely that, as far as the Tellurometers MRA-2 are concerned, the zero error is significant on a measured length of 25 km. For the Tellurometers MRA-3, there is evidence that by measuring with these instruments alternately as master and remote from opposite ends of the line, as was done in this investigation, residual zero error is reduced to less than 1·5 cm (0·6 ppm) which is almost negligible[1].

A total of 893 Tellurometer measurements were completed.

Additionally, measurements were made with a Geodimeter NASM-4 (No. 266, with mercury-vapour lamp) from one or other of the auxiliary stations whenever visibility permitted. A total of fifty-seven measurements were made, forty during the first period and seventeen during the second. During the second period, vertical angles (zenith distances) were observed simultaneously with the measurements. The object of this was to make a reduction of the measured distance in the manner suggested by Saastamoinen.[2] There was a slight doubt about the accuracy of the Geodimeter because of asymmetry in the Kerr cell, produced by progressive oxidization of the nitrobenzine.

4. *Meteorological Instruments*

Preliminary investigations indicated that there was no suitable equipment available commercially for the remote measurement of temperatures. The necessary apparatus was specially made, based on a design developed by the Meteorological Office. The main components were thermistors mounted in silicone rubber inside stainless steel sheaths, with a temperature response time of a few seconds. They were wired to a galvanometer through a switch box, the latter incorporating balancing resistors; several thermistors could be read on one meter. The galvanometer scale was graduated in °C with an upper and lower scale range, each covering 20°C, and could be read to 0·05°C. Initially, some trouble was experienced with the temperature scales since they did not overlap; 10°C was the change-over point, and calibration was critical at this point. The meters were later modified to incorporate a 5°C overlap between scales, and this solved the initial problems.

Calibrations were carried out, using a water bath, both before and after each observing period. The calibration figures show remarkable consistency, indicating very little drift. It is considered that the field temperatures, having been corrected for calibration, can be relied on to be correct to ±0·2°C.

The thermistor probes were mounted in pairs in retaining brackets (see Appendix B). One had a textile sheath fed by a wick from a water bottle to act as a wet bulb. Each was housed centrally inside two concentrically mounted open-ended cylinders, which formed a shield against radiated heat. Each pair was aspirated by an electric fan mounted on the end of the protective cylinders, drawing 30 ft³ of air per minute over the probes. To prevent possible heat transfer from the fan to the probes, 'Tufnol' insulating blocks were inserted between the base of the fan and the cylinders. A single unit thus performed the function of a wet- and dry-bulb psychrometer, with the advantage that the instrument could be placed at some distance from the recording station.

Measurements of atmospheric pressure were made with altimeters operated in pairs. Pressures were read in equivalent metres of altitude and have been converted directly to mm of mercury through the calibration values. The average difference between simultaneous readings on a pair of altimeters is less than 1·0 mm Hg; (1 mm error in pressure represents approximately 0·3 ppm error in the refractive index determination).

5. *Meteorological Observations*

Two meteorological stations were established at each terminal, and three more along the line of the base. In addition, standard meteorological measurements were taken by the Tellurometer observers using aspirated psychrometers and aneroid barometers. The position of all these stations is shown on the diagram in Appendix A. At each terminal a 50-ft (15-m) mast was erected, approximately 14° off line to avoid interference by reflections, such that the top of the mast was level with the line of sight between base terminals. Wet- and dry-bulb thermistor units were placed at the top, and at the 30-ft (9-m) level, of each mast. They were also placed at the top of each terminal tower, i.e. approximately 30 ft (9 m) above the base terminal, and on the tower, level with the instrument on the terminal pillar. At Annfield, approximately 8 km from Spital Hill a 103-ft (30-m) Bilby tower was erected. Wet- and dry-bulb thermistor units were established at the top, at the 60-ft (18-m) and 20-ft (6-m) levels on this tower. At Hillhead, the mid-point of the base line, and at Slickly, further 103-ft towers were erected. At each of these towers, a wet- and dry-bulb unit was attached to a rope and pulley, so that it could be raised and lowered at will to predetermined levels on the tower. All temperatures were read remotely on a meter at some convenient position at each location. Temperatures were read and recorded to 0·05°C.

At the base terminals, and at Annfield, the reading drill was the same. The observer recorded wet- and dry-bulb temperatures from each of his thermistors at five-minute intervals. He also recorded barometric pressures at ten-minute intervals. Owing to the shortage of equipment and man-power it was not possible to man the Hillhead and Slickly stations full time; the observer alternated between these two stations.

At Slickly, he read temperatures at every 20-ft (6-m) level [i.e. at 100, 80, 60, 40 and 20 ft (30, 24, 18, 12 and 6 m)] of the tower, twice in succession, commencing on the hour. He then moved to Hillhead to make similar observations on the half hour. Barometric pressures were not recorded at these stations.

In addition to these readings, each observer made a detailed note every half hour of the prevailing weather conditions, including wind speed and direction.

6. *Preliminary Reductions and Analysis*

6.1 *Distance Measurements—Tellurometer*

Each of the distance measurements has been reduced, on the field sheets, to an air distance. A sample booking sheet is shown in Appendix C, p. 37. Note that the times booked are British Summer Time (subtract one hour for Greenwich Mean Time). The standard meteorological observations taken at the instrument station, and given on the field sheets, have been used in this reduction. All air distances have been further reduced to the spheroidal distance between terminals using the corrections tabulated in Table 1. One surprising fact has emerged. The measurements made with Tellurometers MRA-3 between the terminal pillars are systematically shorter than those made with the same instruments between any other combination of auxiliary stations, no matter which pair of instruments is involved. The reason for this is not apparent, and investigation into it continues. It is not a function of the fixation of the auxiliary stations; these have been checked and are errorless. It will be noted that the same phenomenon does not appear to occur with Tellurometers MRA-2, although the pillar-to-pillar sample is too small for this to be said definitely. The relevant figures are given in Table 3.

Note that 133 measurements made from steel towers are not included in these mean figures as, despite being made in an atmosphere free from possible ground anomaly, these proved very difficult to observe. The results, particularly those from the MRA-2 Tellurometer, indicate gross swings from which no pattern emerges. They give widely divergent answers ranging from 24 826·850 to 24 829·232 m.

6·2 *Distance Measurements—Geodimeter*

All Geodimeter measurements were made between auxiliary stations and reduced to the terminal pillars. There are rather fewer of these measurements (fifty-seven only), but no more detailed analysis than that given below has yet been possible. The spheroidal distances obtained

TABLE 3. *Mean derived distances*

Stations	Period	Instruments	Numbers in mean	Derived distance (m)	s.D. of single observation (m)	s.D. of mean (m)
Pillar to pillar	1st	MRA-3 (308, 311)	237	24 827·958	±0·081	±0·005
	1st	MRA-2 (773, 774)	6	28·111	±0·168	±0·068
	2nd	MRA-3 (270, 274)	240	27·922	±0·081	±0·005
	Both	All instruments	483	27·942	±0·083	±0·004
Peg to peg	1st	MRA-3 (270, 274)	64	28·064	±0·124	±0·016
	1st	MRA-2 (773, 774)	92	28·080	±0·083	±0·009
	2nd	MRA-3 (308, 311)	87	28·026	±0·082	±0·009
	2nd	MRA-2 (773, 774)	34	28·068	±0·080	±0·014
	Both	All MRA-3	151	28·042	±0·102	±0·008
	Both	All MRA-2	126	28·077	±0·082	±0·007
	Both	All instruments	277	28·058	±0·094	±0·006
Overall mean figure			760	24 827·984	±0·087	±0·003

The accepted catenary measurement is 24 828·000 m.

(compare catenary 24 828·000 m) were:

First visit, mean of 40 measurements	24 828·097 m
s.d. of single measure	± 0·026 m
s.d. of mean	± 0·004 m
Second visit, mean of 17 measurements	24 828·029 m
s.d. of single measure	± 0·039 m
s.d. of mean	± 0·009 m

Applying Saastamoinen's[2] method to the results of the second period gives:

Mean of 17 measurements	24 828·006 m
s.d. of single measure	± 0·054 m
s.d. of mean	± 0·013 m

(Note that vertical angles were not observed during the first period.)

6.3 *Temperature and Pressure Observations*

All thermistor temperature measurements and barometric pressures have been corrected for calibration and abstracted on to sheets similar to the samples shown in Appendix C, p. 37. Abstraction has been done by stations with the sheet headed simply Spital, Annfield, Hillhead, Slickly, or Warth. The nominal altitude of the thermistor unit above ground level at the station is given. Where sheets include a column headed Meter Temperature this should be ignored. It was originally included because it was suspected that the calibration procedure had introduced an error by having probes at one temperature and the meter at another. This suspicion was subsequently disproved. Remarks are included where necessary on the occasions when the readings may be suspect. This happened from time to time when the wick for the wet bulb failed to draw up water, when electrical faults occurred, when the battery activating the meter became discharged, or on one occasion when a cow chewed the cable.

7. *Available Data*

Samples of the available data are attached. Those which can be supplied to interested research workers are as follows:

Tellurometer Field Booking Sheets (see Appendix C)

	First period	439 sheets
	Second period	455 sheets

Geodimeter Field Booking Sheets (see Appendix D)

	First period	40 sheets
	Second period	17 sheets

Abstracted Meteorological Data (see Appendix E)

Spital Hill	First period	58 sheets
	Second period	60 sheets
Warth Hill	First period	58 sheets
	Second period	60 sheets
Annfield	First period	52 sheets
	Second period	60 sheets
Hillhead	First period	4 sheets
	Second period	5 sheets
Slickly	First period	4 sheets
	Second period	5 sheets

Requests for these data should be addressed to the Director General, Ordnance Survey, Leatherhead Road, Chessington, Surrey, England, stating briefly the object of the research and giving exact details of data required.

References

[1] BOSSLER, J. D. and LAURILA, S. H., 1965. Zero error of MRA-3 Tellurometer. *Bulletin Géodésique*, No. 76.
[2] SAASTAMOINEN, J., 1962. The effect of path curvature of light waves on the refractive index application to electronic distance measurement. *Canadian Surveyor* (March).

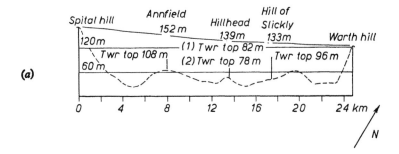

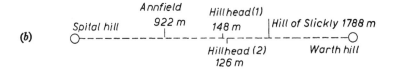

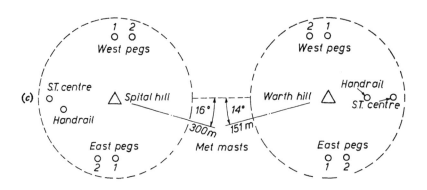

Appendix A. (a) Clearance of Tellurometer ray above top of towers.
(b) Distances of intermediate tower sites from base-line.
(c) Auxiliary stations at terminals.

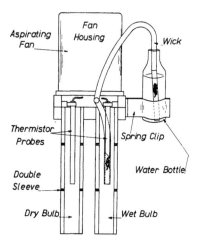

Appendix B. Thermistors mounted for measuring air temperature and humidity.

TELLUROMETER Field Sheet

Block Reg. No.
CAITHNESS BASE REFRACTIVE INDEX INVESTIGATION.

274

		Inst. Height	Inst. No.
Master Station *NORTH HILL (Pillar)*	Operator *V. R. BARKER*	*364 ft.*	*274*
Remote Station *SPITAL HILL (Pillar)*	Operator *E. JOHN*	Inst. Height *3.61 ft*	Inst. No. *270*
	Booker *E.C. LATTER*		

Date _28/9/14_
Time _1758 U_ _1759 .05 T_
Weather *Overcast dull*
dry fresh SW Wind
Heater - On/Off
NOTES:-

Coarse Readings (initial)
A+　A+　A+　A+
B　　C　　D　　A-

```
        8·30
      290
      832
      493
      254
  24828·30
```

Coarse Figure
Coarse Reading (Final)
A+　A+　A+·　A+
B　　C　　D　　A-

Approx. Distance

Coarse Figure

Met. Readings

	Crystal Temp.	Dry Bulb °F	Wet Bulb °F	Dep. °C	Dew Point	Vap. Pres.	Baro. Rdg. ins.	Baro. Correct ins.	Corrected Baro. ins. mm	Baro. Serial No.
Master Initial	R I.	53.0	50·2				29.39	-·08	744·5	**Master:**
		53·2	50·2				745.9	-0·3	45·6	3047
Final	1·000	53·0	50·2				29 38		4·2	
	1221						745·2		4·9	2699
							29.38		4·2	
Remote Initial	C	11·7	10·1	1·6		8 443	745·2		4·9	**Remote**
							-3-		744·9	
		54·2	50·0				29 85	-·16	741·4	70
Final		54·8	49 8				28·80	+·32	39·6	
Sum.	1·000	53·2	48 8				29.84		41·2	A 39
	3153						28·81		39·9	
Mean	C	123	09·7	2·6		7 746	29.32		40·7	
							28·80		740·4	
For Office Use Only	1·000 316·9°									

Air Distance _24818·532 m_

Range 0·76 metres

Appendix C. (*a*) An example of a tellurometer field booking sheet.

	Freq. Dial.	A+ A-	A+R A-R	Mean Diff.	Fine Reading	Master AVC.Reg	Remote AVC.Reg	Pulse A.
1.	22	821	832	8.265	/			
2.	30	833	830	8.315	/			
3.	40	8.44	8.42	8.430	/			
4.	50	8.45	8.44	8.445	/			
5.	60	8.62	8.55	8.585	/			
6.	70	8.72	8.73	8.725	/			
7.	80	8.50	8.60	8.550	/			
8.	90	8.48	8.43	8.455	/			
9.	100	8.03	8.30	8.165	/			
10.	110	8.36	8.27	8.315	/			
11.	120	8.31	8.01	8.160	/			
12.	130	8.23	8.19	8.210	/			

	Freq. Dial.	A+ A-	A+R A-R	Mean Diff.	Fine Reading	Master AVC.Reg.	Remote AVC.Reg.	Pulse A.
13.	140	797	7.97	7.970	/			
14.	150	8.13	7.99	8.060	/			
15.	160	8.40	8.51	8.455	/			
16.	170	8.52	8.38	8.550	/			
17.	180	8.48	8.45	8.465	/			
18.	190	8.48	8.44	8.460	/			
19.	200	8.56	8.56	8.560	/			
20.	Mean of 19 - 8.376 /							
21.								
22.								
23.								
24.								

Appendix C. (*b*) An example of a fine reading booking sheet.

DISTANCE MEASURING   Geodimeter Model 4 serial No. **264**, Date **20.7.64.** Time on **22.56** off **00.27** BST.

Geodimeter station **SPITAL HILL EAST** ... Observer **G. BELLAMY** Recorder **S. DERRICK** ....

Reflector station **WORTH HILL EAST** ... Type of Refl. **12** prisms ... wedges

Approx. dist. **24.829** Area **CAITHNESS BASE** ... Visibility: Good fair poor **Good**

Meteorological and other data:

Geod. eccentr. **NIL** m   Height **4.74 Ft** above W.P. +G   **SPITAL HILL** Temperature **49.2** °F

Plumb corr. **-.003** m   Above W.P.+   Bar. press. **29.15 A39** mmHg **74.90 N°2699**

Reflector eccentr. **NIL** m   Atm. corr. **+21.0** 10 D

Geod. constant **+0.230** m

Reflector constant **-0.030** m   Height **4.08 Ft** above W.P.+R   **WORTH HILL**   TEMP:- **49.1°F**

Atm. con. **+0.521** m   BROM: N° A26 :- **29.24"-05**

Sum of corr. **+0.718** m   N°2698:- **756.0"**

Phase	Frequency f1				Frequency f2				Frequency f3			
	i/o	C	i/o	R	i/o	C	i/o	R	i/o	C	i/o	R
1	S	54	O	114	S	53	S	70	S	58	O	213
2	O	53	S	121	O	52	O	72	O	57	S	215
3	S	55	O	114	S	56	S	71	S	61	O	215
4	O	55	O	117	O	56	O	73	O	60	S	213
Sum of phase 2 and 3		54	235	115.5		54		71.5		59		215
Sum of phase 1 and 4		54.5	231	115.5		54.5		71.5		59		214
Mean of phase 1,2,3,4		54.25		116.5		54.25		71.5		59.0		215.5
Mean metres from table	S	0.767	O	1.395	S	0.752	S	0.952	S	0.553	O	2.437
If C)k add U and change		(+U1)		(+2,500)		(+U2)		(+2,494)		(+U3)		(+2,181)
"1" to "o" or "o" to "1"												
(L) = R-C(+U) (-U)		-C1				-C2				-C3		
If "1o" or "o1" add U	R1 - C1(+U1)		S/o		R2 - C2(+U2)		S/o		R3 - C3(+U3)		S/o	
		(+U1)		(+2,500)		(+U2)		(+2,494)		(+U3)		(+2,181)
If (L2)<L1, (L3)<L1 add		L1 = 3.128				(L2)				(L3)		
2U2 resp. 2U3								(+4,988)				(+4,762)
					L2 = 5.188				L3 = 4.265			
Nearer multiple of 5					-L1 = 3.128				-L1 = 3.128			
(F) = F					A=L2-L1 = 2.060				B=L3-L1 = 1.137			
Nearer hundred					400 A =				+20 B =			
(E) = E					-F =				(F) - 21 B=			
					(E) - 400 A-F =				F =			
					E =							
Formulae (see table)	L1k		L2k		L3k				MEAN			
D' = E + F	L1 = 3.128		L2 = 5.188		L3 = 4.265				D' = E + F = 8.23			
K2 = D' ÷ 0.002493766			-K2 = 2.05		-K3 = 1.190				(L1+L2+L3) = 3.111			
K3 = F ÷ 0.0476190			-2k = 3.131		-3k = 3.075				Sum of corr. = +0.718			
L1k = L1	Air disc Corr								n x 2000 = 24000			
L2k = L1 - K2	As Velocity Change		24828.796 m						Mean dist = 24828.829 m			
L3k = L1 - K3												

Appendix D. An example of a geodimeter field booking sheet.

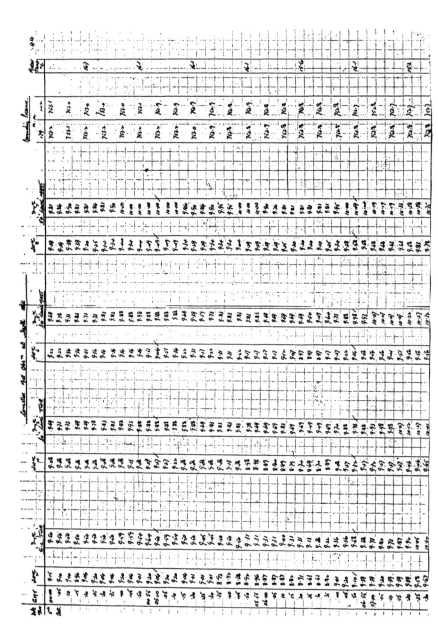

Appendix E. An example of abstracted meteorological data.

DISCUSSION

R. C. A. Edge: The weather was, from the point of view of comfort, not as good as it appeared in the slides shown, though for EDM purposes the conditions were good and so the variation in the meteorological conditions was not as great as had been hoped. Although a very large quantity of data had been accumulated, there had not been time to do a lot with these and it was hoped that some country (with a surfeit of professors) would make use of the material which would be given to anyone interested without charge. The only stipulation made was that any results should be passed back to the O.S.

S. K. Sharma: Were meteorological readings taken along the line in addition to those at the terminals?

M. R. Richards: Owing to the height of the line of sight above the ground it was not feasible to make meteorological readings exactly on the line, but measurements were taken from three towers along the line and the data deduced by extrapolation from the readings on the tower.

J. J. Gervaise: Did you find any change or drift in the calibration of the thermistors?

M. R. Richards: No appreciable change.

A. Marussi: Were balloons considered as a lifting device for the thermistors?

M. R. Richards: Yes, they were considered, but owing to the likelihood of strong winds, we were advised not to use them.

J. J. Gervaise: Was the use of automatic recording apparatus considered?

M. R. Richards: Yes, it was considered but we decided on manual recording as being the easier.

D. A. Rice: Were vertical angles taken during measurement as a means of determining the vertical gradient of refractive index?

M. R. Richards: Seventeen only when measuring with the geodimeter.

M. C. Walsh: Were the Radomes removed during measurement?

M. R. Richards: The Radomes were removed for some measurements, but it appeared to make no noticeable difference.

H. D. Holscher: What was the reason for the 30-ft towers at the terminals?

M. R. Richards: It was thought advisable to try this because of Poder's papers on the subject. Unfortunately, the swing from the towers was worse and the measurements have not yet been analyzed.

I. R. Brook: Is the removal of the rail from the Bilby Tower standard practice and has the O.S. definitely been able to state the effect of it on the fine readings?

M. R. Richards: It is not normal practice; in this case, the measurement appeared to be equally bad whether or not the rail was removed.

F. J. Hewitt: Were measurements made from the ground at the same time as from the towers and, if so, were they equally successful?

M. R. Richards: The tower measurements were not good.

R. Bill: As the Caithness Base was a bad Tellurometer line, were you not making it worse by going up high (without evaluating the optimum height)?

M. R. Richards: Yes, it certainly did make the measurements worse.

R. C. A. Edge: The swing pattern changed very sharply if the instrument was moved quite a short distance.

K. Poder: Did you derive distances using the intermediate as well as the terminal meteorological observations?

M. R. Richards: Yes, this is discussed in another paper (on pages 102–16).

K. Poder: The reason for the elevation on towers is so that the meteorological readings will be more representative of the line. The surface of the ground near the station can be very critical. We have found that a very small amount of movement can possibly alter the measurement. In what form are the records?

M. R. Richards: As in the examples given with this paper.

W. C. J. Burki: Have any comparisons of the standard deviation of the present observations been made with previous measurements of the base?

M. R. Richards: Figures are not available now but I should be happy to write to the questioner.

R. Bill: This was in a report by Wadley in 1957.

O. B. Andersen: Was zero error allowed for?

M. R. Richards: We are doing this but it has not yet been completed.

A. R. Robbins: Was it taken into account that there may be a correlation with absolute humidity as well as with relative humidity?

M. R. Richards: This was not taken into account.

Report on Polish Electromagnetic Distance Measurements Programme

W. KRZEMIŃSKI

(Institute of Geodesy and Cartography, Warsaw)

Polish work on an EDM programme, partly co-ordinated by the Institute of Geodesy and Cartography, has, in the last four years, been concentrated on three main tasks:

1. Construction of microwave distance equipment;
2. Research into EDM errors;
3. Application of EDM in geodetic surveys; the technical and economic factors.

The research into EDM errors consists of the statistical elaboration of standard errors of MDM. During the test measurements of Telemetr OG-1, the routine of standardization of MDM instruments has been elaborated. Special care must be taken on the dependence of the instrument constant on carrier frequency. Research into the influence of other conditions on EDM accuracy are in progress. Good results have been obtained, both technical and economic, with the application of Telemetr OG-1 to construction of triangulation nets of 3rd and 4th order. The use of MDM in 2nd-order nets is appropriate only in special cases. In photogrammetric fixed-point measurements, the EDM instruments are not used on a wide scale in Poland.

1. *Construction Work*

Work began in 1958–9 on the construction of geodetic microwave distance measuring equipment in the Department of Radiolocation of the Warsaw Technical University. The work, directed by Professor S. Slawinski, has been carried out until today. The first prototype, marked KG-1, was constructed in 1962 and has been tested in the field and in the laboratory. The following year, in accordance with obtained results, an improved model was developed, and is now being constructed under the name of Telemetr OG-1. It includes, of course, some changes in detail. Its constructor is Dr K. Holejko.

The Telemetr OG-1 works on a carrier wave of about 3 cm. Its measuring system consists of two non-interchangeable stations: a Master and a Remote. Both stations are of the same dimension, $31 \times 32 \times 21$ cm; the weight of the Master station is 10 kg, while the

Remote weighs 9 kg. The antennae in the form of paraboloids of revolution are screwed to the cover of the instrument and fed by a widely spreading exit tube of the waveguide. The width of the antenna is characteristically 4°·5 at 3 dB. The stations are fed by accumulators of 12 V each, supplying 3·8 A for the Master and 3·2 A for the Remote station. The working temperature range of the instrument is between −30°C and +40°C, with a warm-up time of ten to twenty minutes. Depending on measuring conditions, the working range of the instrument extends from 100 m to between 15 and 20 km. Thus the Telemetr OG-1 is suitable for the measurement of distances of medium length. The carrier frequency of 94·000 MHz is tuned continuously in a range of ±400 MHz. Basic standard frequencies are F = 15 MHz, F10 = 13·5 MHz and F100 = 14·85 MHz, thus permitting the determination of distances up to 1 km with accuracy of the order of 1 cm. The number of full kilometres is determined separately.

Readings are displayed on the screens of an oscillograph tube. Two impulses (direct and reversed) appear simultaneously on the screen, which has a circular time scale, and permit the reduction of the number of switching connections for measuring each frequency to two. Identification of impulses (intervals) is obtained by switching out the reversed impulse in the Remote station following the order given by the Master station. Both Master and Remote stations are provided with suitable lightweight microtelephones (150 g). The constructional system of the instrument is based on valves, but some sub-assemblies (converter) use transistors. The standard error of Telemetr OG-1 is

$$M_s = \pm(3 + 0\cdot5 \times 10^{-5}D) \text{ cm}$$

As a result of investigations and tests carried out in the Institute as well as during the field experiments, the following advantages of the instrument have been found:

   (i) Small weight of both stations;

   (ii) Relatively small energy consumption;

   (iii) Simplicity in servicing both stations, especially the Remote where, owing to the facility for checking its work from the Master station, only slightly experienced personnel can be used;

   (iv) Good stability of the standard frequency; the three years' operation did not reveal changes exceeding ±10 Hz;

   (v) Measuring time of one series using ten to twelve carrier frequencies amounts to 8 to 12 minutes.

The newest programme of constructional work points to the need for:

   (*a*) The reduction of instrument dimensions, weight and energy consumption (general use of transistors);

(*b*) Production of a simplified instrument for work of lower accuracy;
(*c*) Improvement of parameters of some sub-assemblies;
(*d*) Use of a shorter carrier wave.

2. *Investigation of Microwave Distance Measuring Equipment*

In order to establish a homogeneous method of determining the accuracy of the MDM equipment and its corrections, a series of experiments has been carried out in the Institute of Geodesy and Cartography.

(*a*) *Accuracy of MDM*

This accuracy is ordinarily determined by the value of standard error given in the form

$$M_s = \pm (A + B \times 10^{-5} D) \, \text{cm}$$

Coefficients A and B of this formula are determined by the method of least squares on the basis of resultant (true) errors whose absolute values are adopted as free terms in observational equations. As resultant errors are considered, the differences between distances D_{MDM} observed and D_{base} are adopted as standard values. The latter were determined by measurements effected with accuracy surpassing at least three times the average accuracy of measuring with microwaves:

$$\omega = |\Delta D| = [D_{\text{base}} - D_{MDM}]$$

To check the elimination of all systematic errors, the formula

$$\sum \Delta D \cong 0$$

has been used.

Under such assumptions, and from our own investigations as well as on information published in different countries, it has been inferred that the standard error of all types of microwave distance measuring equipment used is the same and can be represented, after some smoothing, by the expression:

$$M_s = \pm [3 + 0 \cdot 5 \times 10^{-5} D] \, \text{cm}$$

In the presence of high standard frequency, generally better than $\pm 10^{-6}$, the value of coefficient B depends, in practice, on the propagation conditions (refraction, reflection). Its magnitude accords well with that of the refraction coefficient ascribed by Ratynski for the case of medium meteorological conditions.

Taking into consideration such a value of standard error and studying the results published by other authors, the most probable distribution of resultant errors in respect to the standard error has been deduced:

$$M \leqslant M_s \text{ about 70 per cent;}$$
$$M_s < M \leqslant 2M_s \text{ about 20 per cent;}$$
$$2M_s < M \leqslant 3M_s \text{ about 10 per cent.}$$

Such a distribution of the error seems to be imposed by external conditions. The instrumental factors cannot, however, be excluded since 10 per cent of observations carried out in most ideal conditions have revealed errors within limits of $3M_s$ (as has been proved by experiments).

An error surpassing $3M_s$ must be considered as a mistake.

(b) Determination of corrections for MDM equipment

Determination of suitable corrections to the microwave distance measuring equipment is of decided significance for the value of measurement of distances. The instrumental correction consists, in principle, of two parts: a constant and a cyclic factor. The latter is a function of reading the oscillograph tube and may be written, e.g., in the form given by Poder and Lilly:

$$V = b_1 + b_2 \sin[7 \cdot 2(A + b_3)]^\circ$$

(The purpose of using an expression of a higher degree has not been tested.) The instrumental correction, as above mentioned, and methods of its determination are universally known.

In the Telemetr OG-1, made in Poland, the constant correction b_1 amounted to 76 to 85 cm and coefficients b_2 and b_3 were practically equal to 1 cm or less, i.e., they represented quantities smaller than the mean error of their determination. The dependence of this correction on the changes of carrier frequency must also be discussed.

As was observed in 1963 during measurements with a Tellurometer MRA-1, the instrumental correction is not constant for the entire range of frequencies but varies systematically with carrier frequency changes. Variation of zero correction for our instrument can be approximately represented by the equation:

$$W = a \cdot C + b \text{ m}\mu\text{s}$$

where $a = 0 \cdot 076$, $b = 0 \cdot 865$ and $C = $ cavity reading.

This result has been obtained, with the use of a swing diagram, as a mean of more than 250 series of measurements effected on several sides of different length and situation.

The results for Telemetr OG-1 have been elaborated by the same method, which later was improved by replacing the numerous and costly measurements by simplified ones along a chosen line (140 m long) practically free of the effect of reflection.

It was stated that, in the instruments of Telemetr-type, variation of instrumental correction depending on carrier frequency changes cannot be approximated by a curve of the first or second degree but requires

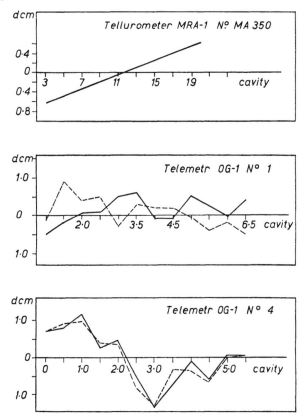

FIG. 1. Diagrams of changing of zero correction with changes of cavity readings. (Increasing cavity readings, full line; decreasing, broken line.)

graphical expression (Fig. 1). It has also been found that in some examples of the same type of instrument the graph is shifted, when observing with the next reading of the cavity, in the direction of results obtained with diminishing readings. Such a dependence is of noticeable significance when measuring with an odd number of series or in the case of using the incomplete tuning range of the instrument. In contrast, when measuring in the range of the recess readings, for which a correction was initially determined, the above-cited phenomenon is of no significance. It must, however, be considered and defined in the case of swings (reflections) analysis under restriction to obtain results charged with a full effect of variation of the instrumental correction.

Besides determination of the above-described corrections for the instruments being used in the state triangulation network measurements, the attestation work carried out at the Institute of Geodesy and Cartography also comprises determination of the tuning range of carrier frequency and checking of standard frequencies. Moreover, a special system for the field control of those frequencies by means of radio signals has been elaborated.

In the domain of MDM, investigations are carried out in order to define the influence of reflections and meteorological conditions upon the accuracy of measurements, but the results obtained do not bring any new ideas into the problems already known and confirm only the achievements published up to now.

3. *Applicability of MDM Equipment to Geodetic Measurements*

These instruments have been applied to the geodetic measurements in Poland in the following cases.

(a) *Measurement of Triangulation Nets*

The possibility of applying the microwave distance measuring equipment in triangulation is limited by the magnitude of standard error and the characteristic distribution of errors larger than the standard. Moreover, some other local factors have a decided influence on the technical and economic effects. Thus it was stated that the MDM can partially replace the angular observations of the net of third and fourth order (according to Polish classification). In some particular cases, as shown in Fig. 2, the weakly constructed angular nets of second order can be completed with MDM.

The manner of replacing angular by linear measurements is shown in Fig. 3. This also presents the ellipses of errors corresponding to the adjustment of an angular net as well as of a mixed angular–linear net with a smaller number of directions.

Results of the adjustment are shown in Table 1.

In nets where both angular and linear measurements are made, it is difficult to assess properly the appropriate set of weights. In the Institute's studies, the weights of sides (when the weights of directions are equal to 1) have been adopted on the basis of standard errors amounting to:

D (km)	1–4	5	6	7	8	9	10	11	12
P	1	0·91	0·83	0·77	0·71	0·67	0·62	0·59	0·59

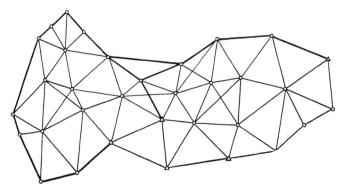

Fig. 2. Example of weakly constructed net of second order connected to five points of first order, completed with microwave distance measurements.

TABLE 1. *Mean errors after different type of adjustment*

Point No.*		Number of sides	Number of redundant observations	M_o^{cc}	M_p (cm)	dx (cm)	dy (cm)	
1	2	3	4	5	6	7	8	9
56	I	—	11	3·89	3·2			single point
	II	2	10	3·92	3·6	+0·2	0·0	
55	I	—	10	2·50	1·4			single point
	II	3	9	2·40	2·3	+4·0	−0·9	
52	I	—	17	2·73	1·3			in a group of points
	II	4	24	2·74	2·2	+0·1	2·5	
51	I	—	4	4·53	2·0			single point
	II	3	1	1·0	0·7	+2·1	+5·2	
53	I	—	17	2·73	1·4			in a group of points
	II	5	24	2·74	1·5	+0·2	+0·8	
54	I	—	17	2·73	2·0			in a group of points
	II	4	24	2·74	2·0	−0·5	+0·3	

* I = angular net, II = combined net.

Economic effects of the application of MDM in triangulation works depend on the degree to which the changes in the net diminish the necessity for the construction of survey beacons.

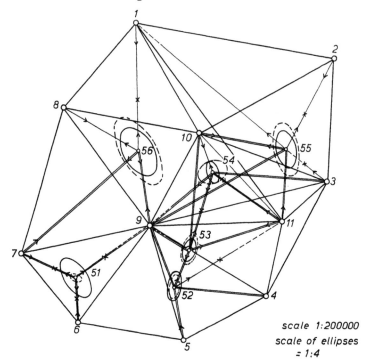

scale 1:200000
scale of ellipses
= 1:4

○   1-11 points of net of 3ʳᵈ order

∘   51-56 determined points of 4ᵗʰ order

→—   Observed directions.

═══   Sights measured with Telemetr OG-1.

—×— Directions not observed in combined measurements.

Ellipses of errors - full line from angular net, broken

line from combined net.

FIG. 3. Example of application of MDM to determination of fourth-
order points.

(b) Photogrammetric Nets

No problem of accuracy arises in the photogrammetric investigations
using EDM because the corresponding sides are measured with an
accuracy much superior to that usually required. Thus, the sole technical considerations and economic effects will decide the application of

a suitable method. In view of the small quantity of control points used in photogrammetric work in Poland, application of EDM is sometimes far from economical or gives an imperceptible effect. As a consequence such work is not being extended.

4. *Measuring Routine, Calculations, Tables and Nomograms*

As for any practical application of MDM, the measuring routine, recording forms, calculation forms as well as some tables and nomograms have been prepared. Of considerable interest is the nomogram for computation of air refractive index for microwave range, by Zofia Majdanowa, first published in *Proceedings of the Institute of Geodesy and Cartography*, **11**, No. 2 [24], 1964.

This nomogram is based on the formula of Essen and Froome, after some transformations written in the form:

$$Q = A + B \cdot p$$

where,

A and B are some functions of t and t',
t is temperature of dry bulb,
t' is temperature of wet bulb,
p is atmospheric pressure.

Nomograms, including two diagrams for terms A and $B \cdot p$, are prepared for the ranges:

$$-10°C \leqslant t \leqslant +40°C$$

$$680 \text{ mm Hg} \leqslant p \leqslant 760 \text{ mm Hg}$$

$$\Delta t = t - t' = 15°C$$

The accuracy of the nomogram is about 1 ppm.

5. *Electro-optical Distance Measurements*

Until now, two models of electro-optical distance measuring equipment have been in operation in Poland. These are SWW-1 from the Soviet Union and Geodimeter NASM-4B from AGA, Sweden, both used in the state triangulation nets of second and third order. The Geodimeter has been used successfully for base measurements.

In Table 2 are the results of series of measurements with Geodimeter 4B on a base of length 6·8 km, first measured with invar wires.

As this Geodimeter is provided with a high-pressure mercury lamp as well as with a standard one, the measurements were made using both lamps. The results indicate that the accuracy obtained is of the same order though operation with mercury lamp takes 50 per cent longer.

3

TABLE 2. *Measurements with the Geodimeter.*

Date	No. of measurements	Diff. = Base − Geod.	
		cm	ppm
2.10.64	5	−0·1	0·1
7.10.64	3	+3·8	5·6
27.10.64	3	−0·5	0·7
19.11.64	2	+1·3	1·9
12.4.65	2	+2·9	4·3
	Mean	+1·7	2·5

DISCUSSION

R. C. A. Edge: Are the accuracies quoted in Section 2 for single measurements or for a complete determination of the distance?

S. Krzemiński: They are for a determination, but I do not know how many measurements.

R. C. A. Edge: It is interesting to see that your figures agree with those of the Study Group. With these accuracies, the instruments appear sufficiently accurate for first-order work. Why are they not used for this?

S. Krzemiński: The problem is the distribution of the standard error, since errors can be up to three times the standard error.

R. C. A. Edge: Is the OG-1 available in the general market?

S. Krzemiński: Maybe, but at present it is in Poland and Czechoslovakia.

P. J. Taylor: Would not interchangeable units be a great advantage?

S. Krzemiński: These instruments are not used for very accurate work. They need to be cheap and interchangeable units might cost up to $1\frac{1}{2}$ times more.

K. Poder: Is the size of the zero error approximately 80 cm?

S. Krzemiński: The b_2 coefficient will need measurements of 1 cm; b_3 can be found by forward and reverse readings. The accuracy of the instrument is about 1 cm. The b_2 coefficient is about 8 cm on our Telemetr. The figures 76 to 85 cm in Section 2(*b*) are a result of the configuration of the antennae.

K. Poder: Is b_2 always negative if you want to apply this correction?

S. Krzemiński: It depends on the instrument.

Electromagnetic Distance Measurement in the US Geological Survey*

EDWARD A. KRAHMER

US Geological Survey

For mapping in difficult terrain, the AirBorne Control (ABC) Survey System has been developed, in which a hovering helicopter with a remote electromagnetic distance measuring unit serves both as a target and as a survey platform. This system includes a Hoversight, to enable the pilot to maintain position above a selected ground point; a height indicator, to measure the vertical distance from the ground to the helicopter; a rotating beacon, for a sighting target; and a retractable antenna. Communications between the helicopter and the ground are maintained by radio.

Field parties at ground stations read both vertical and horizontal angles to the beacon, and measure the distances electromagnetically. The horizontal positions are established with at least one redundancy in angle or distance, and the elevations are based on at least two individual determinations. To establish a K factor (refraction and Earth-curvature correction) for refining elevations, simultaneous reciprocal vertical angles are read hourly between ground or tower stations near the operations area.

Field tests in the States of Arizona, Maine and Alaska have established that the system is economical in remote areas and accurate enough horizontally to control 1:24 000-scale mapping and vertically to permit a contour interval of 20 ft.

Introduction

Although the theme of this Symposium concerns the scientific aspects of electromagnetic distance measurement, this paper is devoted mostly to applications in control surveys in the National Topographic Mapping program. The US Geological Survey has about eighty units of microwave distance measuring equipment of several makes: Tellurometer, Electrotape and Micro-Chain.

These instruments have been used in a wide variety of terrain and under all possible conditions with generally excellent results. Our

* Publication approved by the Director, US Geological Survey.

55

experience agrees with that of other mapping and surveying organizations. Electromagnetic distance measurement is one of the greatest single advances in surveying procedures ever made. Since the introduction of the Model MRA-1 Tellurometer some years ago, the accuracy of our basic horizontal control has improved substantially, and during the same period the cost has been reduced. We consistently obtain second-order closures with third-order methods.

For the most part our field procedures are conventional. They include long-leg traverse, control extension by radial methods and the establishment of elevations by trigonometric methods with electromagnetically measured distances. The one exceptional method, developed by the Geological Survey, is the use of airborne electromagnetic distance-measuring instruments.

The AirBorne Control (ABC) Survey System

In the ABC system, under development for the past six years, a hovering helicopter carries a remote electromagnetic distance measurement unit and serves both as a target for distance and angle measurements and as a survey platform. The specialized instrumentation includes a Hoversight, to enable the pilot to maintain position above a selected ground point; a height indicator, to measure the distance from the ground to the helicopter; a rotating beacon, for a sighting target; and a retractable antenna that can be rotated in azimuth.

The Hoversight, the basic instrument on which the system depends, consists of a damped pendulum with a self-contained light source which projects a collimated light beam through a semi-transparent mirror. With the aid of a second mirror, the observer can see the image of the light source superimposed on the image of the ground below. The ground image is viewed stereoscopically in the same sense that the ground itself is viewed with both eyes, enabling the observer to judge the distance from the ground. The lamp-filament image projected against the ground image defines the vertical below the Hoversight. The pilot views the instrument at a depression angle of about 30°, so that he retains the peripheral vision needed for maintaining aircraft stability.

The height indicator, mounted immediately above the Hoversight, consists of a weighted Dacron line which is lowered to the ground, passing over a calibrated drum geared to a counter reading in feet. A motor-driven magnetic coupling maintains a tension of about 2 lb on the line.

The flashing beacon is mounted above the helicopter cabin, on the vertical axis of the Hoversight to avoid eccentricities. The beacon is the target for vertical and horizontal angle measurements, and also aids observers in spotting and tracking the helicopter from ground base stations.

The electromagnetic distance measuring equipment (DME) component is the Hydrodist, manufactured by the Tellurometer Pty Ltd. A remote unit is carried in the helicopter, and one or more master units are at the ground stations. Instrument readings are in metres, and ambiguities are resolved in terms of 100, 1000 and 10 000 m. Reading patterns permit estimation of distances to about 20 cm. A retractable, rotatable antenna is mounted on the underside of the helicopter to shield the signal from rotor interference. Fully extended, the antenna dish is entirely below the skids; retracted, it is in front of the cockpit bubble.

A vital element of the system is communication between ground parties and the helicopter. A fully duplexed frequency-modulated (FM) system is provided, operating in the very high frequency (VHF) portion of the electromagnetic spectrum. The ground units each embody two transmitters and one receiver. The airborne unit functions as a repeater station having one transmitter and one receiver. With this arrangement, communications between ground stations can either be relayed through the helicopter or sent direct, depending on which transmitter is used. The repeater transmitter can be keyed by either the pilot or the observer. Helicopter personnel hear only relayed messages and therefore are not distracted by communications pertinent only to ground crews. The repeater transmitter in the helicopter has a 1-kc/s beep-tone generator controlled by the pilot, to indicate to the ground crews when the aircraft is in position over a point.

General Operating Procedures

When the pilot signals by radio that he is over the point to be established, horizontal and vertical angles are read to the flashing beacon on the helicopter from the ground stations, and distance measurements are made from master units on each ground station to the remote unit in the aircraft. The airborne engineer determines the hover height and photo-identifies the ground point. He is also responsible for navigation. The rate of production and the accuracy of the results depend largely on his ability. Reliable two-way communication with all party members is essential.

The system in practice has shown itself to be very flexible. Using components together or in various combinations permits positions to be computed by triangulation, trilateration, or a combination of the two. The configuration of stations, therefore, can be selected with considerable freedom. Base stations need not be situated to give strong intersection angles as in conventional triangulation.

Developing AirBorne Electromagnetic Distance Measurement

In the beginning, the most difficult problem was to obtain and maintain a readable display on the DME instruments. Model MRA-1 Tellurometers were used in the first trials. The crude prototype Hoversight, pilot inexperience, and poor antenna configuration combined to give completely negative results. If we had fully realized the extent of these problems at the time, we might have dropped the project. Fortunately, some Hydrodist instruments, designed for shipboard use, became available about this time. With the meter crystals of the Hydrodist, and a somewhat improved Hoversight, we obtained results good enough to encourage us to continue.

With hovering accuracy improved enough to be certain that horizontal movement of the helicopter was not a factor, we still experienced difficulties from rotor interference. A grassy, erratic display with a beat synchronized with the rotor revolution was produced whenever the antenna could 'see' the rotor. We considered some type of shielding and tried a number of locations. The problem was solved satisfactorily when the antenna was mounted in its present position below the cabin, between the skids, and shielded completely from the rotor. In this position, the antenna must be retractable for take-off and landing, and to reduce drag when the helicopter is travelling.

Interference from the metal skids is still troublesome at bearings near 90° on either side, and measurements in these directions must be avoided. The principal problem, other than rotor chop, however, is to maintain an antenna pointing that will transmit full power. It is characteristic of helicopter hovering that the pilot can keep a constant elevation and position, but he cannot at the same time maintain a constant heading, unless atmospheric conditions are unusually stable. He must have one degree of freedom to compensate for wind changes, and a shift of heading is the most convenient.

The resultant change in antenna pointing and consequent loss of signal disrupted the tuning pattern and made it impossible to use the automatic switching for which the Hydrodist was designed. The solution, achieved after a series of modifications, consists of a feedback circuit activated by signal strength, which rotates the antenna in the direction that increases the signal level. An ideal solution would be the use of an omnidirectional antenna which would avoid pointing problems and permit two or more distances to be measured at the same time. But this would operate on an entirely different signal level, and complete redesign of the internal components would be required. The 'hunting' capability based on maximum signal level operates satisfactorily.

The third major problem of airborne DME is protection against

shock and vibration. The helicopter vibration quickly renders components and connections inoperable unless they are properly designed and
protected. We found no simple cure-all for this difficulty, only a gradual
process of trial-and-error improvement, complicated by the limited
space in the cabin. At one stage we mounted the instrument on a pillar
between the pilot and the observer, but this arrangement resulted in a
'whip' action that magnified vibration effects and was therefore quickly
abandoned.

We found that the most effective vibration protection was the human
body, and that carrying the instrument on the lap of the engineer in
the helicopter gave the best results. Unfortunately, this is not practical
from the standpoint of operational efficiency. In the present shock-
mounting arrangement, the instrument is on the skid rack outside the
cabin, supported by four Barry mounts, a type of rubber piston insulator
widely used in supporting various kinds of airborne electronic instruments. In addition, the airborne components are thoroughly tested for
shock and vibration resistance by the manufacturer, following standard
military specifications.

ABC Survey System Operational Projects

Following a series of field tests and considerable equipment modification, the ABC system has been used on a number of operational projects. Some of these surveys with unusual features are described below.

In 1963, a densely wooded area in Maine, about 400 square miles,
was surveyed with the ABC system to 1 : 24 000-scale map accuracy
standards. Some 265 control points were established from ten base
stations on towers. Of these, 250 vertical control points checked within
2 ft, and fifteen horizontal control points checked within 3 ft. For
identification of the horizontal control points, the helicopter was used
as a photographic target. While the helicopter was hovering above a
point, it was photographed with a 6-in. focal length mapping camera from
about 6000 feet. The image of the rotor blade—two small pie-shaped
segments—was used to photo-identify the point on the mapping photographs.

In 1964, the ABC system was used to set and position several hundred
special survey targets in the Pisgah Crater area in California, in terrain
selected for its resemblance to the lunar landscape. The project was
completed in a few days, and the area is being used for studies of moon
landing procedures.

For several years the Bureau of Land Management has used the
ABC system to establish cadastral monuments in Alaska. In this
application, base stations on geodetic positions are used with pre-
computed angles and distances to 'talk' the helicopter into the desired

position. A monument is then dropped and used as a hover target while precise distances and angles are measured. From these, the distance and direction to the true position are computed, and a ground party is landed to move and set the monument. As the subdivision line is not surveyed by the usual cadastral methods, the ABC system has yielded considerable economy in the rugged Alaskan terrain.

We plan to use the ABC system this fall (1965) to establish vertical control in the Okefenokee Swamp, an area in Florida of about 600 square miles which is reputed to have a general slope of less than 15 ft. The swamp is almost wholly wooded, and 80-ft towers will be required for base stations. By holding sightlines to 10 miles or less, we plan to establish a dense pattern of spot elevations to an accuracy of 1 ft. All supplementary data will be obtained by helicopter, eliminating the need for ground surveys in the swamp. Special problems in determining K factors for curvature and refraction corrections are under study prior to the field operation.

The Aeris II Autotape System

The Aeris II Autotape, introduced recently by the Cubic Corporation, is a very accurate helicopter positioning system which will give two simultaneous slant distances, with digital readout. At the present time, one system is on order for the Bureau of Land Management, and two for private firms. We are awaiting with considerable interest reports on experience with this equipment. The Autotape is compatible with the ABC Survey System.

Conclusion

The accuracy of control established by the ABC system, as demonstrated by numerous tests, has been better than expected. Horizontal accuracy is within 1 m at any range, and vertical accuracy is within 1 ft at ranges less than 15 000 ft. We believe that the principal reason for the surprisingly high accuracy is that the helicopter hovering motion is essentially random, both vertically and horizontally. Consequently, observing procedures designed to average out pointing errors and other random errors also average out helicopter motion. The same principle operates in distance measurements. The uncertainty of hovering, whatever it may be, apparently is never cumulative. A second factor contributing to accurate results is that the helicopter end of all lines is above the ground and therefore relatively free from refraction anomalies.

The ABC Survey System is now considered operational and is in use on mapping surveys by the Geological Survey and on cadastral surveys by the Bureau of Land Management, and certain private firms have systems available on contract. It is a practical tool for establishing horizontal and vertical control in difficult terrain.

DISCUSSION

R. Bill: When using a helicopter what methods were used to measure the meteorological parameters in order to determine refractive index?

E. A. Krahmer: Vertical angles (reciprocal) were taken at hourly intervals between ground stations as near to the line being measured as possible and from these observations and the known heights of stations the refractive index was deduced.

R. Bill: Was atmospheric pressure measured inside the helicopter and, if so, what allowance was made for effect of the rotor blades?

E. A. Krahmer: It was not measured as it would not have made sufficient difference to make it worth while on this particular job.

G. J. Strasser: What was the average length of the rays?

E. A. Krahmer: About 10 miles.

R. C. A. Edge: I see from your paper that you used air photographs of the helicopter to locate the photopoints, and that for this purpose you used the photo image of the rotor blades. As the centre of rotation of the blades was not identical with the position of the hoversight did this require some correction?

E. A. Krahmer: Before the use of the hoversight we used to mark a point on the ground and then photograph the helicopter from another plane so that the image of the rotors appeared over the mark. It is true that there was a slight discrepancy between the centre of observation and the hoversight position, but this was too small to make a correction worth while.

E. H. T. Silva: As long as there was optical clearance, did you measure the lines even if grazing or did you confine the measurements to lines of greater clearance?

E. A. Krahmer: The system is not sufficiently precise to worry about grazing rays and, if a line would measure, we would use it. If it was sufficiently grazing to make a difference we could always go up 20 ft or so.

R. C. A. Edge: Was any clearance necessary in order to use these particular frequencies in conjunction with aircraft?

E. A. Krahmer: No, the main trouble was in cutting red tape such as loading figures and number of personnel, etc.

Zur Elektronischen Messung sehr langer Entfernungen (50–120 km) mit dem Electrotape DM-20

(Fehlerquellen und Versuchsergebnisse)

H. SEEGER

Seelze/Hannover

In carrying out a detailed and careful test of the Electrotape DM-20 at the Geodetic Institute of the Hanoverian Technical University from 1962 to 1964, a great number of measurements were made in order to find out the maximum range of this equipment, to investigate the disturbing effects which appear on particularly long lines and to establish the accuracy attainable on distances of 50 to 120 km. These tests have been carried out on thirty-one particularly long lines over terrain of different topographical and morphological characteristics and under various meteorological conditions; all these lines are part of two special I.O. test-fields.

This report gives a detailed discussion of all those errors which may arise on such particularly long distances and a careful analysis of the results. In particular, those disturbing effects which are produced by meteorological factors are illustrated; these are the irregularities of the horizontal and vertical gradients of the refractive index, effects of absorption and reflection caused by atmospheric inhomogeneities. The effects of ground reflections on lines of this region are also discussed; in particular the maximum spread of 10 fine-readings is demonstrated and analysed. In a final section, the results of 381 sets of measurements on thirty-one particularly long lines are summarized; a careful distinction between precision and accuracy was drawn. Concerning accuracy, it is deduced that the root mean square error of a single observation (= mean of two sets of measurements done from the two endpoints of a line) on such particularly long distances amounts to

$$m_s = \pm 5 \text{ mm/km } (1 : 200\ 000)$$

If such a line is measured three times in periods of different meteorological conditions, we get a root mean square error of

$$m_s = \pm 2.9 \text{ mm/km } (1 : 345\ 000)$$

Im Rahmen der in [1] und ausführlich in [2] beschriebenen Erprobung des elektronischen Entfernungsmeßgerätes Electrotape DM-20 wurden —insbesondere auf Anregung militärischer Dienststellen—u.a. auch zahlreiche Versuchsmessungen zur Ermittlung der maximalen Reichweite bzw. der auf 'überlangen' Strecken (50-120 km) erzielbaren Genauigkeit sowie zur Erfassung der in diesem Bereich wirksamen Störeffekte und Fehlerquellen durchgeführt.

In der Praxis können solche überlangen Strecken bei den verschiedensten Anlässen auftreten; drei typische Beispiele seien hier gennant:

(a) Die Überbrückung von Meerengen, wobei wie an einigen Stellen des Ärmel-Kanals, des Skagerraks bzw. Kattegats durchaus mit Entfernungen um 100 km zu rechnen ist; ähnlich lange Strecken treten gelegentlich auf, wenn über sehr ausgedehnte Flußtäler hinweg beobachtet werden soll (z.B. am Oberrheingraben und im südlichen Rhonetal);

(b) Vorwiegend in Entwicklungsländern könnte es bei der Neuanlegung eines Hauptdreiecksnetzes im Hinblick auf die photogrammetrischen Verdichtungsverfahren durchaus wirtschaftlich sein,—sofern die erzeilbare Genauigkeit hinreichend groß ist— besonders großflächige Netze mit minimaler Punktdichte zu schaffen;

(c) Bei Trilaterationsarbeiten zur Ergänzung bzw. Überprüfung klassischer Hauptdreiecksnetze können häufig zur Versteifung des Streckennetzes überlange Diagonalen beobachtet werden. Das Verfahren ist jedoch nur sinnvoll, sofern die auf überlangen Strecken erzielbare relative Genauigkeit den Ergebnissen auf den klassischen 20-50 km langen Seiten entspricht.

Im Zuge der am Geodätischen Institut der TH Hannover durchgeführten Erprobung wurden insgesamt 381 Meßreihen auf 31 topographisch sehr unterschiedlichen 'überlangen' Strecken beobachtet. Bis auf eine Ausnahme, einer 67 km langen Strecke über der Nordsee, sind all diese Seiten Bestandteil zweier Versuchsfelder:

(a) des Testnetzes 'Niedersachsen', das sich von der Lüneburger Heide bis zum Hohen Meißner bei Kassel erstreckt (siehe Abb. 1) und

(b) des erweiterten Zentralsystems München im Testnetz 'Oberbayern' (siehe Abb. 2).

Vom Grundsätzlichen her muß bei der Messung auf überlangen Strecken mit den gleichen Fehlerursachen gerechnet werden wie auf Dreiecksseiten normaler Ausdehnung (20-50 km). Allerdings verschiebt sich—sofern wir den relativen Streckenfehler zum Ausgangspunkt unserer Betrachtungen machen—die Bedeutung der einzelnen

Fehlerfaktoren mehr und mehr zu den meteorologisch bedingten Störeinflüssen hin.

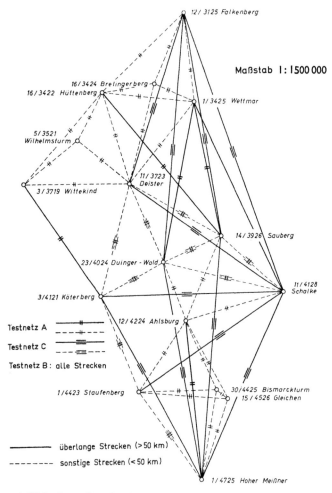

ABB. 1. Ueberlange Strecken (> 50 km) im Testnetz I.O. 'Niedersachsen'.

1. *Instrumentell bedingte Fehler*

1.1 Im Hinblick auf den relativen Streckenfehler verliert der systematisch verfälschende Einfluß einer ungenügend genau erfaßten Additionskonstanten—wobei an Fehler in der Größenordnung von wenigen

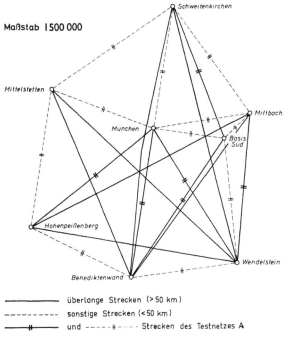

ABB. 2. Ueberlange Strecken im Testnetz 'Oberbayern'—erweitertes Zentralsystem München.

Zentimetern gedacht ist—auf besonders langen Strecken immer mehr an Bedeutung.

1.2 Erhöhte Sorgfalt muß in diesem Zusammenhang der Bestimmung des absoluten Wertes der I_1—Modulationsfrequenz gewidmet werden. Sofern man bestrebt ist, diesen Fehleranteil auf 0,5 mm/km zu beschränken, müßte die Eichung der Modulationsfrequenzen so häufig wiederholt werden—gegebenenfalls im Feldquartier mit einer Anlage wie sie z.b. von Leitz in [3] beschrieben wird—, daß der durch Interpolation für einen bestimmten Zeitpunkt ermittelte Frequenzwert mit Sicherheit genauer als ± 3 Hz ist (7,5 Hz entsprechen beim Electrotape 1 mm/km). In der Regel ist die hier geforderte hohe Genauigkeit gewährleistet, wenn bei Messungen dieser Art einmal wöchentlich eine Frequenzkontrolle durchgeführt wird. Darüber hinaus darf mit einer Meßreihe erst dann begonnen werden, wenn die Modulationsfrequenzen 5–20 Minuten (— bei den einzelnen Geräten stark unterschiedlich —) nach dem Einschalten des Betriebsstromes genügend stabil geworden sind. Da das hier angeschnittene Einlauf-Verhalten der I_1—Kristalle sich als Folge von Alterungserscheinungen verändern —insbesondere verlängern—kann, sollte dieser Vorgang einmal jährlich überprüft werden.

2. *Meteorologisch bedingte Fehlereinflüsse*

Mit wachsender Streckenlänge muß die Erfassung und richtige Beurteilung der verschiedensten meteorologischen Parameter mit noch großerer Aufmerksamkeit betrieben werden als bei den üblichen Entfernungen eines Hauptdreiecksnetzes. Zweifellos resultiert aus diesem Problemkreis—insbesondere aus den Instabilitäten des dreidimensionalen Brechungsindexfeldes—eine Schranke für die maximal erzielbare Reichweite und Genauigkeit. Grundsätzlich müssen in diesem Zusammenhang drei verschiedene Faktoren beachtet werden:

2.1 Aus den meteorologischen Daten, die in klassischer Weise an den Endpunkten einer Strecke registriert werden, kann nur dann ein repräsentativer Mittelwert für das Brechungsindexfeld entlang des gesamten Wellenweges abgeleitet werden, wenn dessen horizontale Gradienten in diesem Bereich gering oder zumindestens konstant sind. Bei überlangen Strecken dürfen jedoch—insbesondere wenn sie sich zwischen Stationen in sehr verschiedenartigen Landschaften erstrecken —derartige Voraussetzungen nicht mehr als gegeben betrachtet werden. Die im Rahmen der hier beschriebenen Versuchsmessungen gewonnenen Erfahrungen lehren, daß die Endpunkte einer überlangen Strecke nicht selten in Gebieten mit sehr unterschiedlichen Witterungs- und Brechungsindexverhältnissen liegen; über Art und Ort des Überganges zwischen solchen Wettertypen können in der Regel keine genauen Angaben gemacht werden. Derartige Gegebenheiten werden vor allem am Mittelgebirgsrand und im Alpenvorland beobachtet.

Ein weiterer Typ horizontaler Gradienten erwächst aus der zeitweise eigenartigen Feinstruktur der Atmosphäre. In diesem Zusammenhang kann unterschieden werden zwischen langzeitigen Schwankungen bzw. Variationen—man spricht von der bubble-Struktur der Atmosphäre—und kurzzeitigen Veränderungen des Brechungsindexes, den sogenannten Fluktuationen.

Beispiele zur bubble-Struktur diskutiert z.B. Brocks in [4]; u.a. beschreibt er Konvektionszellen, deren Durchmesser etliche hundert Meter betragen und an deren Ränder sprunghafte Veränderungen des Brechungsindexes bis zu 30 N-Einheiten auftreten. Ergebnisse von Fluktuationsmessungen werden z.B. von Schünemann und Bull in [5] beschrieben und diskutiert. Aus diesen Registrierungen, bei denen Höhen bis zu 75 m über dem Erdboden erfaßt wurden, folgt zunächst als Mittelwert der Fluktuationsamplituden ein Betrag um 0,5 N-Einheiten; als Maxima treten gelegentlich jedoch Schwankungen von

3-5 *N*-Einheiten auf. Besonders unruhige Fluktuationen werden häufig dann beobachtet, wenn sich die Witterungsverhältnisse verändern.

Eine dritte Gruppe horizontaler Gradienten erwächst aus den besonderen Eigenschaften der bodennächsten Luftschicht, der sogenannten labilen Unterschicht. Wie an Hand ausführlicher Beispiele in [2] gezeigt wird, muß in den untersten 10–40 m der Atmosphäre stets mit außerordentlich großen vertikalen Gradienten des Brechungsindexes gerechnet werden, aus denen—wie leicht einzusehen ist—in unebenem Gelände stets auch horizontale *N*-Gradienten resultieren. Der verfälschende Einfluß der bodennächsten Luftschicht wird besonders dort stark anwachsen, wo in Landschaften mit Mittelgebirgscharakter und kuppenförmig ausgeprägten Bergformen von Bodenpunkten oder nur sehr kleinen Beobachtungsgerüsten aus gemessen wird. Zwischen der auch auf solchen Erhebungen zweifellos vorhandenen labilen Unterschicht und den Verhältnissen entlang des Wellenweges in der freien Atmosphäre werden stets größere Abweichungen auftreten. Eine mathematische Behandlung der aus der labilen Unterschicht resultierenden Probleme ist nach Brocks[6] (—siehe auch Bakkelid in [7]—) zwar grundsätzlich möglich; sie dürfte jedoch für geodätische Zwecke solange wenig erfolgversprechend sein, wie auf ausführliche meteorologische Messungen mit Hilfe von Ballonaufstiegen oder ähnlichen Verfahren verzichtet werden muß.

2.2 Eine weitere Fehlerquelle, die bei den Hauptdreiecksseiten üblicher Länge noch ohne besondere Bedeutung war, resultiert aus der in jedem Fall anzubringenden Krümmungskorrektion

$$K_{Kr_{[mm]}} = (1-k)^2 \cdot \frac{D^3}{24 \cdot R^2} = \frac{D^3}{24 \cdot R^2} \cdot 10^6$$

$$+ \frac{D^3}{12 \cdot R} \cdot \left(\frac{dN}{dh}\right) + \frac{D^3}{24} \cdot \left(\frac{dN}{dh}\right)^2 \cdot 10^{-6}$$

$$(D \text{ und } R \text{ in km})$$

Die an den vertikalen Brechungsindexgradienten zu stellenden Genauigkeitsanforderungen folgen aus:

$$d(K_{Kr})_{[mm]} = \left[\frac{D^3}{12 \cdot R} + \frac{D^3}{12} \cdot \left(\frac{dN}{dh}\right) \cdot 10^{-6}\right] \cdot d\left(\frac{dN}{dh}\right)$$

$$(D \text{ und } R \text{ in km})$$

Eine unten zusammengestellte Übersicht enthält die mit Hilfe dieser Gleichung als Funktion von D und d(K_{Kr}) = 10 mm bzw. 0,5 mm/km errechneten Werte d(d*N*/d*h*).

$$d(K_{Kr}) =$$
$$10 \text{ mm} \quad 0,5 \text{ mm/km}$$

$dN/dh = -40 \,[\text{km}^{-1}]$; $S = 60 \text{ km}$	$d(dN/dh) = 4,7$	$= 14$
$S = 90 \text{ km}$	$= 1,4$	$= 6,3$
$S = 120 \text{ km}$	$= 0,6$	$= 3,6$

Die hier abgeleiteten Genauigkeitsanforderungen sind bereits außerordentlich schwer erfüllbar. Für Versuchszwecke wurde an sämtlichen 74 Tagen, an denen Messungen auf überlangen Strecken erfolgt sind, die 12 Uhr-Terminaufstiege des Deutschen Wetterdienstes an den Flughäfen Hannover-Langenhagen und München-Riem ausgewertet. Die dabei gewonnenen 162 Gradientenwerte streuen zwischen

$$\frac{dN}{dh} = -110 \quad \text{und} \quad \frac{dN}{dh} = 0$$

77 der 162 Gradientenwerte weichen um mehr als ± 10 N-Einheiten vom allgemein benutzten Mittel $dN/dh = -40 \,[\text{km}^{-1}]$ ab. Diese besonders große Streuung der Einzelwerte läßt erkennen, wie außerordentlich schwierig die tatsächlichen Krümmungsverhältnisse durch Mittelwerte zu erfassen sind; der für die Krümmungskorrektion erzielbaren Genauigkeit sind in diesem Entfernungsbereich bereits Grenzen gesetzt.

2.3 Zwei weitere meteorologisch bedingte Vorgänge, aus denen gewisse Störeffekte bei der Streckenmessung mit 3 cm—Trägerwellen erwachsen können, sind atmosphärisch bedingte Dämpfungs—und Reflexionserscheinungen.

Neben der Absorption durch Gase, die im 3 cm—Trägerwellenbereich noch außerordentlich gering ist, können die hier interessierenden Dämpfungsvorgänge zwei verschiedene Ursachen haben:[8]

(*a*) dielektrische Verluste, wobei Strahlungsenergie in eine andere Energieform, insbesondere in Wärme, umgewandelt wird;

(*b*) Streuungen, bei der durch Beugungserscheinungen ein Teil der Strahlungsleistung abgeleitet wird.

Mit erheblichen Dämpfungseinflüssen muß vor allem dann gerechnet werden, wenn Wolken, Nebel, Regen oder andere Niederschläge im Strahlungsraum auftreten; besonders störend können sich in diesem Zusammenhang starke Regenintensitäten auswirken.[8,10,11]

Bei den in den Jahren 1962-4 in Hannover durchgeführten Versuchsmessungen ist bei normalen meteorologischen Bedingungen noch auf einer 116 km langen Strecke gemessen worden; noch längere Strecken standen nicht zur Verfügung. In diesem Entfernungsbereich ist die

Unruhe des Zeigers am Nullindikator bereits so groß und die Signalstärke so gering, daß auf noch längeren Strecken kaum noch mit ver-
wertbaren Ergebnissen zu rechnen ist.

Spürbare Dämpfungserscheinungen sind auf überlangen Strecken
stets dann registriert worden, wenn starke Regenfälle im Profil auftraten.
Häufig verläuft bei solchen Strecken der Wellenweg so hoch über dem
Erdboden, daß der Kernstrahl ausgedehnte Wolkenfelder durchdringen
muß. Insbesondere sind durch heftige Regenfälle und schwere Wolken
bedingte Störungen vielfach dann beobachtet worden, wenn ausges-
prochenes 'April-Wetter' herrschte, d.h. wenn häufig heftige Schauer
das Profil einer Strecke kreuzten. Das Durchziehen eines solchen Regen-
gebietes äußert sich zunächst in einer deutlich spürbaren Abnahme
der Signalstärke und in einer entsprechenden Vergrößerung der
Zeigerunruhe. Dabei kann das Rauschen im Kopfhörer so stark anwachsen, daß der eigentliche 1,5 KHz—Meßton weitgehend unter-
geht; außerdem schwankt die Intensität des Meßtones. Gelegentlich
wachsen diese Störungen so stark an, daß die Durchführung einer
Meßreihe außerordentlich erschwert oder sogar unmöglich wird; obwohl
noch immer eine brauchbare Sprechfunkverbindung zustande kommt,
bricht der Kontakt vielfach dann ab, wenn auf die 7,5 MHz—Modulation umgeschaltet worden ist. Bei Meßreihen, die durch Vorgänge dieser
Art gestört sind, wird stets eine mit $\pm(10–15)$ Skalenteilen außerordentlich heftige Unruhe des Zeigers am Nullindikator registriert.
Offenbar wirken sich hier als Folge der sehr geringen Empfangsleistung
immer stärker werdende Rauschvorgänge aus, die dann eine der
Ursachen für die besonders großen und vor allem unregelmäßigen
Zeigerschwankungen sind.

Weitere Störeffekte, die es hier zu berücksichtigen gilt, erwachsen
aus atmosphärisch bedingten Reflexionserscheinungen. Grundsätzlich können dabei drei verschiedene Reflexionstypen auftreten.[8,9]

(*a*) Reflexionen an niedrig gelegenen Sprungschichten des Brechungsindexes;

(*b*) Gewisse Streureflexionen an Turbulenzkörpern des dreidimensionalen Brechungsindexfeldes (— siehe bubble-Struktur —) in jenem
Bereich der untersten Atmosphäre, der durch den Strahlenkegel erfaßt
wird;

(*c*) Reflexionserscheinungen an niedrig gelegenen Wolken. Der
Reflexionskoeffizient von Wassertropfen ist an sich im 3 cm—Bereich
noch außerordentlich klein. Die ausgesandte Energie reflektiert deshalb
nicht an der Grenzfläche einer Wolke sondern dringt in diese ein;
wirksam wird dann die Summe der im gesamten Volumen reflektierten
und gleichzeitig von der Antenne aufgenommenen Leistung.[8]

In irgendeiner Form werden die drei genannten Reflexionstypen stets in den untersten Schichten der Atmosphäre auftreten; dabei nimmt die reflektierte Leistung mit kürzer werdender Wellenlänge zu. Da die genannten Ursachen offenbar instabil bzw. turbulenzartig sind, darf in ihnen eine weitere Quelle der auf überlangen Strecken immer vorhandenen Zeigerunruhe gesehen werden.

2.4 Aus den vorangegangenen Betrachtungen folgt, daß drei sehr verschiedenartige, meteorologisch bedingte Vorgänge Ursache der vor allem auf überlangen Strecken häufig sehr stark ausgeprägten Zeigerschwankungen sein können, nämlich Geschwindigkeitsvariationen als Folge schneller N-Fluktuationen sowie Reflexions- und Absorptionseigenschaften der Atmosphäre. Störungen dieser Art wurden bei fast allen Versuchsmessungen auf überlangen Strecken beobachtet. Während man bei 30–50 km langen Entfernungen in der Regel Zeigerschwankungen von $\pm(1\text{--}3)$ Skalenteilen registriert, wird dieser Effekt auf überlangen Strecken schnell größer. Bei 50–80 km langen Seiten überwiegt noch eine Zeigerunruhe von $\pm(3\text{--}5)$ Einheiten der Skala, bei Entfernungen um 100 km beträgt sie bereits $\pm(5\text{--}10)$ Einheiten; gelegentlich treten sogar Störungen auf, die ± 10 Skalenteile übertreffen. Wesentlich ist, daß sich sowohl die Amplitude als auch die Frequenz dieser Schwankungen sehr schnell ändern können; innerhalb einer Stunde ist manchmal eine Verschiebung von ± 10 herab zu $\pm(2\text{--}4)$ Einheiten der Skala beobachtet worden. Verursacht durch die zunehmende Unsicherheit beim Koinzidieren des schnell schwankenden Zeigers nimmt der mittlere Fehler einer aus 2 Ablesungen gebildeten Entfernungsinformation bei überlangen Strecken größere Werte an als bei ruhigem Zeiger, also bei kürzeren Entfernungen:

Zeigerunruhe $\pm(0\text{--}2)$ Skalenteile $(0\text{--}50\,\mathrm{km})$ $-\mathrm{m}_{(2-1)}$ $= \pm\ 2\,\mathrm{cm}$
Zeigerunruhe $\pm(3\text{--}10)$ Skalenteile $-\mathrm{m}_{(2-1)}$ $= \pm\ 5\,\mathrm{cm}$
Zeigerunruhe ± 10 Skalenteile $-\mathrm{m}_{(2-1)}$ $= \pm 10\,\mathrm{cm}$

Der hier ermittelte Genauigkeitsabfall ist überraschend gering; er kann durch mehrfaches Koinzidieren weitgehend kompensiert werden.

Grundsätzlich sind die oben diskutierten Zeigerschwankungen nichts anderes als ein Ausdruck für die Instabilität der atmosphärischen Verhältnisse längs des Wellenweges bzw. für Dämpfungserscheinungen; sie sollten deshalb in ihrem negativen Einfluß auf die Güte einer Meßreihe nicht überschätzt werden. Da die vermeintlichen Ursachen dieser Erscheinung alle statistischen Gesetzen unterliegen, ist zu erwarten, daß die aus den Zeigerschwankungen resultierenden Meßfehler sich schon innerhalb einer Meßreihe weitgehend eliminieren. Dies insbesondere, wenn bei jeder Ablesung mehrfach koinzidiert wird, so daß auf jeder Trägerfrequenz bereits eine gewisse Mittelbildung erfolgt. Sofern gelegentlich auf überlangen Strecken einmal nur geringe Zeigerschwankungen auftreten, bestehen meßtechnisch gesehen evtl. ungünstige Verhältnisse. In wenig turbulenten Atmosphären können sich stationäre Schichten ausbilden, aus denen dann systematische Fehlereinflüsse resultieren. Auf überlangen Strecken sollte man deshalb Meßreihen mit einer durchschnittlichen Zeigerunruhe von ±(3–5) Einheiten der Skala höher bewerten als solche, bei denen die Zeigerschwankungen sehr gering sind.

3. *Durch Bodenreflexionen bedingte Meßfehler*

Die als Folge von Bodenreflexionen bei der Messung auf überlangen Strecken denkbaren Meßfehler werden—wie sich aus der Fejer-Gleichung leicht ableiten läßt (siehe [2])—in der Regel nicht stärker ausgeprägt sein als auf Hauptdreiecksseiten üblicher Länge. Zwar muß auch hier noch—insbesondere in Landschaften mit größeren Höhenunterschieden—mit Exzessweglängen bis maximal 2 m gerechnet werden; aus solchen großen Mehrweglängen resultiert jedoch nur eine zunehmende Variationsbreite der Feinablesungen, nicht jedoch—wie aus einer Integration der Fejer-Gleichung über den gesamten Trägerfrequenzbereich und bei Beschränkung auf eine nur im Hinweg wirksame Reflexionsstelle folgt—ein Anwachsen des eigentlichen Meßfehlers. (Als Ergebnis der genannten Integration wird in [2] festgestellt, daß unter den oben angenommenen Bedingungen der größtmögliche Streckenfehler bei einem Exzessweg von $\Delta d = 0,75$ m auftreten und einen Betrag von $\Delta D = 0,19$ m erreichen kann [$R = 0,41$]).

Eine dämpfende Wirkung wird in diesem Zusammenhang insbesondere auch von dem in [2] ausführlich diskutierten Einfluß der Bodenrauhigkeit auf den wirksamen Reflexionskoeffizienten ausgehen. Die durch das Raleigh-Kriterium angedeutete Abhängigkeit des Reflexionskoeffizienten R von der jeweiligen Höhe h der Bodeninhomogenitäten dürfte sich trotz der in der Regel sehr kleinen Erhebungswinkel bei den hier diskutierten überlangen Strecken außerordentlich reflexionsschwächend auswirken, zumal in Profilen dieser Art stets mit sehr mannigfaltigen und vor allem kräftig ausgeprägten Bodeninhomogenitäten gerechnet werden muß. Mit hoher Wahrscheinlichkeit wird daher bei der Messung sehr großer Entfernungen weitgehend nur noch diffuse Zerstreuung auftreten.

Bevor an Hand einiger Beispiele der Verlauf der Feinablesungen charakterisiert wird, sollen heir einige allgemeine Angaben zur Größenordnung der bei den Feinablesungen verzeichneten 'Variationsbreite' erfolgen; wir verstehen darunter die Differenz zwischen dem jeweils größten und kleinsten Wert der auf den einzelnen Trägerfrequenzen erzielten Ergebnisse.

In der unten angefügten Übersicht sind die Ergebnisse von insgesamt 381 Meßreihen zusammengestellt und dabei in drei Entfernungsbereiche aufgegliedert worden. Trotz der auf überlangen Strecken stets sehr stark ausgeprägten Zeigerunruhe ergibt sich auch in diesem Bereich ein ähnliches Bild wie bei den 10–50 km langen Seiten (seihe [1] und [2]). Mit wachsender Entfernung verschiebt sich zwar das Maximum geringfügig von der (11–20) cm- zur (21–30) cm-Spalte; außergewöhnlich große Variationsbreiten sind hier jedoch offenbar noch seltener als bei den 10–50 km langen Strecken. Verursacht wird diese an sich überraschende Erscheinung vermutlich durch eine stetige Abnahme der Bodenreflexionseinflüsse mit wachsender Entfernung. In diesem Bereich wirken sich vorwiegend nur noch diejenigen Störeffekte aus, die oben als Ursache der stets vorhandenen Zeigerschwankungen diskutiert wurden; die dazu im 2. Abschnitt beschriebenen meteorologischen Vorgänge werden sich meistens so wechselseitig variierend überlagern, daß das Messen zwar schwieriger und zeitraubender, die Variationsbreite der Feinablesungen letztlich aber nicht bedeutend größer wird.

Die Variationsbreite von 10 Feinablesungen einer Meßreihe bei jeweils veränderter Trägerfrequenz

	Anz. der Meßr.	0–10 cm	11–20 cm	21–30 cm	31–50 cm	51–80 cm	gr. 80 cm
km							
50– 70	132	10	56	39	22	5	—
70– 90	186	11	79	59	28	8	1
90–116	63	—	17	26	16	3	1
Summe:	381	21	152	124	66	16	2
	= 100%	= 5 5%	= 39,9%	= 32,6%	= 17,3%	= 4,2%	= 0,5%

In den Abb. 3–6 sind alsdann für den hier diskutierten Entfernungsbereich einige typische Beispiele zum Verlauf der 10 Feinablesungen bei jeweils veränderter Trägerfrequenz graphisch dargestellt. Ausgangspunkt der geplanten Wertung muß in jedem Fall diejenige Grundfigur sein, die sich bei der Messung auf vollkommen störungsfreien Strecken ergibt. Sofern eine solche Figur mit genügend großer Sicherheit bestimmt ist, kann aus ihr auf eine evtl. vorhandene Trägerfrequenzabhängigkeit der Additionskonstanten geschlossen werden. Die genannte Grundfigur ist eine Funktion der elektronischen Abstimmung und Justierung der einzelnen Geräte; in den Jahren 1963–4

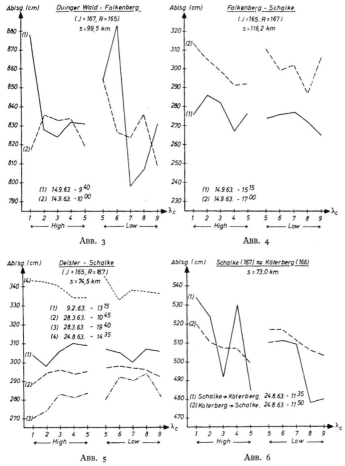

ABB. 3

ABB. 4

ABB. 5

ABB. 6

Der Verlauf der Feinablesungen unter gleichen und unterschiedlichen
Bedingungen auf vier überlangen Strecken (73,0–116,2 km).

setzte sie sich bei den hier benutzten Electrotape-Geräten auf störungsfreien Strecken in der Regel aus zwei stetig abfallenden Geraden
zusammen. Dabei entstand ein erstes Teilstück bei den Trägerfrequenzen 1 High–5 High, ein entsprechendes zweites Teilstück bei
den Frequenzen 5 Low bis 9 Low; die Differenzen zwischen dem
jeweils größten und kleinsten Wert dieser Äste erreichen 10–15 cm.
Sofern Reflexionseinflüsse wirksam sind, werden sie sich dieser Grundfigur überlagern [2].
Durch zahlreiche Versuchsmessungen ist erwiesen, daß die oben

beschriebene Grundfigur nicht nur bei sehr kurzen Entfernungen auftritt, sondern daß auch bei den 10–50 km langen Strecken der abfallende Verlauf der Feinablesungen—durch geringfügige Störungen überlagert—vielfach noch recht gut erhalten bleibt. Dieses Bild ändert sich bei den hier besprochenen überlangen Strecken; die aus der Überlagerung zahlreicher Störeffekte resultierende große Ziegerunruhe führt naturgemäß zu einer unregelmäßigeren und manchmal auch etwas größeren Variation der Einzelwerte. Die oben beschriebene Grundfigur geht dabei häufig, jedoch nicht immer, verloren (siehe Abb. 4 bis 6). Auch bei unmittelbar nacheinander beobachteten Wiederholungsmessungen ergibt sich hier in der Regel schon ein ganz andersartiger Verlauf der Feinablesungen; vermutlich handelt es sich dabei um eine mehr zufällige Streuung, die der im 2. Abschnitt beschriebenen Instabilität der atmosphärischen Verhältnisse in den untersten Luftschichten gerecht wird.

4. Die auf überlangen Strecken erzielte innere und außere Genauigkeit

Bei der nachfolgenden Diskussion der tatsächlich erzielten Ergebnisse soll unterschieden werden zwischen der inneren Genauigkeit, hergeleitet aus Wiederholungsmessungen unter verschiedenen meteorologischen Bedingungen, und der äußeren Genauigkeit, die hier sowohl über Ausgleichungsergebnisse als auch aus einem Vergliech mit Koordinatenstrecken ermittelt werden soll. Zu diesen Betrachtungen werden die Ergebnisse der eingangs erwähnten 381 Meßreihen auf 31 verschiedenen überlangen Strecken herangezogen; im einzelnen werden dabei die gemessenen Strecken in 5 verschiedene Entfernungsbereiche zusammengefaßt (siehe Abb. 7).

4.1 Die innere Genauigkeit

Aus den zur Verfügung stehenden Wiederholungsmessungen (— auf den einzelnen Strecken wurden in der Regel 10–20 Meßreihen beobachtet —), die stets bei unterschiedlichen meteorologischen Verhältnissen und möglichst auch bei veränderten Vegetationsbedingungen erfolgten, ist zunächst als Maß für die innere Genauigkeit für jeden Entfernungsbereich ein mittlerer Fehler m_1 einer einmal im Hin- und Rückweg gemessenen Strecke abgeleitet worden. Die dabei erzielten Werte streuen zwischen $\pm 1{,}1$ und $\pm 1{,}6$ mm/km (siehe Abb. 7); das gewogene Mittel beträgt

$$m_1 = \pm 1{,}4 \text{ mm/km} \, (1 : 715\,000)$$

Von Interesse sind darüber hinaus die Extremwerte der Differenzen zwischen den Beobachtungen unter verschiedenen meteorologischen

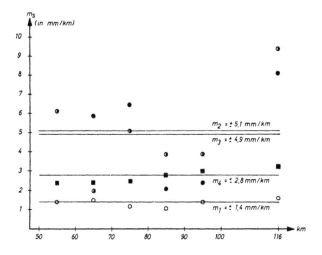

○ m_1 = *innere Genauigkeit aus Wiederholungsmessungen bei unterschiedlichen meteorologischen Bedingungen*

◑ m_2 = *Vergleich mit Strecken der zwangsfreien Ausgleichung des Hauptdreiecks-netzes*

● m_3 = *Vergleich mit Strecken der Testnetze A*

■ m_4 = *Ausgleichungsergebnis (Langstreckennetz)*

ABB. 7. Darstellung der auf überlangen strecken erzietten inneren und äusseren Genauigkeit.

Verhältnissen. Im Mittel betragen diese Werte in den hier diskutierten Entfernungsbereichen 7,0 mm/km; als absolutes Maximum wurden 10,4 mm/km auf einer Strecke in der (70–80) km-Gruppe verzeichnet.

4.2 Die äußere Genauigkeit

Eine Beurteilung der äußeren Genauigkeit elektronisch gemessener Strecken über Koordinatenstrecken des Hauptdreiecksnetzes ist in der Regel problematisch, da die Genauigkeit der Streckenmessung häufig größer oder zumindestens gleich jener des derzeitigen Netzes I.O. ist. Im Zusammenhang mit Untersuchungen über die Genauigkeit 10–50 km langer Electrotape-Strecken ist allerdings festgestellt worden (siehe [2]), daß in Nordwestdeutschland in der Regel eine sehr gute Über-einstimmung zwischen den elektronisch gemessenen Strecken und denen der Zwangsfreien Ausgleichung des Hauptdreiecksnetzes besteht; in der Umgebung von München (Testnetz Oberbayern) ergaben sich im mittleren Entfernungsbercich die kleinsten Differenzen gegenüber den Strecken des Gebrauchsnetzes (BHDrN). Die Koordinatenstrecken der genannten Grundlagen-Netze sollen auch hier für eine Beurteilung der überlangen Strecken herangezogen werden. Die in Abb. 7 für die einzelnen Entfernungsbereiche dargestellten mittleren Fehler m_2—wiederum auf eine einmal in Hin- und Rückweg gemessene Strecke bezogen—streuen zwischen $\pm 2{,}0$ und $\pm 9{,}4$ mm/km; das gewogene Mittel erreicht

$$m_2 = \pm 5{,}1 \text{ mm/km} \, (1 : 196\,000)$$

[Nach einer zweiten bzw. dritten Messung bei verschiedenen meteorologischen Verhältnissen würde man erhalten:

$$(m_2)_{2M} = \pm 3{,}6 \text{ mm/km} \, (1 : 278\,000) \text{ bzw}$$

$$(m_2)_{3M} = \pm 2{,}9 \text{ mm/km} \, (1 : 345\,000)]$$

Wesentlich ist, daß die einzelnen Differenzen wechselndes Vorzeichen tragen; ein systematischer Restanteil beträgt im Testnetz Niedersachsen (—Vergleich mit der Zwangsfreien Ausgleichung des Hauptdreiecksnetzes—) $d = +1{,}5$ mm/km und in Testnetz Oberbayern (—Vergleich mit den Strecken des BHDrN—) $d = -0{,}7$ mm/km. Als größte Abweichung wird für eine Meßreihe auf einer 75 km langen Strecke $d = +12{,}5$ mm/km registriert.

Ein zweites Kriterium für die äußere Genauigkeit der überlangen Electrotape-Strecken kann aus den Koordinaten der in Abb. 1 und Abb. 2 gezeichneten Testnetze A abgeleitet werden; diese zwangsfrei ausgeglichenen Figuren enthalten nur Entfernungen zwischen 10 und 60 km. Sofern man die aus diesen Mittelstreckennetzen abgeleiteten Werte der überlangen Seiten den gemessenen Strecken gegenüberstellt, ergeben sich mittlere Fehler einer einmal im Hin- und Rückweg gemessenen Strecke, die in den einzelnen Entfernungsbereichen zwischen $\pm 2{,}1$ und $\pm 8{,}2$ mm/km variieren; das gewogene Mittel beträgt hier

$$m_3 = \pm 4{,}9 \text{ mm/km} \, (1 : 204\,000)$$

$$[(m_3)_{2M} = \pm 3{,}5 \text{ mm/km} \, (1 : 285\,000) \text{ bzw}$$

$$(m_3)_{3M} = \pm 2{,}8 \text{ mm/km} \, (1 : 357\,000)]$$

Das Maximum aller Differenzen erreicht bei dieser Gegenüberstellung $d = +17{,}6$ mm/km auf einer 78 km langen Strecke des Testnetzes Niedersachsen.

Festgehalten sei hier, daß im Testnetz Oberbayern kaum ein systematischer Restanteil verbleibt ($d = -0{,}3$ mm/km); im Testnetz Niedersachsen dagegen erreicht dieser Anteil jetzt $d = +3{,}5$ mm/km.

Die Ursache dieser Erscheinung muß in solchen systematischen Feh-
lereinflüssen gesucht werden, die aus der labilen Unterschicht der
Atmosphäre erwachsen; im Testnetz Niedersachsen führen mehr als
50 prozent der überlangen Strecken zu Stationen in ebenen Landschaften bzw. zu einem Bodenpunkt im Mittelgebirge.

Schließlich kann ein drittes Fehlermaß aus einer Ausgleichung des
ebenfalls in Abb. 1 skizzierten Langstreckennetzes (Testnetz C) hergeleitet werden; bis auf einige wenige Ausnahmen enthält diese Figur
nur überlange Entfernungen (13 Unbekannte, 7 Überbestimmungen).
Aus dem Gewichtseinheitsfehler dieser Ausgleichung ist über den
eingeführten Gewichtsansatz $p = p_n \cdot p_s$ (siehe [2]) für jeden der
Entfernungsbereiche ebenfalls ein mittlerer Fehler für eine einmal im
Hin- und Rückweg gemessene Strecke errechnet worden; als gewogenes
Mittel folgt

$$m_4 = \pm 2,8 \text{ mm/km} (1 : 357\,000)$$

$$[(m_4)_{2M} = \pm 2,0 \text{ mm/km} (1 : 200\,000) \text{ bzw}$$

$$(m_4)_{3M} = \pm 1,6 \text{ mm/km} (1 : 625\,000)]$$

Die hier abgeleiteten Beträge differieren von den m_3—Werten um ca.
4 mm/km; Ursache dierser großen Unterschiede dürften die bereits im
vorherigen Abschnitt erwähnten systematischen Verfälschungen der
überlangen Strecken im Testnetz Niedersachsen sein, die letztlich
meteorologisch bedingt sind.

Aus den oben dargestellten Ergebnissen folgt, daß eine einmal im
Hin- und Rückweg gemessene überlange Entfernung im allgemeinen
mit einem m.F. von

$$m_s = \pm 5 \text{ mm/km} (1 : 200\,000)$$

behaftet ist; sofern man solche Strecken in drei verschiedenen Witte-
rungsperioden beobachtet, wird mit

$$(m_s)_{3M} = \pm 2,9 \text{ mm/km} (1 : 345\,000)$$

eine relative Genauigkeit erzielt, die den Beobachtungsergebnissen auf
10–50 km langen Stracken entspricht. Sofern man bemüht ıst, die
meteorologischen Daten oberhalb der turbulenten Unterschicht der
Atmosphäre zu ermitteln, können demnach elektronisch gemessene
überlange Entfernungen mit gutem Erfolg zur Lösung der eingangs
erwähnten Aufgaben beitragen. Insbesondere sei festgehalten, daß auch
dort, wo es gilt großräumige bzw. weitmaschige Lagefestpunktfelder
zu bestimmen, mit Hilfe sogenannter Langstreckennetze die geforderte Genauigkeit erreicht werden kann.

Schrifttum

[1] SEEGER, H., 1965. Die Ergebnisse einer Erprobung des Electrotape DM-20 *Z.f.V.*, **7**, 222–31.

[2] SEEGER, H., 1965. Ein Beitrag zur elektronischen Streckenmessung mit 3 cm—Trägerwellen (10 GHz), insbesondere mit dem Electrotape DM-20 (Dissertation TH Hannover).

[3] LEITZ, H., 1965 Eine zweckmäßige Einrichtung zur Kontrolle der Modulationsfrequenzen bei elektronischen Entfernungsmessern, *Z.f.V.*, **4**.

[4] BROCKS, K., FENGLER, G. and JESKE, H., 1963. Models of the troposphere derived from direct measurements of the atmospheric refractive index (Institut für Radiometeorologie an der Universität Hamburg, Bericht Nr. 7).

[5] SCHÜNEMANN, R. und BULL, G., 1961. Über die Abhängigkeit der Fluktuationen des atmosphärischen Brechungsindexes von den meteorologischen Parametern. Ergebnisse von Messungen mit einem Mikrowellen-Refraktometer.(Mitteilung aus dem Heinrich-Hertz-Institut der Deutschen Akademie der Wissenschaften zu Berlin. Hochfrequenztechnik und Elektroakustik.)

[6] BROCKS, K., 1948. Über den täglichen und jährlichen Gang der Höhenabhängigkeit der Temperatur in den untersten 300 m der Atmosphäre und ihren Zusammenhang mit der Konvektion. (Berichte des Deutschen Wetterdienstes in der US-Zone, Nr. 5.)

[7] BAKKELID, S., 1964. Preliminary Results of Studies of some Tellurometer Problems. (Norges Geografiske Oppmaling, Geodetic Publication No. 14.)

[8] KLEEGREWE, C., 1959. Bau eines Wolkenradargerätes zur gleichzeitigen Messung bei 3,2 cm und 0,86 cm Wellenlänge. (DVL-Bericht Nr. 96.)

[9] BROCKS, K., 1959. Feldstärkeregistrierungen auf Richtfunkstrecken im cm- bis m-Band über der Nordsee und ihre Beziehungen zu meteorologischen Vorgängen. (Radiometeorologische Forschungsgruppe des Geophys. Instituts der Universität Hamburg, Bericht, Nr. 2.)

[10] RYDE, J. W., 1946. The Attenuation and Radar Echoes Produced at Centimetre Wave-Lengths by Various Meteorological Phenomena (*Meteorological Factors in Radio-Wave-Propagation*, London, The Physical Society).

[11] ROBERTSON, S. D. and KING, A. P., 1946. The Effect of Rain upon the Propagation of Waves in the 1- and 3-Centimetre Regions (Proceedings the IRE and Waves and Electrons, April 1946).

Propagation Problems

Microwave Reflection Problems

K. PODER and O. B. ANDERSEN

Danish Geodetic Institute

The paper shows the approaches used for deriving formulae describing the effects of reflections upon microwave distance measurements.

The approach is a geodetic one, and no deeper physical description of the reflection phenomenon is given, as the only aim is a reduction of distance measurements suffering from reflection errors.

The paper has five sections dealing with geometry, effective reflection coefficient, reflection error for simple and complex surfaces, optimum carrier wavelength and modulation wavelength, and examples.

Acknowledgement

This paper is published with the permission of the Director of the Royal Danish Geodetic Institute, Professor E. Andersen, D.Sc. We express our deep gratitude to Professor Andersen and State Geodesist E. Kejlsø, M.Sc., for their support of our work. Likewise, we would like to thank foreign and Danish colleagues for contributions by work and discussion.

1. *Introduction*

When microwaves are propagated in a beam over the surface of the Earth, part of the waves will be reflected from the surface. The resulting field therefore will be a compound one.

This is important for microwave distance measurements, where the distance is derived from the phase relations of propagated waves between two co-operating microwave transmitters. The basic assumption here is that the phase shifts by 2π for each path element of one wavelength.

If the phase of the direct waves is changed by the addition of reflected waves, the conclusions concerning the distance obviously become erroneous. Fortunately it is possible to produce a variation of the relative phase of the reflected waves. This circumstance and the fact that the reflected waves are smaller in amplitude than the direct ones will permit a conclusion concerning the phase of the direct signal. As the phase measurement takes place as a phase measurement of the modulation sidebands of a microwave carrier, there will be a complication because the mean value of these sideband phases will not be zero even if the mean value of the phase error of the carriers is zero, except for the case where the reflected waves have a very small amplitude.

The variation of the distance indication is thus periodic. With a crude approximation it is a sine curve, frequently termed the ground swing curve. Unfortunately, almost any variation of the indication versus the carrier frequency has frequently been interpreted as resulting from ground reflections.

We shall here show that ground reflections at least in the more severe cases can be revealed by a much more characteristic ground swing curve, and that the procedure of taking the mean value of the readings frequently will lead to an erroneous result because the reflection, in the more complex cases, will have a periodic variation which systematically is shifted away from the value corresponding to an idealized direct propagation.

The approach for the problem thus becomes as follows:

1. Determine the excess path length of the reflected signal. From this length, the phase of the reflected signal is known when the wavelength is known. This is not an easy problem, but as the variations of the wavelength are known, it will be easy to ensure that the phase has been varied one full period.

2. Find the reflection coefficient of the surface. From this, the amplitude of the reflected signal is known. The phase will mostly be shifted by *ca. π* at the reflection.

3. Find the compound signal and express its phase relation to the idealized direct signal. This phase relation is the reflection error.

The method will in most cases deliver satisfactory results for simple surfaces such as water, where the errors are felt most strongly. Examples of both success and failure of the method have been given by Gardiner-Hill (1963). For land surfaces, the value probably is of a more illustrative nature, because the theory shows that a strong regular variation of the readings presupposes certain excess path lengths and reflecting surfaces of a certain size and physical properties.

Finally it should be added that a far better approach to the whole problem seems to be a reduction of the wavelengths in order to obtain more narrow beams and mainly to obtain a diffuse reflection from the ground.

2. *The Geometry*

Fig. 1 shows the geometry of the reflection problem. The direct path is P_1P_2, and the indirect one is $P_1P_rP_2$. The excess length is the difference of these paths. The grazing angle g is an important quantity for the

reflection coefficient and besides it is a convenient intermediate parameter for computing the excess path length.

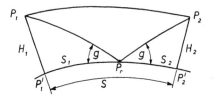

FIG. 1. Geometry of paths on a curved surface.

Let the heights of the terminals be H_1 and H_2, the ellipsoidal distance
between the stations be s, and the corresponding two distances for
$P'_1 P_r$ and $P'_2 P_r$ be s_1 and s_2, respectively. Then obviously

$$s = s_1 + s_2 \qquad (2.1)$$

If the two paths could be observed separately, the indications of the
instrument would be cT for the direct path and $c(T_1 + T_2)$ for the
reflected one, where $c = $ a standard propagation velocity of electromagnetic waves, and T, T_1, and $T_2 = $ the transit times for the paths
$P_1 P_2$, $P_1 P_r$, and $P_2 P_r$.

The expression for $\mathrm{d}l$ thus becomes

$$\mathrm{d}l = c(T_1 + T_2 - T) \qquad (2.2)$$

Reversing the general reduction formula from transit time to ellipsoidal distance (see e.g. Poder and Andersen 1963), the expression for
the transit time may be inserted in (2.2). For this purpose it will be
necessary to extrapolate a value of the refractive index at the formal terminal P_r. This requires that the vertical gradient of the refractive index
is determined or fixed at some empirical value. The value of this gradient can for most geodetic purposes be fixed to:

$$\mathrm{d}n/\mathrm{d}H = -0.040 \times 10^{-6} \qquad (2.3)$$

and this value is accepted here too.

By this means $\mathrm{d}l$ can be found with s as a parameter. Using the law
that the incident angle must equal the reflected angle and the simplest
possible formula for trigonometric levelling fictitiously carried out in
P_r, the expression for $\mathrm{d}l$ becomes

$$\mathrm{d}l = 2[(H_1 - s_1^2/2R_e)(H_2 - s_2^2/2R_e)]/s \qquad (2.4)$$

and the grazing angle is found by

$$\tan g = H_1/s_1 - s_1/2R_e = H_2/s_2 - s_2/2R_e \qquad (2.5)$$

where R_e is the effective radius of curvature of the Earth defined by

$$1/R_e = 1/R + dn/dH \qquad (2.6)$$

where R is the radius of curvature of the Earth.

The following algorithm will give dl and g in a simple way.

real procedure $dl(s, H_1, H_2, R_e, g)$;
   value s, H_1, H_2, R_e;
   real s, H_1, H_2, R_e, g;
   begin
   real $D, d, d\tan$;
         $s := d := s/2$;
         $D := 0$;
   for $d\tan := H_1/(s+D) - H_2/(s-D) - D/R_e$ **while** $\text{abs}(d\tan) > 10^{-6}$ **do**
   begin
         $d := d/2$;
         $D := D + d \times \text{sign}(d\tan)$
   end
         $H_1 := H_1 - (s+D) \uparrow 2/(2 \times R_e)$;
         $H_2 := H_2 - (s-D) \uparrow 2/(2 \times R_e)$;
         $g := \arctan\{[H_1/(s+D) + H2/(s-D)]/2\}$;
         $dl := H_1 \times H_2/s$
end

The procedure starts from the middle of the line and compares the tangent of the two angles at the ground. The point is then shifted in decreasing steps until the angles are equal within 10^{-6}.

3. *The Effective Reflection Coefficient*

The reflection coefficient for specular reflection is easily derived from Fresnel's equations for the reflection coefficients of polarized electromagnetic waves.

The incident polarized signal is split into vertical and horizontal components, which are multiplied by the respective reflection coefficients to give the components of the reflected signal. The resulting component of these in the plane of polarization will then be the reflected signal.

For an inclination of 45° of the plane of polarization of the incident signal, the reflection coefficient will be the mean value of the reflection coefficients (regarded as complex quantities).

If the dielectric constant of the reflecting surface, ϵ, has a modulus greatly exceeding unity, the reflection coefficient can be approximated by

$$a_t \exp jr = -1/(1 + \sqrt{\epsilon} \sin g) \qquad (3.1)$$

where $\epsilon = 69 - j39$ for sea water (j is the imaginary unit). In this case, the modulus and argument of the reflection coefficient become

$$a_t = 1/\sqrt{(1 + 17 \cdot 2 \sin g + 79 \sin^2 g)} \qquad (3.2)$$

$$r = \pi - 2 \cdot 5 \sin g$$

For fresh water, the dielectric constant has a slightly smaller imaginary term and for ordinary ground both terms mostly will be smaller. For very dry ground, the assumption of a large modulus is not satisfied and (3.2) must be replaced by more elaborate expressions.

The divergence factor introduced by Bremmer and Van der Pol (1939) is used for taking into account that the Earth and the paths are curved. This will give an increase of the divergence of the rays describing the propagation of the signal. The net effect will be a reduction of the amplitude of the reflected signal.

The divergence factor can be found by a comparison of the areas at one terminal corresponding to a given spatial angular element at the other terminal in the two cases of curved Earth with curved paths and plane Earth with straight paths.

The divergence factor can be expressed as:

$$D = \sqrt{\{1/[1 + 2s_1 s_2/R_e(H_1 - s_1^2/2R_e + H_2 - s_2^2/2R_e)]\}} \qquad (3.3)$$

$$= \sqrt{\{1/[1 + dl/R_e g^3]\}}$$

The effective reflection coefficient thus becomes:

$$a = a_t D \qquad (3.4)$$

and the phase shift, at reflection, is approximately π.

Roughness of the surface will reduce the reflection considerably if the surface irregularities are sufficiently large, because there will then be a significant phase difference of the waves reflected from the top of the irregularities and those reflected from the bottom. It is evident that very small grazing angles will mean that the difference will be small. Rayleigh's criterion for surface roughness suggests that a surface may be considered rough when the phase difference between top and bottom

4

is $\pi/2$. This means that the irregularity times the sine of the grazing angle must exceed one eighth of a wavelength. It has been found that this criterion may be too severe and that somewhat smaller irregularities are sufficient (see e.g., Kerr 1951, pp. 411–6).

4. *The Reflection Formula*

In section 2, the approach for finding the phase relations of the direct signal and the reflected one is given, and in section 3 it is shown how the relative amplitude of the reflected signal to the direct signal is found. The sum signal $e = e_d + e_r$ will be derived here.

4.1 *The Direct Signal*

The direct signal from the transmitter can be written as:

$$\exp(jwt) + m \exp[j(w+W)t] + m_1 \exp[j(w-W)t] \qquad (4.1)$$

where

$\quad j =$ the imaginary unit
$\quad w =$ the angular carrier frequency
$\quad W =$ the angular measuring frequency (modulated on the carrier)
$\quad t =$ the time
$\quad m =$ half the generalized modulation coefficient of the upper sideband
$\quad m_1 =$ half the generalized modulation coefficient of the lower sideband

If $m_1 = -m$, the modulation is frequency modulation. In this case, higher-order sidebands will exist, but for small values of m they may be neglected, not least because they do not participate in the measuring process. If $m_1 = m$, the modulation is amplitude modulation.

In (4.1), the voltages are assumed normalized and only the real part of the expression is used.

After propagation to the other terminal along the direct path with a transit time of T the signal becomes

$$\begin{aligned} e_d &= \exp jw(t-T) + m \exp j(w+W)(t-T) \\ &\quad + m_1 \exp j(w-W)(t-T) \\ &= \exp jw(t-T)[1 + m \exp jW(t-T) \\ &\quad + m_1 \exp -jW(t-T)] \end{aligned} \qquad (4.2)$$

The desired phase information to be used is the relative phase of the modulation $-WT$. It is the function of the receiver to utilize this information and produce the indication at the master and the reference signal at the remote.

4.2 *The Reflected Signal*

The reflected signal has an increased transit time of:

$$Q = \mathrm{d}l/c \qquad (4.3)$$

and also an additional phase shift of r given by (3.2). The amplitude is multiplied by a, which is the product of the modulus of the reflection coefficient and the divergence factor.

The expression for the reflected signal thus is:

$$
\begin{aligned}
e_r &= a \exp j[w(t-T-Q)-r] + am \exp j[(w+W)(t-T-Q)-r] \\
&\qquad + am_1 \exp j[(w-W)(t-T-Q)-r] \\
&= a\{\exp j[w(t-T)-z] + m \exp j[(w+W)(t-T)-z-Z] \\
&\qquad + m_1 \exp j[(w-W)(t-T)-z+Z]\}
\end{aligned}
\qquad (4.4)
$$

where

$$ z = wQ + r \quad \text{and} \quad Z = WQ \qquad (4.5) $$

As z and Z always will be found as arguments of sine, cosine, or the complex exponential function they can be regarded modulo 2π.

4.3 *The Sum Signal*

The sum signal is the vector sum of e_d and e_r found by adding and taking the real part:

$$
\begin{aligned}
e &= \exp jw(t-T)(1 + a \exp -jz) \\
&\quad + m \exp j(w+W)(t-T)[1 + a \exp -j(z+Z)] \\
&\quad + m_1 \exp j(w-W)(t-T)[1 + a \exp -j(z-Z)] \\
&= \exp jw(t-T)(1 + a \exp -jz)\{1 + m \exp jW(t-T) \\
&\quad \times [1 + a \exp -j(z+Z)]/(1 + a \exp -jz) \\
&\quad + m_1 \exp -jW(t-T) \\
&\quad \times [1 + a \exp -j(z-Z)]/(1 + a \exp -jz)\}
\end{aligned}
\qquad (4.6)
$$

Introducing

$$
\begin{aligned}
n_1 + n_2 &= m[1 + a \exp -j(z+Z)]/(1 + a \exp -jz) \\
n_1{}^* - n_2{}^* &= m_1[1 + a \exp -j(z-Z)]/(1 + a \exp -jz)
\end{aligned}
\qquad (4.7)
$$

where the asterisk denotes the complex conjugate, the sum signal may be written as

$$
\begin{aligned}
e &= \exp jw(t-T)(1 + a \exp -jz) \\
&\quad [1 + n_1 \exp jW(t-T) + n_1{}^* \exp -jW(t-T) \\
&\quad + n_2 \exp jW(t-T) - n_2{}^* \exp -jW(t-T)]
\end{aligned}
\qquad (4.8)
$$

As n_1 and n_2 may be written as

$$n_1 = \bmod n_1 \exp (j \arg n_1)$$
$$n_2 = \bmod n_2 \exp (j \arg n_2) \tag{4.9}$$

where mod n means the modulus of n and arg n its argument, it is seen that the signal now is simultaneously amplitude modulated with n_1 as half the modulation degree and frequency modulated with n_2 as half the modulation index (approximately). The phase errors are arg n_1 and arg n_2, respectively.

Let the intended modulation be frequency modulation. This means that this modulation will produce the desired internal signals for phase indication and reference signal. Furthermore, it will be assumed that the unintended modulation will not be detected by the second detector inside the instrument. This is rather well satisfied because detectors intended for one type of modulation only will mostly be poor detectors for the other type.

In the case of frequency modulation $m_1 = -m$, and n_1 and n_2 become:

$$n_1 = (m/2)[\,\mathrm{1} + a \exp -j(z+Z)]/(\mathrm{1} + a \exp -jz)$$
$$\quad - (m/2)[\,\mathrm{1} + a \exp j(z-Z)]/(\mathrm{1} + a \exp jz)$$
$$n_2 = (m/2)[(\mathrm{1} + a \exp -j(z+Z)]/(\mathrm{1} + a \exp -jz)$$
$$\quad + (m/2)[(\mathrm{1} + a \exp j(z-Z)]/(\mathrm{1} + a \exp jz) \tag{4.10}$$

which may be reduced to:

$$n_1 = jm(\mathrm{1} - \exp -jZ)B$$
$$n_2 = m[\mathrm{1} + \exp (-jZ - \mathrm{1})]A, \tag{4.11}$$

where

$$A = A(z,a) = a(a + \cos z)/(\mathrm{1} + a^2 + 2a \cos z) \quad \text{and}$$
$$B = B(z,a) = a \sin z/(\mathrm{1} + a^2 + 2a \cos z) \tag{4.12}$$

From this follows:

$$\arg n_1 = \operatorname{arc} \tan (\mathrm{1} - \cos Z)/\sin Z = Z/2 \quad \text{and}$$
$$\arg n_2 = -\operatorname{arc} \tan A \sin Z/[\mathrm{1} - A(\mathrm{1} - \cos Z)] \tag{4.13}$$

Writing arg n_2 as $-WX$, it is seen that the actual phase information is not the desired $-WT$, but $-W(T+X)$. The parameters of X are Z (which depends on the ratio of excess path length to modulation wavelength) and A (which depends on the ratio of excess path to carrier wavelength and on the effective reflection coefficient).

The variation of X is sinusoidal only for very small values of a, where A is roughly sinusoidal. Table 1 (values of A) shows how the variations are in the case of greater reflection coefficients. The curve tends to be almost flat for the greater part of the cycle and only for values of z near π is a large excursion found. The values of a exceeding unity are not physically realizable with a single reflector, but, as will be shown in section 4.5, the presence of more than one reflecting surface may effectively increase the reflected signal in such a way that the reflected signal exceeds the direct signal in strength.

The amount of unintentional modulation is characterized by n_1. The phase of this signal is permanently changed by $Z/2$, but for z equal to any integer multiple of π, this modulation disappears completely. A perusal of the deductions above will show that if the intended modulation was amplitude modulation, then this modulation would have the error of arg n_2, and arg n_1 would give the error of the now undesired frequency modulation.

The phase errors constitute, of course, the main effect of the reflections but, apart from this, the amount of information also is affected. The moduli of the modulation coefficients are:

$$\text{mod } n_1 = 2mB \sin Z/2$$
$$\text{mod } n_2 = m\sqrt{[1 - 4A(1 - A)\sin^2(Z/2)]} \qquad (4.14)$$

The content of unintended modulation is thus rather insignificant for shorter excess path lengths. For the intended modulation, the information for small values of A is almost unchanged. For greater excess paths, it will be realized that the modulation will depend strongly on the actual values of A and B.

4.4 *The Reflection Error for a Set of Instruments*

The reflection error of the phase of a primary signal can be found from (4.13). This means that the phase error is determined as the angle

$$\text{arg } n_2 = -WX \qquad (4.15)$$

where W may be taken as the modulation frequency of the master in all cases, because the remote frequency is very little different from the former. This does not imply that the errors are equal for master and remote, as the carrier frequencies are different, resulting in different values of A being used in (4.13).

As the carrier frequencies are maintained to have a constant difference equal to the intermediate frequency, it means that the two values of z must satisfy

$$\text{abs } (z_R - z_M) = bQ \qquad (4.16)$$

where b is the angular intermediate frequency, as will be seen from (4.4). The arguments

$$z - bQ/2 \quad \text{and} \quad z + bQ/2 \qquad (4.17)$$

will give corresponding values of A to be used for finding the two phase errors.

The display of phase at the master is without reflection errors

$$S = \pm(W_R T + W_M T) + \text{abs}(W_M - W_R)T \qquad (4.18)$$

where W_M and W_R are the modulation frequencies of master and remote, respectively. The plus sign is valid when the remote frequency is smaller than the master frequency and conversely for the minus sign. When reflection errors are present, all three values of T in (4.18) are changed and (4.18) is replaced by:

$$S = \pm(W_R T + W X_R + W_M T + W X_M)$$
$$+ \text{abs}(W_M - W_R)(T + X_{\text{ref}}) \qquad (4.19)$$

where X_R and X_M are the errors for remote and master, respectively, found from (4.13) and (4.15) with the arguments from (4.17). As the difference of a positive and a negative display is used finally, it is seen that the error of the reference signal X_{ref} is cancelled. The difference then is

$$4 W_M[T + (X_M + X_R)/2] \qquad (4.20)$$

as the mean value of the remote frequencies equals the master frequency and W is made equal to W_M.

The displays of different types of instruments are rather different, and it is therefore natural to express the reflection error in units of length. As (4.13) gives WX directly in angular units, the reflection error in length units is

$$E = c(W X_M + W X_R)/2W = (W X_M + W X_R)L/4\pi \qquad (4.21)$$

where L is the modulation wavelength.

The reflection error for a set of instruments thus will be the mean value of the reflection error for each instrument. This means that the curve representing the reflection error is rather flat for the greater part of a full cycle but with two (mostly) negative excursions, as shown in Fig. 2. The negative excursions are separated in relation to the intermediate frequency, as will be seen from (4.17).

4.5 *The Reflection Formula for two Reflecting Surfaces*

If more than a single reflecting surface is present, the sum signal will be the sum of the direct signal and all the reflected signals. It is easy to derive an expression for several reflected signals from (4.10) by replacing the reflected signal terms with sums.

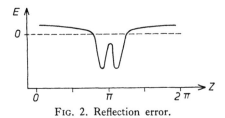

FIG. 2. Reflection error.

The special case with one surface with a high reflection coefficient a and the parameters z and Z, and one surface with low reflection coefficient a_1 and the parameters z_1 and Z_1 should be considered here. This will be found in practice with a water surface as the main reflector and the ground near one of the stations as the other reflecting surface. If the excess path of the latter is short, z_1 will vary only slowly compared with z, and for Z_1 the approximation

$$\exp(-jZ_1) = 1 - j\sin Z_1 \qquad (4.22)$$

may be used.

The conditions may be illustrated in two typical cases of z_1, viz. $z_1 = 0$ and π. For $z_1 = 0$, n_2 becomes

$$n_2 = m\{1 + [\exp(-jZ) - 1]A(z,a^1)$$
$$- ja_1 \sin Z_1(1 + a_1 + a\cos z)\}/(1 + a_1)^2(1 + a^2 + 2a\cos z) \qquad (4.23)$$

where

$$a^1 = a/(1 + a_1) \qquad (4.24)$$

When $\sin Z_1$ and a_1 are sufficiently small, the term with the factor j may be neglected, and the result becomes that the reflection effect appears as due to the main reflector with a reflection coefficient equal to the real coefficient divided by the factor $1 + a_1$. In the case of $z_1 = \pi$, the effective reflection coefficient is a divided by $1 - a_1$. This means that the reflection coefficient may exceed unity even for a weak reflection from a surface near one terminal of the line.

From this it will appear that when severe reflection effects may be expected, it will be important for a simple reduction that the instruments are well clear of the nearby ground. Attempts to use screening are more likely to give complications, because there will be a considerable diffraction round the edges of the screen. The estimation of the amount

TABLE 1. *Values of* $A = a[\cos(z)+a]/[1+2a\cos(z)+a^2]$

z *in fractions of* π

a	0·0	0·1	0·2	0·3	0·4	0·5	0·6	0·7	0·8	0·9	1·0
0·1	0·09	0·09	0·08	0·06	0·04	0·01	−0·02	−0·05	−0·08	−0·10	−0·11
0·2	0·17	0·16	0·15	0·12	0·09	0·04	−0·02	−0·10	−0·17	−0·23	−0·25
0·3	0·23	0·23	0·21	0·18	0·14	0·08	0·00	−0·12	−0·25	−0·38	−0·43
0·4	0·29	0·28	0·27	0·24	0·20	0·14	0·04	−0·11	−0·32	−0·55	−0·67
0·5	0·33	0·33	0·32	0·30	0·26	0·20	0·10	−0·07	−0·35	−0·75	−1·00
0·6	0·37	0·37	0·36	0·35	0·32	0·26	0·18	0·01	−0·32	−0·96	−1·50
0·7	0·41	0·41	0·40	0·39	0·37	0·33	0·26	0·12	−0·21	−1·11	−2·33
0·8	0·44	0·44	0·44	0·43	0·42	0·39	0·34	0·24	−0·02	−1·02	−4·00
0·9	0·47	0·47	0·47	0·47	0·46	0·45	0·42	0·37	0·23	−0·47	−9·00
1·0	0·50	0·50	0·50	0·50	0·50	0·50	0·50	0·50	0·50	0·50	undef.
1·1	0·52	0·52	0·53	0·53	0·54	0·55	0·57	0·61	0·74	1·39	11·00
1·2	0·55	0·55	0·55	0·56	0·57	0·59	0·63	0·71	0·94	1·90	6·00
1·3	0·57	0·57	0·57	0·58	0·60	0·63	0·68	0·80	1·09	2·09	4·33

TABLE 2. *Value of $B = a \sin(x)/[1 + 2a \cos(x) + a^2]$*

x in fractions of π

a	0·0	0·1	0·2	0·3	0·4	0·5	0·6	0·7	0·8	0·9	1·0
0·1	0·00	0·03	0·05	0·07	0·09	0·10	0·10	0·09	0·07	0·04	0·00
0·2	0·00	0·04	0·09	0·13	0·16	0·19	0·21	0·20	0·16	0·09	0·00
0·3	0·00	0·06	0·11	0·17	0·22	0·28	0·32	0·33	0·29	0·18	0·00
0·4	0·00	0·06	0·13	0·20	0·27	0·34	0·42	0·47	0·46	0·31	0·00
0·5	0·00	0·07	0·14	0·22	0·31	0·40	0·51	0·61	0·67	0·52	0·00
0·6	0·00	0·07	0·15	0·24	0·33	0·44	0·58	0·74	0·91	0·85	0·00
0·7	0·00	0·08	0·16	0·24	0·35	0·47	0·63	0·85	1·15	1·36	0·00
0·8	0·00	0·08	0·16	0·25	0·36	0·49	0·66	0·93	1·36	2·09	0·00
0·9	0·00	0·08	0·16	0·25	0·36	0·50	0·68	0·97	1·50	2·84	0·00
1·0	0·00	0·08	0·16	0·25	0·36	0·50	0·69	0·98	1·54	3·16	undef.
1·1	0·00	0·08	0·16	0·25	0·36	0·50	0·68	0·97	1·50	2·89	0·00
1·2	0·00	0·08	0·16	0·25	0·36	0·49	0·67	0·94	1·42	2·35	0·00
1·3	0·00	0·08	0·16	0·25	0·35	0·48	0·66	0·91	1·30	1·85	0·00

of energy arriving in this way will be more difficult, and the additional signal reflected from the screen will be a further complication.

4.6 *The Reduction of the Readings*

The reduction of the readings is easy in the simple cases. Knowing the excess path length and the effective reflection coefficient, the reflection error is computed for one full period of the error for two instruments. The computed error is then subtracted from the portions of the observed curve where it is almost flat and steady. This means that readings in the peak areas should be avoided, whether they are positive or negative. A reduction programme is available in GIER–ALGOL, based upon the algorithms given by Poder and Andersen (1963). Gardiner-Hill (1963) gives several examples and valuable advice for the reduction work.

5. *Geodetic Precautions against Reflections*

The effect of reflections may—as mentioned here—be accounted for by corrections. However, the network and its determination may be planned with regard to reflections.

Firstly, the layout of the network may be designed in such a way that lines over water with long excess paths are avoided. A reasonably short excess path and a mainly low divergence factor may be obtained by means of auxiliary stations situated at suitable heights. The azimuth is then carried directly between the desired stations and the scale is carried via the auxiliary stations. The over-water lines may even be avoided by bridging the water with one or more quadrilaterals and measuring the lengths of the inland sides only.

Secondly, if over-water lines cannot be avoided, the net form must be such that several sides are measured to give a high redundancy of the scale determination permitting the worst cases of reflection to be detected and if necessary rejected.

References

Bremmer, H. and van der Pol, B., 1939. Further note on the propagation of radio waves over a finitely conducting spherical earth. *Phil. Mag.* (7th Ser.), **27**, No. 182.

Gardiner-Hill, R. C., 1963. Refinements of precise distance measurement by Tellurometer, *Conference of Commonwealth Survey Officers*, Cambridge.

Kerr, D. E., 1951. *Propagation of Short Radio Waves*, London and New York.

Poder, K., 1962. Reflections. *Tellurometer Symp.*, London.

Poder, K. and Andersen, O. B., 1962. Results and experiences of electronic distance measurements. IUGG 13, Gen. Ass. Berkeley, Copenhagen.

DISCUSSION

R. C. A. Edge: Does this represent a fundamental change in your theories on this subject?

K. Poder: Not really. My advice is to keep the instrument clear of the near ground, because the ground may change the effective reflection coefficient.

R. C. A. Edge: Also your advice is not to use the positive hump in the swing curve?

K. Poder: To some extent, yes. You should subtract from the flat portion of the swing curve the excess path length.

R. C. A. Edge: You are assuming this half excess path length from the geometry of the swing curve rather than the geometry of the ground?

K. Poder: Yes, but only if the geometry of the ground cannot be determined.

A. Marussi: If we are given a point source, and a distant misty surface which provides reflection and refraction, we can look at the problem of trying to describe the phenomenon by considering the virtual images of the source signalled by the distant misty surface itself.

If the surface be a plane, we would only have to consider a point-like image, but in general the correspondence between the point source and its virtual images will not be astigmatic, and we will have to deal with a continuous distribution of sources or a source-density function. This distribution will be, in general, two-dimensional. The study of reflections and refractions can therefore be reduced to the study of the density function of this distribution.

K. Poder: I had not considered these points. They do represent a useful way of approaching the problem.

A. Marussi: Also dn/dH is strictly related to the curvature of the path, that is to say it is a measurement of the zenith distance of the distant station. Is such a value of importance in determining the refractive index?

K. Poder: If one acquires a value for dn/dH by zenith distances, to this must be applied a correction for humidity effects. Such a value might be of use for electro-optical systems, but would be of limited value for microwave systems.

S. K. Sharma: The point source described by Prof. Marussi and Dr Poder is in fact a source on a spherical surface. Does this not affect Prof. Marussi's suggestion of a series of points along a vertical line?

A. Marussi: The distribution of sources can be imagined on whatever surface we like; in practice, there will be an obvious advantage considering the distribution over a surface containing the original source, and in particular the vertical plane normal to the horizontal direction of the line under consideration.

K. Poder: Yes, I agree with this suggestion.

J. Kelsey: Has Dr Poder put his theories to a practical use in correcting observations made in the field, or has he restricted his work to deciding which lines to observe?

K. Poder: I have only corrected one line in Denmark which was over water. This line had strong reflection characteristics. Otherwise we have used it for observations made in Greenland.

Some Aspects of the Meteorology and Refractive Index of the Air near the Earth's Surface

G. D. ROBINSON

Meteorological Office, Bracknell, Berkshire

The factors which determine the density and water content of the air very near the ground and their change with height are discussed. Examples are shown of mean conditions and fluctuations in various meteorological situations, including the effect of discontinuities in the nature of the surface.

Electromagnetic wave propagation in the atmosphere is affected by air density and water content in the transmission path. In the lower layers of air, these are controlled by the transfer of heat and water vapour between the atmosphere and the Earth's surface. A sensitive instrument near the Earth's surface records fluctuations of temperature, humidity and wind, and to obtain a smooth mapping of average conditions observations are usually meaned, instrumentally or arithmetically, for a period generally between 5 and 30 minutes.

During daylight, the ground absorbs solar radiation and the long-wave radiation from the atmosphere, and emits long-wave radiation appropriate to its temperature (its emissivity is always high and usually near 1.0). At night, it exchanges long-wave radiation with the atmosphere and space. In general, it gains heat by radiation from a time a little after sunrise to one a little before sunset, and loses heat at other times. Neglecting the effects of storage in the ground—which is not very significant in the present context—the surface must lose heat to the atmosphere by convective processes during the day, and gain heat by these processes at night. Convective heat transfer may occur by transfer of 'sensible' heat—associated with temperature fluctuations and potential temperature gradients—or by transfer of latent heat, i.e., by evaporation of water, associated with humidity fluctuations and humidity gradients.

The flux of 'sensible' heat through a constant pressure surface may be written as

$$F_H^{\bullet} = \overline{c_p \rho w T} \simeq c_p \rho \overline{w' T'} \tag{1}$$

where ρ is air density,

c_p is specific heat at constant pressure,

w is the 'instantaneous' vertical wind component at a point,

T is the 'instantaneous' temperature at the same point at the same time; the bar denotes an average over space and time,

w', T' are deviations of the 'instantaneous' values from an average value over space and time.

Similarly the flux of water vapour may be written as

$$F_e \simeq \overline{\rho w' e'} \tag{2}$$

where e is the concentration of water vapour. The relations between the fluxes and the corresponding gradients are usually written

$$F_H = c_p \rho K_H(z) \frac{\mathrm{d}\theta}{\mathrm{d}z} \tag{3}$$

$$F_e = \rho K_e(z) \frac{\mathrm{d}e}{\mathrm{d}z} \tag{4}$$

the K terms being eddy diffusion coefficients for the appropriate quantity, so defined, θ potential temperature and z height.

The problem of specifying the temperature and humidity distribution near the ground is thus, most significantly, a problem in free and forced convection, of which there is no satisfying theoretical treatment. Although certain useful relations, applicable when time-mean conditions are horizontally homogeneous, have been established by dimensional methods, we must rely on empirical generalization from the numerous observations which have been made in special cases. Meteorologists investigating near-surface conditions usually try to find horizontal homogeneity, which means a surface whose thermal and mechanical properties are constant over distances many times (some observations suggest hundreds of times) the height at which observations are being made. In practice, electromagnetic distance measurements will very rarely be made over a meteorologically homogeneous surface—the only condition which would allow the refractive index along the path to be specified, even by empirical methods, from observations at one or two points. Thus the meteorological correction can be dealt with only by understanding it sufficiently to work in weather conditions which minimize it, or by choosing methods which reduce it to a second order small quantity involving dispersion rather than refractive index itself.

Fig. 1 illustrates some aspects of the mean profiles of temperature and humidity near the ground (Rider, Philip & Bradley 1963). The series of measurements was made over irrigated mown grass at, and upwind of, the edge of a substantial area of tarmac. Over the dry area, there is no evaporation and zero humidity gradient, with a strong lapse of temperature. Over the grass, evaporation is so strong that the surface temperature cools below free air temperature, 'sensible heat' is transported downwards by convection, and a temperature inversion is created. The approximate changes in refractive index in 16 m of horizontal path are:

	At 5 cm	At 50 cm
Light	5 N units	1 N unit
Cm waves	25 N units	6 N units

At normal meteorological screen height, the change is about half that at 50 cm. Similar measurements on a larger scale have been made by Dyer and Crawford (1965) who observed upwind of a sharp transition

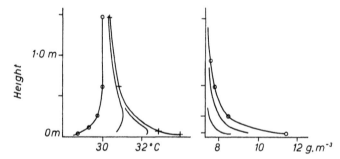

FIG. 1. Changes of mean temperature and humidity at a sharp surface discontinuity between tarmac and irrigated lawn. Readings at 0 m (+), 1 m, 4 m and 16 m (o) upwind of the boundary. (Rider, Philip and Bradley, Canberra, Australia, 13.30 hr, 11 January 1961.)

between arid and irrigated ground. After 200 m of horizontal path, they found temperature changes of up to 1°C at a height of 5 m—approximately a linear scaling of Rider, Philip and Bradley's results. The surface changes in these observations are extreme, being effectively equivalent to the changes from desert to open water at noon. More usual causes of horizontal inhomogeneity in the air near the surface are differences in radiative properties of the surface or crop—mainly in the reflectivity for solar radiation—the heat capacity and thermal conductivity of soils, and varying restriction of the availability of water for evaporation or transpiration, short of complete aridity.

At night, these surface differences have less effect, because the total heat transport involved is less, so long as the wind is not too light, but if the mean wind measured at 1 m falls below about 1 m sec^{-1} and the sky clears to allow maximum radiative heat loss, considerable horizontal inhomogeneities—of several °C in near-saturated air—may develop in consequence of differences in heat capacity and thermal conductivity of the surface layers of the Earth, which in turn are very dependent on its water content.

At heights above about 100 m local differences in vertical structure of the atmosphere are small, except for coastal effects (which may be carried many miles from the actual coastline in the form of 'sea-breeze fronts'). Fig. 2, the mean of several observations in similar weather con-

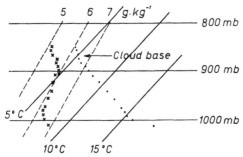

FIG. 2. Mean values of aircraft observations of temperature and humidity below small cumulus clouds. (Grant; Pershore, Worcestershire, May–July 1962.)

ditions reported by Grant (1965), illustrates conditions inland on a sunny summer day in southern England. The lowest aircraft observation is at a height of about 150 m; above this, the atmosphere has the temperature structure appropriate to complete mixing (the 'dry adiabatic lapse rate') and a slowly decreasing water vapour concentration. The local surface effects are made evident by the 'superadiabatic lapse' and large humidity gradient below the lowest aircraft observation. Conditions in the region between about 10 m and about 100 m have not been extensively observed. Some aspects have recently been discussed by Webb (1964).

We turn now from consideration of mean structure to consideration of fluctuations from mean conditions, which we have seen to be an inescapable consequence of transfer of heat and water. Fig. 3 shows how large the effect can be close to the ground; the extreme range of temperature in observations made 25 cm above an extensive lawn during a 30-minute period near noon on a cloudless summer day in southern England is as much as 6°C. This is by no means an unusual case.

Equation (1) may be written

$$F_H = c_p \rho r_w \, T \cdot \sigma_w \sigma_T \tag{5}$$

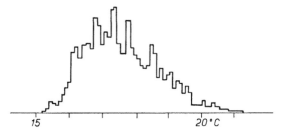

FIG. 3. Distribution of 1200 instantaneous readings of temperature sampled over a 30-minute interval 0·25 m above a lawn. (Lander and Robinson; Kew, Surrey, 2 June 1951.)

the σ terms being standard deviations and r a correlation coefficient, with a similar expression for equation (2). Observation has shown the correlation coefficients to have values of order $\frac{1}{3}$; reasonable values for heat flux and evaporation at summer noon in sunshine in the British Isles then lead to values of σ_r of about 1°C and of σ_e of about 1 g kg^{-1}. These correspond to standard deviations of refractive index of order 1 N unit for light and 5 N units for cm waves. The variation with height of the standard deviation is small, at least up to heights of order 100 m, but the dominant scale of the fluctuation changes with height.

Scale is often specified in terms of a spectral function. If we assume we have an instantaneous map of the temperature field at a given height over a large area, we may determine the variance of temperature. We may also subject the field to a form of Fourier analysis and determine the amount of variance at any given wave number (k) over the whole range of wave numbers. This is the spectrum function $S(k)$, the total variance being $\int S(k) \, dk$. Observations are usually made in terms of time-series of temperature readings at a point, and there are conceptual and practical difficulties, which we must ignore here, in relating these to a synoptic mapping of observations. We find in practice that the value of $kS(k)$ has a maximum which changes with height above the ground, and that, in the first 100–200 m, the dominant scale of disturbance has dimensions roughly equal to the height above the ground.

Most of the observations on which this generalization is based have been made in winds of order a few m sec^{-1} or more. It may not hold in calm conditions near the surface, nor at heights greater than about 100 m when there is a very large heat flux. These 'free convection' conditions form the main subject of Webb's (1964) article, but knowledge of their nature is still rather speculative, and generalization may be dangerous. Fig. 4 is an illustration taken from Grant (1965) of the nature of the fluctuation of temperature and humidity in 'free convection' conditions up to cloud base. It shows the magnitude of the refractive index gradients which may occur, and illustrates the danger of applying ideas on the spectrum of turbulent fluctuations which have emerged in the context of observations in forced convection conditions near the surface. Recent observations in extensive temperature inversions have emphasized the same point (Lane 1964).

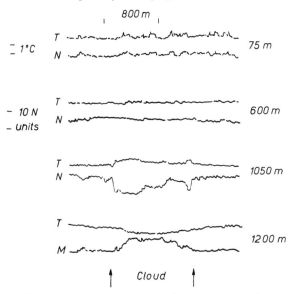

FIG. 4. Instantaneous measurements of temperature and refractive index during aircraft traverses at different heights below a small cumulus cloud. (Grant; Pershore, Worcestershire, Noon, 30 May 1962.)

This incomplete and rapid survey does not lead to any comforting conclusions. The refractive index of the air will be least liable to variation in time and space when the vertical flux of heat and water vapour is small. Near certain coastlines and irrigated areas, these conditions can never be found. Elsewhere they occur with overcast skies or near sunrise and sunset, and are least predictable in light winds. The most extreme surface effects can be avoided by working as little as 10 cm away from the surface and by making meteorological observations at this height. Meteorologists would not in general care to estimate mean refractive index along a transmission path to an accuracy of one or two N units, even given quite extensive observations, and are surprised when consistent distance measurements show that this can sometimes be done.

References

Dyer, A. J. and Crawford, T. V., 1965. *Quart. J.R. met. Soc.*, **91**, 345–8.
Grant, D. R., 1965. *Quart. J.R. met. Soc.*, **91**, 268–81.
Lane, J. A., 1964. *Nature, Lond.*, **204**, 438–40.
Rider, N. E., Philip, J. R. and Bradley, E. F., 1963. *Quart. J.R. met. Soc.*, **89**, 507–31.
Webb, E. K., 1964. *Appl. Optics*, **3**, 1329–36.

DISCUSSION

A. R. Robbins: Is it fair to say that you recommend surveyors to take EDM measurements at night when the wind strength is not less than 1 m sec^{-1} and also to ensure that the ray path is never less than 10 m from the ground?

G. D. Robinson: I would recommend conditions in which the geostrophic wind strength exceeded 20 miles per hour. An overcast day is perfectly suitable and much better than a sunny day.

R. C. A. Edge: The Caithness Base investigation [see pp. 25–44], showed that if the measuring instrument was placed on a 30-ft tower, there were undesirable reflection effects. Would you recommend elevating only the meteorological instruments to a height of say 10 m?

G. D. Robinson: Yes, the important thing is to obtain the meteorological conditions at a height of 10 m. The ray path is not likely to be always near the ground.

R. C. A. Edge: As this implies the higher the ray path the better, might not a systematic effect be introduced?

G. D. Robinson: The best way of measuring refractive index is to shine a light through the atmosphere. I would rather give a mean value for a ray path at 1000 ft above the ground than for a ray path at 10 ft.

R. C. A. Edge: From meteorological observations at ground level terminals?

G. D. Robinson: No; I am very suspicious of observations at or near ground level. 10 m up is a reasonable height to reduce the errors to 1°C in temperature and 1 millibar in pressure.

R. C. A. Edge: We seek an accuracy of 1 part per million. This requires a knowledge of the refractive index to about 1 N unit. At the moment it seems we are nearer 5 N units.

J. C. Owens: Is the figure of 5 N units the average over the path or the value at a point? How are we to get accurate refractive index at a point? What method would the meteorologist use to get humidity?

G. D. Robinson: The best measures of humidity are made with radio refractometers. A single measure is then much better than 5 N units. I could get 5 N with a selection of meteorological data at the ends of a line and a weather chart.

A. Marussi: We as geodesists put our faith in statistics. We hope that, by taking the average, errors become harmless. In former times we were concerned with the gradient of refractive index which affected our angular work. Here we relied on statistics and averaged our measurements. Frequently I found that when observing conditions seemed ideal the results were poor.

R. J. Hewitt: What criteria would you suggest for good observing? Should the temperature be as low as possible to reduce absolute humidity?

G. D. Robinson: I was expressing criteria for estimating refractive index. I would not dare to express criteria for good observing. Overcast conditions with a fairly good wind enable a good estimate of refractive index to be made.

K. Poder: If we can get refractive index to within $5 N$ to $10 N$ this is encouraging because this error will only act on small portions of the path.

G. D. Robinson: There is always a danger in estimating refractive index from local observations.

J. C. Owens: Where the conditions of measurement are mountainous there can be no certainty that the systematic errors are eliminated by repeated observations.

Multiple Meteorological Observations Applied to Microwave Distance Measurement

M. R. RICHARDS

Ordnance Survey of Great Britain

The general aspects of the investigation on which this study is based are contained in the paper on the Caithness Base Investigation.

A detailed study is made of two relatively short periods of the investigation. One period was by day, mostly overcast, wind varying from light to moderate and visibility less than 30 km. The second was by night with a distinct temperature inversion, no cloud, little wind and excellent visibility. Various combinations of observed meteorological information are applied to the observed distances and the results obtained are discussed.

General

The Caithness Base Investigation by the Ordnance Survey of Great Britain, described on pages 25–44, produced a great deal of field data. In addition to 900 distance measurements by Tellurometer and 57 by Geodimeter, with associated normal (or routine) meteorological observations, over 150 000 other meteorological events were recorded.

It would be well beyond the range of a single paper to analyse and discuss every aspect of the investigation; it would certainly prove to be most indigestible. For the purpose of stimulating discussion, two relatively short periods have been chosen for analysis. Each of the two periods is treated separately, and some tentative conclusions have been drawn from the results obtained.

It must be emphasized that the intention of this analysis is to try to assess the combination or combinations of meteorological observations which, when applied to the observed Tellurometer distances, are most likely to give the best mean value for the length of the base. This mean length must of course have a standard deviation which shows it to be a significant improvement on other values when compared with the accepted catenary value.

The meteorological instruments used have already been described, but for convenience are here briefly outlined. The Tellurometer operators observed temperature and pressure in the normal way. That is, they employed standard electrically-aspirated wet- and dry-bulb mercury thermometers, and aneroid barometers. Thermometers were suspended in free air about 2·5 m from the observer and about 1·2 m above the ground. Barometers were placed beside the instrument.

Temperatures at other specially established stations were observed using thermistors. These could be remotely read, and were emplaced at some distance from the recording station. Particular care was taken over the calibration, and it would appear probable that the recorded temperatures are correct to ±0·2°C. This of course is within the capabilities of a mercury thermometer, but the advantage lies in the ability to read the thermistor remotely and thus save time. Also, possible effects of the observer's body on temperature and humidity are avoided. Barometric pressures were recorded using altimeters in pairs. The difference in barometric readings from a pair of altimeters, as corrected for calibration, rarely exceeded 1 mm of mercury; this represents approximately 0·3 ppm in refractive index determination.

The relative locations of all the observing stations are shown in Appendix A to the paper on the Caithness Base (p. 37).

Constants and Formulae

The velocity of light in vacuo accepted for this analysis is 299 792·5 km sec^{-1}.

All refractive index computations have been made with the Essen and Froome formula recommended by Special Study Group No. 19 of the IAG. Each determination was made directly from the formula, and not interpolated from any precalculated refractive index tables. It has also been independently computed by two individuals.

First Period

General

This period was chosen because the weather conditions were those that are normally considered to be acceptable for microwave measurements. Thirteen distance measurements, between 0850 and 1450 GMT, on 5 October 1964, have been subjected to analysis.

Weather

The weather was cool and cloudy. The sun shone briefly at the start, but then did not appear during the rest of the period. There was a fresh to strong wind blowing initially, but this tended to drop away towards the end. Visibility was just sufficient to see between the terminals, but there were no indications of low cloud or fog patches that could cause meteorological uncertainties.

Distance Measurements

The thirteen measurements were all made with Tellurometers MRA-3. Six were made with sets (numbers 270 and 274) between the terminal pillars of the Caithness base line; the other seven were with sets (numbers 308 and 311) from the auxiliary stations to the west of the line. The overall path length measured from the auxiliary stations was some 3 m shorter than the actual base line. All measures have been reduced to the base for purposes of analysis. Reference must be made here to the previous paper, where it was stated that the measurements made between the terminal pillars were systematically shorter than those from auxiliary stations. The overall mean of the measurements agreed closely with the catenary value. The measurements now being considered are divided almost equally between pillars and auxiliary stations. It is reasonable to suppose therefore that the mean value is free from any systematic effect of this kind.

The ground swing patterns for the measurements made from the terminal pillars showed good correlation, irrespective of the time of measurement. The ground swing patterns for the measurements made from the auxiliary stations were also correlated with each other, but were quite different from the pillar patterns. Three specimen patterns of each type are given in Figs. 1 and 2.

The mean readings are derived, with one exception, from at least seventeen frequency settings covering the full cavity range. The odd one had twelve readings only, extending a little beyond halfway on the dial, but it was accepted as it had a negligible effect on the mean.

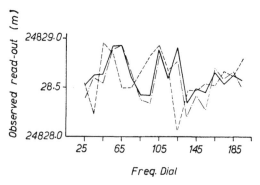

Fig. 1. Measurements between terminal pillars (first period).

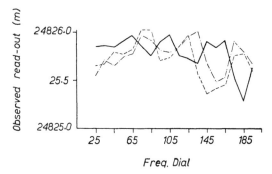

FIG. 2. Measurements between auxiliary stations.

The range of fine readings varied from 1·0 m (7 mμ sec) to 0·4 m (3 mμ sec) with a mean range of 0·6 m (4 mμ sec). Under normal circumstances, all would be considered acceptable within the recommendations of Special Study Group No. 19.

Refractive Index Determinations

This paper deals essentially with variations of refractive index, and for this reason no detailed analysis of variations of temperature, pressure and humidity has been made. However, it may be mentioned that the difference in dry-bulb reading between the 20-ft and 100-ft level of the tower at Annfield remained fairly steady at about 0·3°C throughout the period. Very approximately this is 1·2°C in 100 m and only slightly in excess of the generally accepted adiabatic lapse rate of 1°C in 100 m for an atmosphere in neutral equilibrium.[1] The wet bulb depression was marginally smaller (0·2 to 0·3°C) at the lower level throughout the period. The depression increased gradually between 0850 and 1215 and then began slowly to decrease once more. Similar conditions were observed at the other tower stations.

The refractive index determinations have been made by meaning the temperature measurements contained within the time of distance observation. The observed, or directly interpolated, atmospheric pressure was then applied. Thus each determination is, in itself, a mean of several measurements; for the more complex of the combinations of refractive index applied, forty or more separate observations of wet and dry bulb temperatures have been taken into consideration.

A study was made of the way in which refractive index varied throughout the period according to the measurements taken at the top of the intermediate towers; Fig. 3 indicates the values obtained. There is some correlation between the values, notably a rising trend during the day.

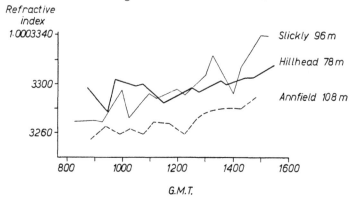

Refractive
index

Fig. 3. Variation of refractive index (determined at top of 100-ft
towers) with time; 5 October 1964.

However, it is perhaps surprising that the station located at the lowest altitude does not give a higher refractive index throughout. It is quite apparent that there will almost certainly be minor fluctuations in atmospheric conditions, and hence refractive index, during the course of any measurement. This indicates that temperature should be measured as frequently as possible. Pressure will vary much less rapidly.

The tower station at Annfield was used as a control for the extrapolation of refractive indices. The main reason for this is that temperatures were read virtually simultaneously at top and bottom of the tower. The thermistor units were also fixed at predetermined and accurately known levels, and were not subject to minor variations in altitudes as experienced at Hillhead and Slickly where they were moved on ropes and pulleys. Refractive index at the upper level was smaller than that at the lower level, with an even gradient over the tower confirmed by thermistors at the 60-ft level. The lapse rate of refractive index varied only slightly over the period, but at a mean figure of $-0\cdot100$ parts per million (ppm) per metre of altitude was two and a half times greater than would be expected of a standard atmosphere which has been calculated as $-0\cdot040$ ppm per metre. This is accounted for by the observed decrease in water vapour pressure with altitude which was more rapid than might be expected.

For the determination of refractive index at ray path level, the observed lapse rate of refractive index at Annfield was applied to the index determined from the observations at the top levels of each of the intermediate towers. This was considered to be more satisfactory than applying a mean lapse rate, and more likely to give a correct answer than a theoretical lapse rate derived from a standard atmosphere.

A theoretical lapse rate of temperature was applied in one instance only. The observations at the top of the terminal (30-ft) towers were designed to be taken in free air, clear of any ground anomaly. One of the determinations of refractive index has involved the observed temperatures at the 30-ft level being reduced by 1°C per 100 m to the level of the terminal pillar.

Results

The combinations of observations that could be used are almost limitless, but common sense rules that only those that might be deemed representative should be used. Nine only have been taken, each of which could be defended as being a reasonable practical solution. Having computed the refractive index for the time of each individual distance measurement at each of the meteorological stations involved, Table 1 was drawn up. The description in the 'Combination' column represents the position of the thermistor units from which the refractive index was derived.

It was found that the mean values varied only slightly with each additional combination tried, and further ones have not been included.

Discussion of Results

It becomes immediately apparent from Table 1 that, in terms of standard deviation, there is little if any significant difference between any of the values derived from the Tellurometer measurements. In view of this, the only real significance that might be found is in the differences between the derived values and the catenary measurement. To a certain extent also the differences within the derived values may be significant.

Taking a mean standard deviation of a derived value as ±0·027 m, the standard deviation of the difference between a derived value and the catenary measurement will be ±0·034 m. Similarly the standard deviation of the difference between derived values will be ±0·039 m.

Accepting confidence limits of twice the standard deviation, i.e. ±0·068, only Serial 7 becomes significantly different when considered together with the catenary value. This may be taken to indicate that the extrapolation has over-corrected for the generally low values obtained at Serials 1 and 2. These themselves are very much borderline cases which lend argument to the suggestion that refractive index as determined from terminal measurement will be too high.

With the same criterion of twice the standard deviation, i.e. ±0·078, for differences within the derived values, many more of the figures become significant. Serial 2, when considered with Serial 3 (and 4), supports the argument about too high a refractive index being obtained by terminal measurements. However, Serial 5 goes against this argument. Although the values at Serials 6–9, derived by extrapolation,

TABLE 1. *Results of applying various combinations of refractive index, day-time period*

Serial	Combination	Mean derived length (m)	Standard deviation of single obs. (m)	Standard deviation of mean (m)
	Catenary measurement	24 828·000		±0·020
1	Normal determination using standard observations	24 827·941	±0·102	±0·028
2	Instrument level at terminals	27·937	±0·087	±0·024
3	30-ft level of terminal towers uncorrected for lapse rate	28·016	±0·100	±0·028
4	30-ft level of terminal towers corrected for lapse rate to instrument level	28·009	±0·100	±0·028
5	Upper (50-ft) level of masts near terminals, i.e., at ray path height	27·939	±0·103	±0·029
6	Instrument level at terminals and extrapolated values at intermediates	28·042	±0·098	±0·027
7	30-ft level at terminals and extrapolated values at intermediates	28·074	±0·101	±0·028
8	50-ft masts near terminals and extrapolated values at intermediates	28·043	±0·103	±0·029
9	Lapse rate corrected 30-ft terminal levels, 50-ft masts and extrapolated intermediates	28·035	±0·100	±0·028

do not seriously conflict with Serials 3 and 4, the indicated tendency is again for an over-correction by application of a derived lapse rate.

Conclusions—First Period

It must be stressed that, because of the generally small differences that are being handled, it is most difficult to draw any really satisfactory conclusions. This is no doubt due to a great extent to the favourable meteorological conditions under which the observations were made. It will be seen that Serial 2 has the lowest standard deviation, and is also within 3 ppm of the catenary value. This would appear to indicate that, in the prevailing conditions, precise measurements of temperature and pressure, at instrument level at the terminals, should give an answer that is not seriously in error. However, the tendency seems to be to derive a refractive index that is, if anything, slightly too high.

The intermediate meteorological observations have not produced any great improvement in the derived distance, although the tendency appears to be to counter the positive error in refractive index deduced from the terminals, and to over-correct it if extrapolation is used. Finally, the application of a lapse rate of refractive index, either theoretically derived or directly observed, is a difficult and misleading operation. It is certainly liable to error.

Second Period

General

This period was chosen since a distinct temperature inversion existed. Seven distance measurements are considered, between 2300 and 0500 GMT on the night of 1/2 October 1964.

Weather

The general weather conditions throughout the period remained stable. The sky was clear, with the moon in its last quarter rising at about 0130. There was a light wind blowing from a west to south-west direction. Visibility was excellent. A temperature inversion persisted over the whole period. It was, in fact, a time when a good measurement was not expected without the facility for sampling meteorological conditions all along the ray path.

Distance Measurements

The seven measurements were all made between the terminal pillars with Tellurometer MRA-3 numbers 270 and 274. These measurements were interspersed with others made from the handrail of the 30-ft terminal towers, but the latter were discounted in the analysis.

The ground swing patterns of the individual measurements do not display any discernible correlation. Specimens of the 'swings' derived from three of the measurements are given in Fig. 4.

The range of the fine readings varies from 1·3 m (9 mμ sec) to 0·4 m (3 mμ sec). The mean range is 0·9 m (6 mμ sec) which would, exceptionally, be just acceptable within the recommendations of Special Study Group No. 19. The mean readings are derived, in every case, from at least eighteen frequency settings covering the full cavity range.

Refractive Index Determinations

The same method of determination was followed as for the first observational period. Again, variations in refractive index have been studied rather than variations in absolute temperature. However, it is interesting to note that the mean dry-bulb temperature at the 100-ft level of the Annfield tower was 1·3°C higher than that at the 20-ft level. Also, the wet-bulb depression at 20-ft, almost constant at 1·0°C, was only half that at the upper level, which averaged just over 2·0°C. The level at which the inversion reversed is not known. If, as is possible, it continued well above the ray path, extrapolation to that height might be valid. Clearly this assumption cannot be made with any confidence, and therefore extrapolated values must be suspect.

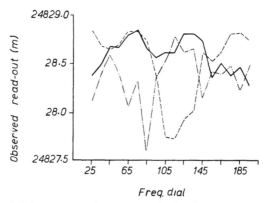

FIG. 4. Measurements between terminal pillars (second period).

A study of the way in which refractive index varied at the top of the intermediate towers proved interesting. As will be seen in Fig. 5, there was good correlation between the derived values at the three stations. The inference is that although conditions throughout the period were not stable, there was little differential variation between the intermediate stations.

For extrapolation of refractive index, the Annfield tower was again used as control. The lapse rate of refractive index remained fairly steady over the period, with a mean figure of approximately −0·189 ppm per metre of altitude. The lapse rate varies in the expected direction, but is five times greater than would be expected from a standard atmosphere. Despite being relatively constant over the height of the tower, it must obviously be treated with some caution. However, this was the best available figure for the conditions prevailing, and was applied to the refractive index as determined at the upper position on each of the inter-mediate towers. Computation of lapse rates at the other towers is hampered by non-simultaneity of observations, but the indications are that the applied figure was representative of the conditions prevailing at those towers.

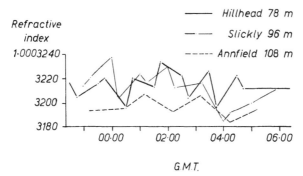

FIG. 5. Variation of refractive index (determined at top of 100-ft towers) with time; 2 October 1964.

Results

The same number of combinations of refractive index have been applied to the measurements in this period. However, immediately the intermediate stations were brought into the argument the results appeared to fall off badly. Table 2 indicates the values obtained.

Discussion of Results

The first thing that appears from Table 2 is the much wider spread of standard deviations. These, unlike the results from the day-time period, are tending to be significant. It is obvious that attempts at extra-polation have over-corrected, and the derived values are not of the same order of accuracy as those obtained from terminal measurements.

Accepting a mean standard deviation of a derived value as ± 0.033 m, the standard deviation of the difference between a derived value and the catenary measurement is ± 0.039 m. Similarly the standard deviation of the difference between derived values is ± 0.047 m. With confidence limits of twice these figures, i.e. ± 0.078 m and ± 0.094 m respec-tively, a number of the differences become significant.

Serial 1 must be suspect. A possible reason for the low value lies in the fact that the mercury thermometers were in a more sheltered position than the thermistors, and have been affected adversely because of it. However, this cannot be the full reason as, being sheltered, a higher temperature and hence longer distance would be expected.

TABLE 2. *Results of applying various combinations of refractive index, night-time period*

Serial	Combination	Mean derived length (m)	Standard deviation of single obs. (m)	Standard deviation of mean (m)
	Catenary measurement	24 828·000		±0·020
1	Normal determination using standard observations	27·904	±0·095	±0·036
2	Instrument level at terminals	27·944	±0·085	±0·032
3	30-ft level of terminal towers	27·935	±0·089	±0·034
4	50-ft level of masts near terminals (ray-path level)	27·999	±0·085	±0·032
5	Mean of instrument level and 50-ft mast near terminals	27·971	±0·084	±0·032
6	Mean of 30-ft tower and 50-ft mast near terminals	27·966	±0·086	±0·033
7	Instrument level at terminals and extrapolated values at intermediates	28·099	±0·114	±0·043
8	30-ft level at terminals and extrapolated values at intermediates	28·095	±0·116	±0·044
9	50-ft masts near terminals and extrapolated values at intermediates	28·120	±0·115	±0·043

There is very little to choose between Serials 2 to 6. All have produced what appears to be a perfectly acceptable measurement. Serial 4 in particular is worthy of note, but the close agreement with the catenary value may be fortuitous (cf. Serial 5 of Table 1).

Serials 7 to 9 must, like Serial 1, be suspect. The high values and comparatively poor standard deviations must indicate that the observed lapse rate of refractive index was too great for the extrapolation up to ray path level.

Reference is again made to the previous paper. All the measurements here considered were made between the terminal pillars, and the indications are that these are systematically shorter than the measurements between auxiliary stations. To make them directly comparable with the catenary value, it may be considered that one should add to the derived distance half of the systematic difference, which for Tellurometer MRA-3, is 0·101 m. Thus 0·050 m might be added to each of the measurements in Table 2. The effect of this on the mean derived values would be very significant, improving a few, but impairing the majority. The effect is not further considered in the context of this analysis.

Conclusions—Second Period

This second period under analysis has reinforced the view that care and accuracy applied to the observation of meteorological conditions will be well repaid. Free air determinations at line of sight near the terminals seem to be of some value, but can hardly be relied upon to produce significant improvements. In the weather conditions prevailing, lapse rates of refractive index appear impossible to predict from the intermediate towers.

Conclusions—Final

Any conclusions drawn here can only be applied with any confidence to microwave measurements of the Caithness Base. However, the line may be typical of many lines likely to be measured in temperate climates, and these results may help to illuminate the general problem of refractive index in other parts of the world.

The intermediate towers were erected for experimental purposes only. It is not suggested that it is practicable, in any but the most isolated cases, to erect towers in order to obtain intermediate meteorological readings during normal production surveys. The indications are that the effort involved will outweigh the benefits obtained from them. Certainly the application of lapse rates should be avoided, as the variation of refractive index with altitude appears impossible to predict with sufficient accuracy.

The determination of temperature values in free air (that is at some distance above ground level) at the terminals should not harm the results appreciably, and may possibly improve them.

The careful and accurate observation of meteorological conditions at instrument level at the terminals has given results which compare favourably with any other derived values. Provided that adequate precautions are taken, results appear unlikely to be seriously in error. Thermometers must be away from any local influence of observer or distance measuring instruments. It would be advantageous to employ a remote reading or recording device.

It would appear desirable to observe meteorological conditions as frequently as possible during the course of each measurement, in order to obtain a truly representative value of refractive index. However these observations should not be permitted to interfere with and protract the distance measurements, which should always be accomplished as speedily as possible.

Before any firm conclusions can be drawn it will be necessary to analyze a great many more readings. This will have to include a breakdown of the separate meteorological quantities of temperature, pressure and humidity, and also a study of the way that refractive index varies along the length of the line during measurement. The invitation previously expressed, to make all the observed data available for research work, is repeated.

Acknowledgements

The opportunity afforded to me by the Director General of the Ordnance Survey for the preparation and presentation of this paper is gratefully acknowledged. Thanks are due to the computing staff of the Ordnance Survey for the help they have given in the compilation of the necessary material, most particularly to Mr W. G. S. Mably who has undertaken the majority of the computations, and to Mr J. K. Holt for his valuable advice and comments.

Reference

[1] O'Connor, D., 1960. Microclimatology and its effects on the accuracy of surveying measurements. *Empire Survey Review*, No. 118.

DISCUSSION

A. R. Robbins: Would you confirm that a 'measurement' means a full run through the cavities?

M. R. Richards: Yes.

A. R. Robbins: What is the difference in measurement conditions between the Serials 1 and 2 in Table 2?

M. R. Richards: Serial 1 is a reduction using the normal meteorological observations made by the field surveyors. Serial 2 is a reduction using thermistor units located at instrument level.

A. D. N. Smith: Perhaps the reason why the standard deviation for a single reading remains the same for good and bad conditions and throughout the trial is related to humidity. I have recently examined some results obtained for MRA-3 Tellurometers by the Director of Inspection of Armaments in which twenty instruments were used to measure the same base line of about 17 km. The results showed a strong correlation between the error and the relative humidity, a dry day giving a bad result. Perhaps in Scotland it is either always raining or just about to rain!

M. R. Richards: Relative humidity was high throughout the whole of the measuring period.

H. D. Hölscher: Have any measurements been made using finer carrier frequency intervals since on such a poor line one expects the exact shape of the swing curve to be strongly dependent on carrier frequency?

M. R. Richards: The measurements were made using steps of twenty units in the cavity tuning scale.

J. E. De Munck: What corrections were applied to obtain the values quoted?

M. R. Richards: The corrections applied to reduce the observed values to the spheroid were (1) Slope correction, (2) Reduction to sea level, (3) Reduction from chord to arc.

J. E. De Munck: What was the precision of the original catenary base measurement?

M. R. Richards: A standard deviation of ±2 cm.

C. F. Trigg: Could the instability of the towers have affected the results?

M. R. Richards: I think not, the towers were short and quite rigid and stable.

R. C. A. Edge: First, I would like to stress again that the Caithness Base has a homogeneous surface and does not really give rise to variations in refractive index. Conclusions drawn from the small differences it has in fact given must be of very dubious value. Secondly, it may seem that the thermistors used were excessively sensitive. This is so, but they had other advantages particularly in the ability to separate the transistor unit and the reading device. Thereby purely local disturbances of the meteorological conditions such as those caused by the presence of human bodies did not contaminate the readings.

R. H. Bradsell: I notice that 30-ft towers were used; this is a quarter of a normal wavelength. Could this have affected the results?

M. R. Richards: I think not.

G. J. Strasser: I have just computed the mean out of the two series of nine in order to see if there was any systematic deviation between or within the results. I found from each series that they deviated by 0·004 m from the taped length. That meant that all eighteen measurements showed a normal distribution. Often if we surveyors cannot explain a scatter we just sum up and average.

M. R. Richards: I am not sure that the method Dr Strasser has used to obtain a mean value is altogether valid.

5

R. C. A. Edge: Nevertheless, I think that what Dr Strasser has said has a great deal of validity. If the conditions are good and if a number of measurements are made it is perfectly reasonable to mean these and forget about the details of the micrometeorology of the line.

Some Results of Microwave Refraction Measurements over Snow and Ice Fields

K. NOTTARP

Institut für Angewandte Geodesie, Frankfurt-am-Main

As a result of the Ross Ice-shelf Survey Expedition 1962–3 and the Quattara Depression Survey Expedition 1964, I had the opportunity to do some research on the vertical and horizontal gradients of the microwave refraction at 3000 Mc/s.

Sometimes a very high gradient near the surface caused complete reflection of the Tellurometer beam so that measurements were impossible. To escape from the space of this high gradient, the antenna system was lifted up to 6 m with good success.

Furthermore, the horizontal gradient of the microwave refraction was measured between two stations, one placed near the sea shore and the other at the farthest point in the desert. Typical values with their large regular daily variations were used to control the often less accurate meteorological observation during the Tellurometer measurements.

In polar regions, Tellurometer measurements over flat ground are sometimes somewhat difficult. For example, the distance to be measured might be some 4 to 6 km and the stations might be just high enough for the ray path to overcome the Earth's curvature. Even if there are no visible obstructions, the signal often proves very weak or even completely disappears although under such conditions the instrument has proved to be very satisfactory.

I think there are two possible reasons for this phenomenon which has puzzled several observers including myself.

Under fine weather conditions during the Ross Ice-shelf Survey Expedition 1962–3 (Hofmann *et al.*, 1964), we found on occasions that there are boundary layers in the atmosphere with high refraction gradients where the carrier wave is reflected upwards. In such a case, broadly speaking, the boundary layer takes the place of the Earth's surface and curvature but at a higher level with respect to the location of the stations. To overcome this disadvantage, the distance between the two stations has to be decreased or at least one of the stations has to be elevated so that the carrier wave path becomes clear of the boundary layer. This is just a problem of geometry and of the available equipment.

During a period of strong katabatic winds we learnt of another reason for the loss of signal between the stations. The very cold and dense layer some metres in height overflows both stations. Owing to the presumably high vertical refraction gradient, the carrier wave seemed to be refracted down to the ground and absorbed in the snow before it could reach the other station. To make measurements possible in this particular case, it was necessary to lift both stations above the wind layer, when the measurements became easy and there was no swing from ground reflections.

As we did not have the technical means or the time to build towers, we used the MRA-2 Tellurometer with a specially designed antenna extension of aluminium tubes from 5 to 7 m high.

To get more and better numerical values of the refraction gradient under different conditions we intend to carry out a special programme in Antarctica during the 1965-6 season.

In another experiment, we made measurements in the Libyan desert in 1964. We had to carry out Tellurometer measurements to establish a framework and numerous control points in a nearly square area of 100 km side with the seashore as the northern limit.

Since barometric levelling was carried out as well as distance measurements, we set up three weather recording stations—two near the seashore at the north-west and north-east corners of the areas, and one at the farthest point south in the desert. After a time, it was apparent that the changes of humidity, temperature and, in particular, air pressure were very regular and reproducible with respect to sunrise, sunset and mid-day. It was thus possible to interpolate from the data provided by the weather stations to obtain the values at the actual geodetic points used (wet- and dry-bulb measurements are sometimes difficult to make and not very accurate under desert conditions).

In the main, the interpolated and the carefully taken weather readings at the geodetic points compared to within ± 2 units of the refraction unit N. I think this was just such an isolated case of meteorological regularity in a reasonable area comparable with the case reported by G. D. Robinson (see pages 96-103).

Reference

Hofmann, W., Dorrer, E. and Nottarp, K., 1964. The Ross Ice-shelf Survey RISS 1962-3. *Antarctic Research Series*, Vol. IV, pp. 83-118. Washington, D.C. American Geophysical Union. In *Antarctic Snow and Ice Studies*, ed. M. Mellor.

DISCUSSION

R. C. A. Edge: I understand from Poder that his view is that when measuring over snow and ice the refraction coefficient becomes unity and if the excess path-length is negligible, there is a reversal of phase and consequent cancellation of the signal. Do you agree?

K. Nottarp: I found no AVC reading, all the energy having been reflected.

K. Poder: If you consider conditions for an ice-cap path and the reflection coefficient is near unity, there will be almost anti-phasing and there will be an extremely weak signal.

K. Nottarp: We have measured with the Tellurometer below the snow in an effort to find the refractive index. We found we could reach a distance of 1 km. When observing at instrument height above the snow and conditions were such that the katabatic wind layer overflowed the station, then the refraction gradient was so great that the ray was sent into the ground. Our programme is to attempt to measure from the surface to 7 m above the ground. We will measure temperature and humidity just to find out what is happening in detail.

J. A. Webley: Were you using 3 cm or 10 cm Tellurometers?

K. Nottarp: 10 cm.

G. D. Robinson: You seem to have experienced a radio mirage effect. Was there also an optical mirage?

K. Nottarp: First of all we could see no targets, then later two targets, one above the other.

Note later by K. Nottarp: It is suggested by Dr Poder that this is a case of single interference. The experiment together with optical measurements is described in *Antarctic Research Series*, Vol. IV, published by the American Geophysical Union in 1964. After setting up, we awaited at one end of the line the reception of the optical and microwave signals. After a time, near-ground meteorological conditions changed and both optical and microwave signals were received at nearly the same time. This leads me to believe that this is a case of reflection and not interference.

The Meteorological Conditions of Electromagnetic Wave Propagation above the Sea

K. BROCKS and H. JESKE

Meteorological Institute of the University of Hamburg

The shape of the vertical refractive-index profile is shown to be approximately logarithmic with a 'profile-coefficient' which is a function of atmospheric stability. New experimental values of this coefficient derived from measurements in the undisturbed sea atmosphere are given, together with some statistical results of the evaporation duct in the German Bight.

To show the influence of the evaporation duct which, for 80 per cent of the time, is the dominating propagation mechanism above the sea for cm- and dm-waves, recent results of propagation experiments with wavelengths from 4 cm to 2 m on radio-paths not too far behind the horizon are discussed: the correlation between field strength and duct thickness, the wavelength dependence on duct propagation, height gain measurements of field strength, fading properties, etc. Also some examples of optical refraction effects are given, especially an optical analogy of the radio duct.

Finally, some measurements with an airborne refractometer will be demonstrated to show the important fine structure of the refractive-index field in both vertical and horizontal directions.

1. *Introduction*

Electromagnetic wave propagation above the sea is of specific peculiarity because of the strong refractive index gradients formed in consequence of evaporation and mixing processes. There are remarkable deviations compared with the propagation properties over land where the conditions of standard refraction are often realized. In dealing with propagation phenomena above the sea, account must be taken of field strength variations of radio links, radar ranges, or bending problems. Also in the field of electromagnetic distance measurements over the sea, we have to note some peculiarities. The existence of an atmosphere wave guide (duct) makes it possible to receive signals far beyond the optical horizon. Furthermore, the curvature correction due to refraction, which has to be considered when measuring over long paths, is essentially smaller above the sea than over land, as the curvature of the rays is nearly the magnitude of the Earth's curvature.[1,2]

A method is given to derive the characteristic refraction parameters (gradients, duct thickness) from simple routine meteorological measurements near the sea surface.[3,4,5] Moreover, some results of transmission measurements are discussed.[6,7,8]

2. *The Refractive Index of the Air and its Gradient*

The refractive index n or the refractive modulus N for radio-optical frequencies is given by the well-known expression:

$$N = (n-1)10^6 = a_1 p/T - a_2 e/T + a_3 e/T^2 \qquad (1)$$

where

p = total atmospheric pressure (mb)
e = water vapour pressure (mb)
T = air temperature (°K).

For optical frequencies (neglecting dispersion effects), the same relationship is valid without the third term. The different influences of water vapour are caused by the unlike reaction of the polar water molecules on the frequencies in the radio-optical and optical part of the electromagnetic spectrum.

With $p = 1013$ mb and $T = 288°$K, we obtain, in the radio-optical range:

$$n = 1 \cdot 000273 + 44 \cdot 4 \cdot 10^{-7} e \qquad (1a)$$

and in the optical range:

$$n = 1 \cdot 000273 - 0 \cdot 45 \cdot 10^{-7} e \qquad (1b)$$

As constants a_1, a_2 and a_3, we use the values of Essen and Froome (1951):

$$a_1 = 77 \cdot 6 \, (°K/mb); \qquad a_2 = 12 \cdot 96 \, (°K/mb);$$
$$a_3 = 3 \cdot 72 \cdot 10^5 \, (°K^2/mb)$$

In refraction problems, the vertical gradient of the refractive index field is important and is given by:

$$dN/dz = -c_1 + c_2 \, de/dz - c_3 \, dT/dz \qquad (2)$$

with

$$c_1 = \partial N/\partial p = (a_1/T)\, \mathrm{d}p/\mathrm{d}z = a_1 g(p - 0.377e)/RT^2 \qquad (2a)$$

$$c_2 = \partial N/\partial e = (a_3 - a_2 T)/T^2 \qquad (2b)$$

$$c_3 = \partial N/\partial T = (a_1 p - a_2 e + 2a_3 e/T)/T^2 \qquad (2c)$$

where

z = height above the ground (m)
g = gravity acceleration, 9·81 (m/sec^2)
R = gas constant, 287·0 (m^2/sec^2 grad)

Table 1 represents the factors c_1, c_2, and c_3 in the radio-optical range with $p = 1013$ mb for several values of air temperature and for the relative humidities o per cent, 60 per cent and 100 per cent. The values for the optical range (corresponding to the dielectric part of the re-fractive index in the radio-optical range) are written in brackets. The factors c_1 and c_3 for o per cent humidity are the same in both ranges.

TABLE 1. *The factors c_1, c_2, c_3 of equation (2)*

T	c_1			c_2	c_3		
	o%	60%	100%		o%	60%	100%
−20°C	0·042	0·042	0·042	5·75 (−0·51)	1·23	1·26 (1·23)	1·28 (1·23)
−10	0·039	0·039	0·039	5·32 (−0·49)	1·14	1·20 (1·14)	1·25 (1·14)
o	0·036	0·036	0·036	4·93 (−0·47)	1·05	1·19 (1·05)	1·28 (1·05)
10	0·034	0·033	0·033	4·59 (−0·46)	0·98	1·22 (0·98)	1·38 (0·98)
20	0·031	0·031	0·031	4·28 (−0·44)	0·92	1·33 (0·91)	1·60 (0·91)
30	0·029	0·029	0·029	4·00 (−0·43)	0·86	1·53 (0·85)	1·98 (0·85)

From the practical viewpoint, the question is how to find the values of the gradients [see equation (2)] and the shape of the profiles without the considerable expenditure on a special measurement set-up.

3. Refractive Index Profiles in the Atmospheric Surface Layer above the Sea and their Determinations

3.1 General Remarks

Immediately above the sea surface, very strong vertical water vapour gradients are usually observed due to evaporation. By this means a corresponding intensive decrease of radio refractive index follows. Simultaneously, the interaction between sea and atmosphere also causes upward or downward fluxes of sensitive heat connected with temperature lapse-rates or inversions. These temperature gradients weaken or strengthen the influence of the water vapour on the refractive index profile. The intensity of the vertical gradients and the shape of the profiles in the maritime boundary-layer are governed primarily by turbulent mixing processes.

3.2 Neutral Case

The similarity theory of boundary turbulence (expounded by Prandtl and v. Kármán) predicts logarithmic profiles for the neutral case, which was verified in several experiments carried out in laboratories and in the atmosphere. For a quantity i we can write:[8]

$$\mathrm{d}i/\mathrm{d}z = \Gamma_1 . \Delta i_1/z + z_0 \tag{3}$$

with the vertical difference $\Delta i = i_1 - i_0$ between the height z_1 and the water surface, the so-called profile coefficient Γ_1 and the roughness parameter z_0. Γ_1 gives the relative logarithmic shape of the profiles, and it is connected with the roughness parameter z_0:

$$\Gamma_1 = (\mathrm{d}i/\mathrm{d}\ln z)/(i_1 - i_0) = 1/\ln[(z_1 + z_0)/z_0] \tag{4}$$

During the years 1957–61, we carried out [9,8] numerous measurements of wind, temperature and water vapour profiles in the undisturbed sea atmosphere (above the North Sea and the Baltic Sea) to find Γ_1 and z_0 [see equation (4)] which are known by turbulence theory in the case of the so-called hydrodynamic smooth surface only. Fig. 1 gives some average neutral profiles of water vapour pressure which are largely logarithmic.

The several hundred wind and humidity profiles obtained up to 15 m gave us the information that the adiabatic profile coefficient does not depend on wind velocity (up to 15 m/s, which was the upper limit of our measurements) and that it has a value of

$$\Gamma_1 = 0.10 \quad \text{for a reference level} \quad z_1 = 4 \text{ m} \tag{4a}$$
i.e. $z_0 = 0.018$ cm

3.3 Diabatic Case

Furthermore, the measurements show also that, under diabatic conditions (i.e. deviation from the neutral equilibrium), the profiles can be sufficiently approximated by logarithmic height functions. But under diabatic conditions, the profile coefficient varies with the stability of atmospheric stratification. Our experiments prove that the profile coefficient is a function of the Richardson number

5*

$$Ri = g(\mathrm{d}\theta/\mathrm{d}z)/T(\mathrm{d}u/\mathrm{d}z)^2 \qquad (5)$$

the well-known stability parameter of the atmospheric turbulence (θ = potential temperature, u = wind velocity). Fig. 2 gives the experimental relation between the profile coefficient of humidity (reference level 4 m) and the so-called bulk-Richardson number $Ri_{(\text{bulk})}$; it is a condensed representation of all our profile measurements.[8,9]

$$Ri_{(\text{bulk})} = g(\theta_1 - \theta_w)z_1/\text{0.1}.Tu_1^2 \qquad (5a)$$

The bulk-Richardson number can be derived from given pairs of wind velocity and air–sea temperature differences. The profile coefficient

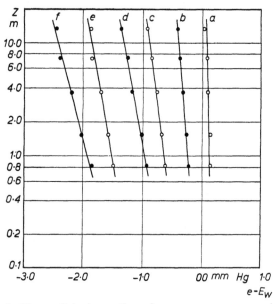

FIG. 1. Mean adiabatic profiles of water vapour, arranged after ΔE ($= e - E_w$.)

	$\overline{u_4}$ m/sec	n	$\overline{\Delta E}$ mm Hg
a	3·79	12	+0·10
b	3·55	10	−0·32
c	3·92	15	−0·78
d	3·28	26	−1·18
e	3·91	13	−1·72
f	3·23	13	−2·22
	($-$0·3 $\leqslant \Delta T \leqslant +$0·2°C)		

as a function of this parameter can be obtained from the diagram of Fig. 2 or by an adequate table. For convenience, the diabatic profile coefficient can also be given as a function of $T_1 - T_w$ and u_1.

With these experimental results and equation (3), the problem of determining the gradients is largely solved. The great advantage of this method is that without any difficulties, and by means of very simple routine meteorological measurements, the shape of the profile and the gradients are known. Consequently expensive special measurements are no longer necessary.

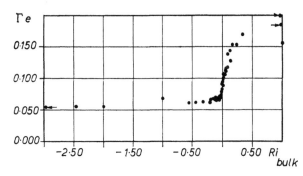

FIG. 2. Profile coefficient Γ_e of water vapour as function of the bulk Richardson number (reference level $z_1 = 4$ m).

The profile of refractive index is given by the following expression:

$$\mathrm{d}N/\mathrm{d}z = -c_1 + [c_2\Gamma_e(e_1 - e_w)/z] - [c_3\Gamma_T(T_1 - T_w)/z] \qquad (6)$$

Provided the profile coefficients for temperature and humidity are identical (which is a good assumption for most purposes) relation (6) can be rewritten:

$$\mathrm{d}N/\mathrm{d}z = -c_1 + \Gamma_1[c_2(e_1 - e_w) - c_3(T_1 - T_w)]/z \qquad (6a)$$

This form gives also an insight into the different influences of the humidity field and the temperature field upon the refractive index gradient.

For practical purposes, we can better write:

$$\mathrm{d}N/\mathrm{d}z = -c_1 + \Gamma_1(N_1 - N_w)/z \qquad (6b)$$

(Here, strictly speaking, N stands for the potential refractive index.[5])

Integration of (6b) gives the refractive index as a function of height. The value of N can be simply derived from a graph as functions of temperature and wet-bulb temperature as parameters.

3.4 *Duct Thickness, Mean n-gradients*

Concerning wave propagation, the ground-based oceanic duct capable of guiding microwave energy around the bulge of the Earth plays the dominating role. The relations (6a) or (6b) give the possibility of deriving the duct thickness which is defined as the height of the layer with an N lapse-rate greater than the Earth's curvature:

$$- dN/dz\ 10^6/r \geqslant 0\cdot157/m$$

With (6b) this height is given by:

$$z^* = \Gamma_1(N_w - N_1)/(0\cdot157 - c_1) \tag{7}$$

The mean n-gradients between the levels z_b and z_a are obtained by integration of (6b):

$$- \frac{\overline{dN}}{dz} = \frac{\Gamma_1 \Delta N_1}{(z_b - z_a)} \ln \frac{z_b}{z_a} \quad \text{or} \quad \frac{(0\cdot157 - c_1)z^*}{(z_b - z_a)} \ln \frac{z_b}{z_a} \tag{8}$$

If we evaluate, for example, the mean gradients between 0·5 m and the height of the duct z^* we obtain for various duct-thicknesses:

z^*:	5 m	10 m	15 m	20 m	30 m
$\overline{dN}/dz$:	$-0\cdot32/m$	$-0\cdot39/m$	$-0\cdot44/m$	$-0\cdot47/m$	$-0\cdot52/m$

The gradients above the duct also show super-refraction conditions.

3.5 Some Statistics of the Evaporation Duct

Heights of the evaporation duct larger than 20 m are observed very seldom in the area of the German Bight. In other climates, for instance in the tropical or trade-wind regions, a thickness up to about 50 m or more is possible.

In 1962 on board of lightship *Elbe 1* a thickness of 7 m was exceeded in 50 per cent of all cases, of 14 m in 10 per cent of all cases and of 19 m in 1 per cent of all cases. Sub-refraction conditions ($z^* < 0$) were observed in 3 per cent of the time. Table 2 gives a detailed cumulative distribution of the thickness of evaporation duct observed in 1962.

A special experiment of simultaneous meteorological measurements on lightships, lighthouses and ships in the German Bight[9,10] shows a great homogeneity of the atmospheric conditions in the maritime sub-layer. Observations carried out on one point are representative for an area of more than 60 km.

4. Experimental Results

4.1 Optical Measurements

It is of interest that the dry component of the radio-optical duct (given by equation 6a for $e_1 = e_w = 0$ and the corresponding factors for the optical frequencies in Table 1) can be determined from optical measurements.

TABLE 2. *Cumulative distribution of duct thickness* z^* (m), 1962

Month	90%	50%	10%	1%
January	−0·5	2·2	7·2	9·0
February	0·6	4·4	7·5	9·5
March	2·0	4·4	7·1	11·0
April	−0·4	3·0	7·6	13·2
May	1·4	6·0	11·0	20·0
June	3·4	8·8	13·5	22·5
July	3·4	8·9	13·7	19·6
August	4·9	10·1	15·5	20·4
September	5·0	11·4	15·0	20·5
October	3·5	8·6	14·0	16·5
November	1·5	8·4	11·4	13·3
December	2·8	5·7	9·0	12·8
Year	1·8	7·2	13·6	19·1

This relation is given by Fig. 3 where angles of optical refraction are plotted against the simultaneously observed temperature differences, air−sea. These angles were measured along rays on paths of 17·4 km length with different heights above the North Sea. We see that also in this case there is a good correlation between the vertical gradients of refractive index, given by the refraction angle, and the simply measured temperature difference, air−sea.[7,9]

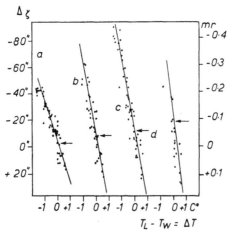

FIG. 3. Angle of optical refraction above the sea. $\Delta\zeta$ difference against adiabatic case; ΔT temperature difference (air−sea). Heights: (a) 6·0, (b) 3·9, (c) 4·3, (d) 2·4 m; distance 17·4 km.

The thickness of the 'dry duct' can be derived from observations of the dip of horizon. The so-called 'elevation of horizon', up to now not explained, [11,9] is an effect of this dry duct, the surface of which is seen as the horizon.

Fig. 4[11] shows the thus derived duct height as a function of the potential refractive index difference air–sea (here called $\Delta \nu$). Its negative value multiplied by 10^6 is round about the positive temperature excess

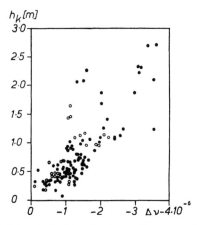

FIG. 4. The height of the 'dry duct' as a function of the potential refractive-index difference (6 m — sea surface).
Height of observation about 24·5 m   O
about   7·9 m   ●

air–sea $(T_1 - T_w)$. At a rough estimation a dry duct with a thickness of about 10 m is caused by a temperature excess of $+10°C$. But commonly the low-level oceanic duct above the ocean is a genuine evaporation duct by water vapour lapse-rate.

4.2   *Radar Ranges and Field Strength Measurements*

The electromagnetic wave theory of ducting predicts an increase of radar ranges or received field strength of radio links with duct thickness. This effect is strongly dependent on the wavelength. In Fig. 5 radar ranges of ships (defined by a blip-scan relation of 0·3) are plotted versus the duct thickness derived with the given method. We see the expected dependence.

Fig. 6 shows a scatter diagram of our example of both duct situations. In the ordinate hourly median values of field strength of the 4-cm link and in the abscissa simultaneous observations of duct thickness are shown. The full points belong to the evaporation duct indirectly

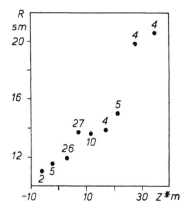

FIG. 5. Radar ranges as a function of the duct width z^* (Ship targets: North Sea, $\lambda = 3$ cm).

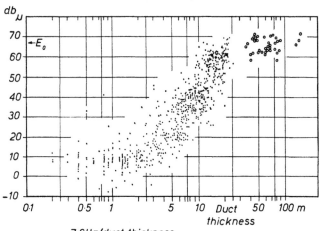

7 GHz/duct thickness

Fading type
- • A. B. (evaporation ducts)
- ○ C. D. (other ground ducts)

FIG. 6. Field strength values as a function of duct thickness.

derived from measurements at the sea surface; the values of bigger ground-based ducts which were observed by refractometer or sonde profiles are characterized by circles. The gap between the heights of 20 and 35 m is caused by the lack of thicker evaporation ducts (during the measuring period) and also as we are not allowed to fly at these low levels.

Fig. 7 shows an example of field strength measurements with one-way radio paths of about 80 km length (the highest antenna heights are about 30 m) with several wavelengths (4·4 cm; 13 cm; 53 cm; 187 cm). Simple meteorological measurements to derive the duct heights were carried out on board of the lightship *Elbe 1* situated under the centre of the paths. We see clearly the correlation between the predicted duct (first file, the strong line representing a duct height derived from a profile coefficient formed in literature out of date) and the field strength of the four wavelengths in use. Evidently the relationship is the more marked the shorter the wavelength. The 187-cm wave does not show any reaction. The observed dependence on wavelength is in good agreement with the theory of ducting.[12]   Our transmission measurements show that in about 80 per cent of all cases situations exist which can be described by the predicted low-level evaporation duct. These situations observed in 80 per cent of the time are connected with a scintillation fading type (indicated by A, B or E in the illustrations).

But sometimes all wavelengths show strong reaction with high values of signal strength and quite another type of slower fading (C,D). In these periods there are acting mostly thicker ground ducts formed by advection or subsidence. They cannot be predicted by measurements at the sea surface; we found them with the aid of airborne refractometers or balloon sondes.[6,7,9]

4.3   *Height Gain Measurements*

Further insights into the tropospheric propagation mechanism can be derived from height gain measurements of field strength. This problem is important also from the practical viewpoint, for instance, to find the best antenna heights.

In Fig. 8 the correlations of signal strengths of several wavelengths for various antenna-height combinations are presented.[7,8] The left side of the figure shows the relation of a 7-GHz link 'SO–EO' (4.4-cm wavelength) with transmitting and receiving antenna at 30 m height versus a 7-GHz link 'SO–EU' with antennas at heights of 30 and 5 m (upper part of the diagram) and versus a 7-GHz link 'SU–EU' with both antennas at 5 m height. The right-hand side gives the simultaneous field strength of the combination 'SO–EO' and 'SO–EU' for the 2 GHz (13 cm) and 600 MHz (53 cm) links.

In most cases the received field strength is stronger for the paths with higher antenna heights. But occasionally we observe a higher field strength level for the links with lower antennas. These situations are primarily connected with a slow fading, i.e., the dominating propagation mode is not the evaporation duct. Such cases are, however, not observed for the 53-cm path.

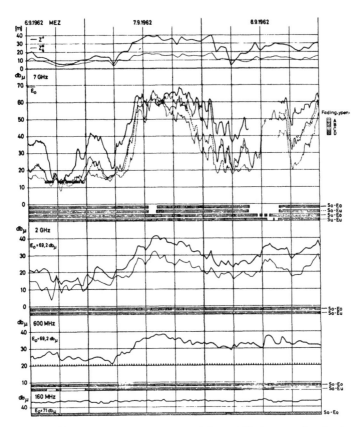

FIG. 7. Simultaneous measurements of duct width (evaporation duct)
and field strengths at paths of various wavelength.

Most of the measured results can be explained by the theory of
ducting. In our case the duct theory of Booker and Walkinshaw[12] with
the fifth root power-law model of refractive index gives a sufficient
theoretical basis. In Fig. 8 the curves present the expected dependence
corresponding to this theory. Evidently there exists quite a good agreement between theory and observations. The large scattering of the values
for very high field strength values connected with fading types C,D. is
mostly caused by interference of several modes. Characteristic deviations from the law of ducting are observed only during periods of low
field strength levels. Then the mechanism of scatter propagation is
acting.[7,8]

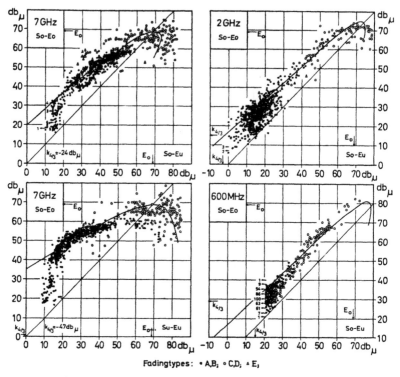

Fadingtypes: • A,B; ○ C,D; ▲ E,

FIG. 8. Simultaneous measurements of field strengths at paths of various heights of transmitting and receiving antennas.

5. Concluding Remarks

The given simple method of predicting the duct thickness or the refractive-index gradient for diabatic cases by approximating the N-profile with a logarithmic height function using a varying profile coefficient can be improved. More realistic modern theories (combining the adiabatic case with similar supplementary considerations) predict more complicated profiles within the surface layer of the atmosphere. They all can be reduced to the following general form:[6]

$$\frac{di}{dz} = \frac{\Delta i_1}{(z + z_0)} \cdot \frac{[1 + (\alpha/n)Ri]^n}{\int_0^{z_1} \frac{[1 + (\alpha/n)Ri]^n}{(z + z_0)} dz}$$

which can be handled easily with electronic computers.[5] With the exponent $n = 1$ we have the model of Monin–Obukhov, with $n = -\frac{1}{4}$

the model of Ellison–Panofsky and with $n = \frac{1}{2}$ the model of Rossby–Montgomery. A widely accepted average value of α is 4·5; for z_0 we can take [see equation $(4a)$] 0·018 cm.

Under extreme deviations from the neutral equilibrium this expression gives better results, but in our climate the given method (Section 2) is sufficient for most practical cases.

Another model used in former times[13] approximates the profiles by means of power-laws: $di/dz \sim z^{-\beta}$. Here β is the stability parameter which is given experimentally as a function of Richardson number.[14] But there are only very few sets of β.

However, it should be emphasized once again that the calculation of duct widths or gradients on the basis of the turbulent mixing theory from the meteorological measurements near the surface is restricted to the determination of propagation conditions within the lowest decameters above the sea surface. Influence of layers on higher levels, especially reflections at higher inversion layers, are generally covered only by the aerological soundings.

References

[1] Höpcke, W., 1964. Uber die Bahnkrümmung elektromagnetischer Wellen und ihren Einfluss auf die Streckenmessungen. *Zeitschr. f. Vermessungswesen*, 89 Jahrg., No. 6.

[2] Höpcke, W., 1964. *Streckenmessungen über See*, 89 Jahrg., No. 11.

[3] Brocks, K., 1955. Der Brechungsindex für elektromagnetische Wellen (Cm- bis M-Band) in der maritimen Grenzschicht der Atmosphäre. *Dt. hydrogr. Z.*, **8**, 186–94.

[4] Brocks, K. and Hasse, L., 1956. Der Brechungsindex für elektromagnetische Wellen (Cm- bis M-Band) in der maritimen Grenzschicht der Atmosphäre, 2 Beitrag. *Dt. hydrogr. Z.*, **9**, 217–21.

[5] Jeske, H., 1965. Die Ausbreitung elektromagnetischer Wellen im Centimeter- bis Meter-Band über dem Meer unter besonderer Berücksichtigung der meteorologischen Bedingungen in der maritimen Grenzschicht. *Hamburger Geophysikalische Einzelschriften*. No. 6, Cram, de Gruyter & Co. (to be published).

[6] Brocks, K., *et al.*, 1963. Radiometeorological Papers, Institut für Radiometeorologie und Maritime Meteorologie an der Universität Hamburg. *Bericht*, No. 7.

[7] Fengler, G., *et al.*, 1964. Radiometeorological Papers II, Institut für Radiometeorologie und Maritime Meteorologie an der Universität Hamburg (Institut der Fraunhofer-Gesellschaft). *Bericht*, No. 9.

[8] Brocks, K., Duct Propagation in the Maritime Surface Layer of the Atmosphere, presented at the Advanced Study Institute on Radio Meteorology, Lagonissi, Greece, 31 August–12 September, 1964.

[9] Brocks, K., 1963. Probleme der maritimen Grenzschicht der Atmosphäre. *Ber. d. Deutsch. Wetterd.*, No. 91, 34–46.

[10] Beuck, G., 1958. Örtliche Unterschiede und zeitliche Schwankungen des meteorologischen Feldes vor der Küste im Wattenmeer, Diplomarbeit, Geophysikalisches Institut der Universität Hamburg.

[11] Hasse, L., 1960. Über den Zusammenhang der Kimmtiefe mit meteorologischen Grössen. *Dt. hydrogr. Z.*, **13**, No. 4.
[12] Booker, H. G. and Walkinshaw, W., 1946. The mode theory of tropospheric refraction and its relation to wave-guides and diffraction, in *Meteorological Factors in Radio Wave Propagation*, pp. 80-127, published by the Physical Society and the Royal Meteorological Society, London.
[13] Anderson, L. J. & Gossard, E. E., 1953. The effect of the oceanic duct on microwave propagation. *Trans. Amer. Geophys. Union*, **34**, (5), 695-700.
[14] Deacon, E. L., 1949. Vertical diffusion in the lowest layers of the atmosphere. *Quart Journ. Roy. Meteorol. Soc.*, **75**, 89-103.

DISCUSSION

R. C. A. Edge: We are very interested in your work owing to our recent measurements across the Channel. We only took meteorological observations at the terminals. Could we usefully re-analyse these observations?

H. Jeske: It would have been better to have had meteorological data along the path.

A. R. Robbins: You may be interested to know that Captain Aslakson of the US Coast and Geodetic Survey made measurements in the Persian Gulf with early model Tellurometers to stations below the optical horizon.

H. Jeske: In measuring over the open sea you need meteorological observations at one point only.

On the Path Curvature of Electromagnetic Waves

J. SAASTAMOINEN

Division of Applied Physics, National Research Council, Ottawa, Canada

To facilitate the calculation of atmospheric refraction from radio-meteorological sounding observations, the coefficients of refraction for light and microwaves have been expressed in the general form

$$k = Ap + BU + (Cp + DU)\ \mathrm{d}t/\mathrm{d}z + E\mathrm{d}U/\mathrm{d}z$$

where p is total pressure, U is relative humidity and A, B, C, D and E are tabulated functions of temperature t. Three-place tables are given in metric units for the temperature range $-10°$C to $+30°$C.

The paper includes a general discussion of the variation of refraction with weather, and outlines computation procedures for electromagnetic measurement of long lines on the basis of radio-sonde data.

The Coefficient of Refraction–Curvature Tables

The path curvature of electromagnetic waves is a function of refractive index n, vertical gradient $\mathrm{d}n/\mathrm{d}z$ and slope angle α of the ray path:

$$1/\rho = -(\cos \alpha/n)\ \mathrm{d}n/\mathrm{d}z \tag{1}$$

A common measure for the curvature is the coefficient of refraction, usually defined as the number expressing the ratio of the path curvature to the curvature of the Earth. Since this ratio varies with the slope angle, it is convenient to refer the refraction coefficient to a horizontal ray path, so that

$$1/\rho = k \cos \alpha/R \tag{2}$$

where ρ is the curvature radius of the ray path, R is the curvature radius of the Earth and k is the coefficient of refraction.

Equating the right sides of (1) and (2) and solving for k gives

$$k = -(R/n)\ \mathrm{d}n/\mathrm{d}z \cong -R\ \mathrm{d}n/\mathrm{d}z \tag{3}$$

the approximate form of which is adequate for practical purposes, considering that $1 < n < 1 \cdot 0005$ within the meteorological range of pressures and temperatures. By the substitution of the total derivative,

$$\frac{\mathrm{d}n}{\mathrm{d}z} = \frac{\partial n}{\partial p} \cdot \frac{\mathrm{d}p}{\mathrm{d}z} + \frac{\partial n}{\partial e} \cdot \frac{\mathrm{d}e}{\mathrm{d}z} + \frac{\partial n}{\partial T} \cdot \frac{\mathrm{d}T}{\mathrm{d}z} \tag{4}$$

Formula (3) may be used for the calculation of the coefficient of refraction when the magnitudes and the vertical gradients of atmospheric pressure p, vapour pressure e and temperature T are known (Höpcke 1964).

The partial derivatives appearing in (4) can easily be formed from the refractive index formula proper. Of the remaining derivatives, the vertical gradients of temperature and humidity are extremely variable and must be either assumed or measured, whereas the lapse rate of pressure can be calculated.

Expressions for the pressure gradient and the partial derivatives of the refractive indices for light and microwaves are given in Appendices I and II. Substituting them in (4), the functional relationship becomes established between coefficient of refraction k and meteorological elements p, e, T, $\mathrm{d}e/\mathrm{d}z$ and $\mathrm{d}T/\mathrm{d}z$.

However, since meteorological recording instruments indicate the relative humidity rather than the vapour pressure, the former quantity should be preferred in the final formulation. For the purpose of conversion, relative humidity may be defined as the percentage ratio of the existing vapour pressure to that of saturation; consequently,

$$e = \frac{Ue'}{100} \tag{5}$$

and

$$\frac{\mathrm{d}e}{\mathrm{d}z} = \frac{e'}{100} \cdot \frac{\mathrm{d}U}{\mathrm{d}z} + \frac{U}{100} \cdot \frac{\mathrm{d}e'}{\mathrm{d}T} \cdot \frac{\mathrm{d}T}{\mathrm{d}z} \tag{6}$$

where U denotes the relative humidity and e' is the saturation vapour pressure of existing temperature T.

The equation of the refraction coefficient may now be written, in metric units:

$$k = 0 \cdot 001 \, p \, (\mathrm{I}) + 0 \cdot 01 \, U \, (\mathrm{II})$$
$$+ [0 \cdot 001 p \, (\mathrm{III}) + 0 \cdot 01 \, U \, (\mathrm{IV})] \mathrm{d}t_{100\mathrm{m}} + (\mathrm{V}) \, \mathrm{d}U_{100\mathrm{m}} \tag{7}$$

where p is total pressure in millibars, t is temperature in °C, U is relative humidity in per cent and

$$\text{(I)} = \frac{9 \cdot 209 R(n_0 - 1)}{T^2}\left(\frac{g}{G}\right)$$

$$\text{(II)} = -\frac{0 \cdot 0035 R(n_0 - 1)e'}{T^2}$$

$$\text{(III)} = \frac{R(n_0 - 1)}{0 \cdot 371\, T^2}$$

$$\text{(IV)} = \frac{1 \cdot 13 R}{10^7 T}\left(\frac{de'}{dT} - \frac{e'}{T}\right)$$

$$\text{(V)} = \frac{1 \cdot 13 Re'}{10^9 T}$$

} Light   (8a)

or

$$\text{(I)} = \frac{2 \cdot 6516 R}{10^3 T^2}\left(\frac{g}{G}\right)$$

$$\text{(II)} = -\frac{Re'}{10^6 T^2}$$

$$\text{(III)} = \frac{776 \cdot 24 R}{10^6 T^2}$$

$$\text{(IV)} = \frac{R}{10^8 T}\left[\frac{e'}{T}\left(\frac{743\,800}{T} - 12 \cdot 92\right) - \frac{de'}{dT}\left(\frac{371\,900}{T} - 12 \cdot 92\right)\right]$$

$$\text{(V)} = \frac{12 \cdot 92 Re'}{10^{10} T}\left(1 - \frac{28\,777}{T}\right)$$

} Microwaves   (8b)

are functions of temperature alone. Their numerical values, which are given in Tables 1a and 1b, were computed for every whole degree of temperature ranging from -10 to $+30°$C and taking the values for e' and de'/dT from *Smithsonian Meteorological Tables* (1951), Nos. 94 and 103, respectively.

To attain the best accuracy in the use of the curvature tables, computations should be carried out to more decimal places than are retained in the final result. In addition, the following correction factors should be observed:

(a) The curvature radius of the Earth changes with azimuth and latitude up to roughly $\pm$ 30 km about the tabular constant, $R = 6370$ km, which will cause an error in k of 0·5 per cent, at the most. Hence the computed refraction coefficients should be multiplied by a factor of $R/6370$, where R is the true curvature radius in km.

(b) Tabular function (I) depends to some extent on the local value of gravity. It was computed for the standard gravity, $G = 980·665$ cm/sec^{-2}, which is nearly equal to the normal gravity (sea-level) at latitude $\phi = 45°$, the latter being given by the international gravity formula:

$$g_0 = 978·049\,(1 + 0·0052884 \sin^2\phi - 0·0000059 \sin^2 2\phi) \quad (9)$$

Thus, accepting a value of $-0·3086$ mgal/m for the vertical gradient of gravity,

$$g/G = 1 - 0·0026 \cos 2\phi - 0·00031 z_{km} \quad (10)$$

expresses the approximate ratio of local gravity to standard gravity.

Correction factor g/G is applied to the first term of (7) only. Since, however, the decrease of pressure with elevation will render the z-term negligible at all heights, it can be omitted and the correction factor can be computed from latitude alone:

$$g_0/G = 1 - 0·0026 \cos 2\phi \quad (11)$$

The resulting corrections of curvature are small, or even negligible.

(c) Table 1a is valid for yellowish-green light of wavelength 0·560 μ. For other colours the coefficient of refraction is obtained by multiplying the tabular value of k_L by the ratio

$$(n_0 - 1)_\lambda/(n_0 - 1)_{0.560} \approx 1 - 0·067(\lambda - 0·560) \quad (12)$$

The approximate value for (12) has been derived from the formula of Barrell and Sears, equation 36 in Appendix II.

One last remark might be added for explanation of notations dt_{100m} and dU_{100m} in formula (7). By them are meant vertical gradients dt/dz and dU/dz where z is expressed in units of 100 m, and they should not be confused with the actual increase or decrease of temperature and humidity within this height interval. A value of $dt_{100m} = +1000°$C, for example, is not at all impossible in a short vertical section of the atmosphere.

Variation of Atmospheric Refraction

By the introduction of the curvature tables the following aspects of the atmospheric refraction become at once obvious.

(a) The curvature of a light ray varies with pressure, temperature and lapse rate. High pressure, low temperature and small lapse rate contribute to a strong curvature; the coefficient of refraction is small at low pressure, high temperature and steep lapse rate.

(*b*) In an extremely dry atmosphere the path curvature of microwaves is approximately equal to that of light; in a moist atmosphere under constant specific humidity the coefficient of refraction is slightly greater for microwaves than for light.

(*c*) The path curvature of microwaves is markedly affected by any variation of humidity (mixing ratio) with height. Decrease of specific humidity with height will increase the curvature; moist air aloft and dry air below will decrease it.

It may seem strange that a significant difference between the path curvatures of light and microwaves occurs only in the case of an unequal mass distribution of water vapour; that is, when the net result of terms $+0.01 U(\text{IV})\, dt_{100\text{m}}$ and $+(\text{V})\, dU_{100\text{m}}$ importantly modifies the curvature of microwaves. As can be easily verified by computation, those terms largely cancel each other out in the case of a constant humidity mixing ratio, or specific humidity. But the primary source of water vapour in the atmosphere is by evaporation from the surface of the earth and the specific humidity normally decreases with height. Fig. 1

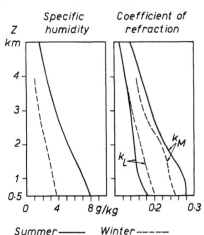

FIG. 1. Average gradients of specific humidity and refraction, computed from meteorological data by Humphreys (1940), pp. 61 and 80, of 416 weather balloon soundings in Central Europe, 1900–12.

shows the effect on the path curvature of microwaves. It is found that, owing to the vapour distribution in the air, the average refraction of microwaves is notably stronger than the average refraction of light, reaching a maximum strength in summer when the rapid rate of evaporation from the ground is favourable to the establishment of a steep humidity gradient.

A great variety of ways exist by which water vapour is removed from or added to the atmosphere, and there is nothing to prevent that occasionally, instead of decreasing, the specific humidity is either constant or increases with height. Therefore the measurement of the humidity gradient must be specifically included in any determination of the curvature of microwaves by meteorological means.

Many of the physical processes that compose the weather are closely associated with the state of vertical stability in the atmosphere as determined by the prevailing lapse rate of temperature. Because the lapse rate also plays an important role in the generation of refraction, it is useful to consider the characteristics of the different degrees of vertical equilibrium so as to learn the effect of the typical weather phenomena on the magnitude of refraction. Appendix III gives a brief account on this topic, the variation of refraction through the various stages being illustrated by numerical samples which were computed assuming in each case a homogeneous distribution of specific humidity. For a detailed discussion of the thermodynamical aspects involved, and to enlarge upon the given description of weather phenomena any textbook in physical meteorology, among others the works of Brunt (1952), Byers (1959) and Humphreys (1940), can be consulted.

The conclusion cannot be avoided—and could presumably be confirmed by a statistical survey of the regional, seasonal and diurnal variations of refraction—that, far from being a constant, the coefficient of refraction varies greatly even under normal weather conditions, particularly so where microwaves are concerned.

The Effect of Path Curvature on Electromagnetic Distance Measurement

Let us suppose that we have at our disposal the record from a meteorological sounding made in connection with an electromagnetic measurement of a long distance. By having recourse to the curvature tables we can readily calculate the vertical distribution of the prevailing refraction, but we still have the problem of evaluating the shape of the ray path in order to assess the due corrections to the measured distance. This subject matter is most conveniently treated beginning with the case of constant refraction (Saastamoinen 1964), and modifying the calculation procedure later on so as to correspond to a refraction coefficient changing with height.

Constant Coefficient of Refraction

The ray path is an arc of a circle whose radius is R/k. Its elevation at any point is given by the formula for trigonometric heights:

$$z = z_0 + \tan \alpha_0 x + \left(\frac{1-k}{2R} \right) x^2 \tag{13}$$

where z_0 and α_0 refer to the origin of the line and x is the horizontal distance from the origin to the point considered. The initial slope may be computed by formula

$$\tan \alpha_0 = \frac{h}{s} - \left(\frac{1-k}{2R} \right) s \tag{14a}$$

if elevation difference h corresponding to total horizontal length s is known. The first derivative,

$$\frac{dz}{dx} = \tan \alpha = \tan \alpha_0 + \left(\frac{1-k}{R} \right) x \tag{15}$$

then gives the slope angles for all the other points along the ray path; particularly

$$\tan \alpha_s = \frac{h}{s} + \left(\frac{1-k}{2R} \right) s \tag{14b}$$

is the slope at the end point of the line.

In the case of a constant refraction coefficient the refractive index changes linearly with height:

$$\Delta n = - \frac{k}{R} \Delta z \tag{3'}$$

It is customary to compute electromagnetic distance measurements using the arithmetic mean of the refractive indices observed at the terminal stations; that is, applying a refractive index that occurs at mean altitude z_m between the terminal points. The average refractive index along the ray path, however, occurs at average height $\bar{z}$ of the path which can be determined by integrating (13):

$$\bar{z} = \frac{1}{s} \int_0^s z \, dx = z_m - \left(\frac{1-k}{12R} \right) s^2 \tag{16}$$

Substituting $\Delta z = \bar{z} - z_m$ from (16) in (3') we obtain a differential correction to the applied refractive index that is equivalent to correction

$$K_v = - \frac{k(1-k)}{12R^2} s^3 \tag{17}$$

to the measured distance.

Applying refractive index or velocity correction (17) will give the true measured length of the ray path. To obtain the slope distance between the terminal points we must also include a geometric correction,

$$K_g = -\frac{s^3}{24(R/k)^2} + \frac{s^3}{24R^2} = \left(\frac{1-k^2}{24R^2}\right)s^3 \qquad (18)$$

in which the first term reduces the measured arc to the chord distance, and the second term is the correction from the chord to an arc with radius R. The total correction to the measured distance finally combines into the formula for *path curvature correction*:

$$K = K_g + K_v = \frac{(1-k)^2}{24R^2}s^3 \qquad (19)$$

The rapid growth of the magnitude of correction (19) with distance is illustrated in Fig. 2.

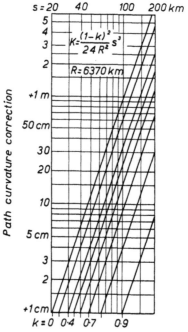

FIG. 2. Nomogram for path curvature correction.

Instead of presenting a complete theory we have in the foregoing treatment tacitly accepted certain approximations which have simplified the derivation of the correction formula. Of them the radii of curvature, R/k and R, asserted to the ray path and the slope line, respectively, can be considered adequate for nearly horizontal lines only. In the airborne systems of distance measurement, however, the measured lines often have slope angles of several degrees. We shall next verify the accuracy of formula (19) in the reduction of such lines, and consider practical ways for the computation of horizontal lengths from given slope distances.

In schematic Fig. 3, let O represent the centre of curvature of the

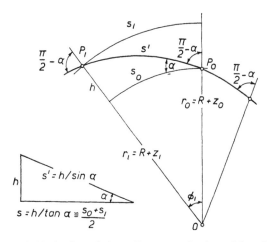

FIG. 3. Reduction of slope distance to horizontal length.

Earth corresponding to the azimuth and mean latitude of slope line s' between stations P_0 and P_1. Let O also be the pole and OP_0 the initial line for polar co-ordinates (r, ϕ). Within the limits of sector P_0OP_1, concentric circles about O $(r = \text{const.})$ and radii vectors $(\phi = \text{const.})$ represent horizontal and vertical lines, respectively.

Line P_0P_1, along which the slope does not change, is an arc of logarithmic spiral which curve possesses the property of cutting radii vectors under a constant angle. Its polar equation is

$$r = r_0 e^{a\phi}; \quad \text{or} \quad \phi = \frac{1}{a}\ln\frac{r}{r_0} \qquad (20)$$

where a stands for tan α. The length and the curvature radius of arc

P_0P_1 are derived from (20) by calculus:

$$\frac{dr}{d\phi} = ar; \qquad \frac{d^2r}{d\phi^2} = a^2r$$

$$s' = (r_1 - r_0) \operatorname{cosec} \alpha = h/\sin \alpha \tag{21}$$

$$\rho' = r/\cos \alpha \cong R/\cos \alpha \tag{22}$$

The approximate value of (22) is sufficient for practical computation.

From equations (2) and (22) we find immediately that geometric correction (18) must be reduced by a factor of $\cos^2\alpha$ if the ray path is inclined. As to the velocity correction, we must consider that formula (13) refers the coefficient of refraction to the inclined ray path, whereas we have expressed k for a horizontal path. To arrive at a proper velocity correction we therefore substitute $k \cos \alpha$ for k in equation (16). The total path curvature correction is consequently obtained as

$$K = K_g + K_v = \frac{\cos^2\alpha(1 - k^2)}{24R^2} s^3 - \frac{k(1 - k\cos\alpha)}{12R^2} s^3$$

Since α is a small angle we can write, to sufficient accuracy, $\cos \alpha = 1 - \frac{1}{2}\alpha^2$ and $\cos^2\alpha = 1 - \alpha^2$, where α is expressed in radians. This gives

$$K = \frac{[(1 - k)^2 - \alpha^2]}{24R^2} s^3$$

Thus we have derived a first reduction term,

$$\Delta K_1 = - \frac{(h/s)^2}{24R^2} s^3 \tag{23}$$

for path curvature correction (19) in the case of inclined measurements.

As only a rough approximation of the measured distance is required for the computation of the path curvature correction, we have so far paid no attention to the variation of horizontal length s with height. In view of equation (21), we shall define horizontal length specifically in such a way that rectified arcs s' and s, together with height difference h, will exactly fit the sides of a rightangled triangle, as shown in Fig. 3. In other words, we select the horizontal length at height z such that condition

$$s = (R + z)\phi_1 = h/\tan \alpha$$

is exactly satisfied. Replacing ϕ in the above equation by its series expansion

$$\phi_1 = \frac{1}{a} \ln \frac{r_1}{r_0} = \frac{1}{a} \ln \left(1 + \frac{h}{r_0} \right) = \frac{h}{ar_0} \left(1 - \frac{h}{2r_0} + \frac{h^2}{3r_0^2} - \dots \right)$$

and solving for z gives

$$z = z_0 + \frac{1}{2} h - \frac{h^2}{3R} + \dots \simeq \frac{z_0 + z_1}{2} \tag{24}$$

The obtained result implies that the conventional slope correction,

$$\kappa = s - s' = (s'^2 - h^2)^{\frac{1}{2}} - s' \tag{25a}$$

which may also be written

$$\left. \begin{array}{c} \kappa = \kappa_1 + \kappa_2 + \dots \\[4pt] \kappa_1 = -\dfrac{h^2}{2s'} \\[6pt] \kappa_2 = -\dfrac{\kappa_1^2}{2s'} \end{array} \right\} \tag{25b}$$

will reduce the slope distance to its equivalent horizontal length at the *mean altitude* between the terminal points, providing a second correction term,

$$\Delta K_2 = + \frac{h^2}{3R} \cdot \frac{s}{R} \tag{26}$$

is included in the curvature correction, which compensates for the error involved in neglecting the third term of expansion (24). Combining (23) and (26) gives a total correction term,

$$\Delta K = \Delta K_1 + \Delta K_2 = \frac{7(h/s)^2}{24R^2} s^3 \tag{27}$$

to be added algebraically to path curvature correction (19) in the case of inclined measurements.

The magnitude of correction ΔK is extremely small. For $s = 150$ km, for example, Fig. 2 gives $s^3/(24R^2) = 3 \cdot 5$ m. A slope of $5/150$ would increase the curvature correction by $2 \cdot 7$ cm only.

To the various reduction formulae discussed above we shall finally add a well-known expression for the sea-level reduction:

$$S - s = -\frac{z_m}{R} \left(1 - \frac{z_m}{R} \right) s \tag{28}$$

by which horizontal length s is reduced from mean altitude z_m of the terminal stations to sea-level distance S.

To conclude, we have shown that providing (*a*) the coefficient of refraction can be treated as constant, and (*b*) the arithmetic mean of the refractive indices at the terminal points has been used in the computation of an electromagnetic distance measurement, the measured distance can be reduced to sea-level arc by applying consecutively the following three reductions:

1. *Path curvature correction*, formulas (19) and (27);
2. *Slope correction*, formula (25*a*) or (25*b*);
3. *Sea-level reduction*, formula (28).

This basic reduction procedure will also be essentially retained in the treatment of the more complex forms of ray path.

Variable Coefficient of Refraction

In the general case of refraction that changes with height, the key problem lies in the computation of the ray path. Once established, the ray path can be divided into sections of, say, 25 km or longer depending on the accuracy desired, each of which is treated separately using the reduction methods discussed previously. In this way the sea-level arc equivalent to the total line is obtained as the sum of the reduced individual sections.

Let us first assume that the coefficient of refraction varies linearly with height, which condition is frequently approximated in the lower atmosphere. In that case we have

$$k = k_0 + (\mathrm{d}k/\mathrm{d}z)(z - z_0) \tag{29}$$

where, again, the subscript refers to the origin of the line, and $\mathrm{d}k/\mathrm{d}z$ is constant. Substituting (29) for k in formula (13), the latter written as

$$z - z_0 = \frac{h}{s}x - \left(\frac{1-k}{2R}\right)(sx - x^2) \tag{13'}$$

gives

$$z = z_0 + \left[\frac{\tan \alpha_0 x + \left(\dfrac{1-k_0}{2R}\right)x^2}{1 - \dfrac{\mathrm{d}k/\mathrm{d}z}{2R}(s-x)x}\right] \tag{30}$$

which is the sought equation of the ray path. The slopes at the terminal points are

$$\tan \alpha_0 = \frac{h}{s} - \left(\frac{1 - k_0}{2R}\right) s$$

and

$$\tan \alpha_s = \frac{h}{s} + \left(\frac{1 - k_s}{2R}\right) s \qquad (31)$$

where k_s refers to the coefficient of refraction at height $z_0 + h$.

The curve represented by equation (30) is a cubic parabola which intersects, from level to level, the corresponding ray paths for constant refraction, as illustrated in Fig. 4. The angles of intersection are, in fact, zero; in other words, the intersecting curves have a common tangent at a triple point.

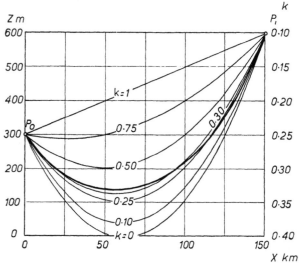

FIG. 4. Example of a compound ray path ($dk/dz = -0.0005$) shown in relation to paths under constant refraction. Plotting of equations (30) and (13').

For most measurements—excepting those where the ray path passes through an inversion—the vertical gradient of refraction can be assumed constant and the path can be divided into sections, computing the required intermediary points by equation (30) above.

A ray path passing through the boundary surface of a sharp inversion suffers a sudden rather than a gradual change of curvature. Using the notation in Fig. 5, and applying formulas (14) or (31), the slope of the path at the boundary is

$$\tan \alpha = \frac{h_1}{x} + \left(\frac{1 - k_1}{2R}\right) x = \frac{h_2}{s - x} - \left(\frac{1 - k_2}{2R}\right)(s - x) \qquad (32)$$

providing the coefficients of refraction are either constant or change linearly in the layers below and above the boundary. Equation (32), which is best solved graphically, gives the position of point P on the boundary. Once that has been determined, the previous methods can be applied to sections P_1P and PP_2 of the ray path.

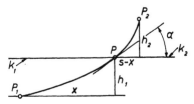

FIG. 5. Passage of ray path through inversion.

With equal values of k_1 and k_2, equation (32) may also be applied to a height level where the vertical gradient of refraction changes abruptly, without there being a discontinuity of the refraction coefficient. In this particular case the solution of (32) coincides with the point at which the ray path under constant refraction, $k = k_1 = k_2$, would intersect the boundary surface.

Generally speaking, whatever the vertical distribution of the prevailing refraction, we can always divide the atmosphere into layers such that through each of them the vertical gradient of the refraction coefficient can be treated as constant. In the computation of the ray path, iterative methods, or calculation by trial and error, can then be used if more than a single passage from one layer to another is involved.

TABLE 1a. *Coefficient of Atmospheric Refraction*

(Metric Units)   Light, $\lambda = 0 \cdot 560\ \mu$ $R = 6370$ km

$g = 980 \cdot 665$ cm sec^{-2}

$k_{\mathrm{L}} = 0 \cdot 001 p\,(\mathrm{I}) + 0 \cdot 01 U\,(\mathrm{II}) + [0 \cdot 001 p\,(\mathrm{III}) + 0 \cdot 01 U\,(\mathrm{IV})]\,dt_{100\mathrm{m}} + (\mathrm{V})\,dU_{100\mathrm{m}}$

$t°\mathrm{C}$	(I)	(II)	(III)	(IV)	(V)
− 10	0·248		0·073	0·001	
− 9	0·246		0·072	0·001	
− 8	0·244		0·072	0·001	
− 7	0·243		0·071	0·001	
− 6	0·241		0·070	0·001	
− 5	0·239		0·070	0·001	
− 4	0·237		0·069	0·001	
− 3	0·235		0·069	0·001	
− 2	0·234		0·068	0·001	
− 1	0·232		0·068	0·001	
0	0·230	−0·001	0·067	0·001	
1	0·229	−0·001	0·067	0·001	
2	0·227	−0·001	0·066	0·001	
3	0·225	−0·001	0·066	0·001	
4	0·224	−0·001	0·065	0·001	
5	0·222	−0·001	0·065	0·001	
6	0·221	−0·001	0·065	0·002	
7	0·219	−0·001	0·064	0·002	
8	0·217	−0·001	0·064	0·002	
9	0·216	−0·001	0·063	0·002	
10	0·214	−0·001	0·063	0·002	
11	0·213	−0·001	0·062	0·002	
12	0·211	−0·001	0·062	0·002	
13	0·210	−0·001	0·061	0·002	
14	0·208	−0·001	0·061	0·002	
15	0·207	−0·001	0·061	0·003	
16	0·206	−0·001	0·060	0·003	
17	0·204	−0·001	0·060	0·003	
18	0·203	−0·002	0·059	0·003	0·001
19	0·201	−0·002	0·059	0·003	0·001
20	0·200	−0·002	0·059	0·003	0·001
21	0·199	−0·002	0·058	0·004	0·001
22	0·197	−0·002	0·058	0·004	0·001
23	0·196	−0·002	0·057	0·004	0·001
24	0·195	−0·002	0·057	0·004	0·001
25	0·193	−0·002	0·057	0·004	0·001
26	0·192	−0·002	0·056	0·005	0·001
27	0·191	−0·003	0·056	0·005	0·001
28	0·189	−0·003	0·055	0·005	0·001
29	0·188	−0·003	0·055	0·005	0·001
30	0·187	−0·003	0·055	0·005	0·001

TABLE 1*b*. *Coefficient of Atmospheric Refraction*

(Metric Units) Microwaves $R = 6370$ km

$g = 980 \cdot 665$ cm sec^{-2}

$k_M = 0 \cdot 001 p \text{ (I)} + 0 \cdot 01 U \text{ (II)} + [0 \cdot 001 p \text{ (III)} + 0 \cdot 01 U \text{ (IV)}] \, dt_{100m} + \text{(V)} \, dU_{100m}$

$t°C$	(I)	(II)	(III)	(IV)	(V)
− 10	0·244		0·071	− 0·069	− 0.010
− 9	0·242		0·071	− 0·074	− 0·010
− 8	0·240		0·070	− 0·078	− 0·011
− 7	0·238		0·070	− 0·083	− 0·012
− 6	0·237		0·069	− 0·088	− 0·013
− 5	0·235		0·069	− 0·094	− 0·014
− 4	0·233		0·068	− 0·100	− 0·015
− 3	0·231		0·068	− 0·106	− 0·016
− 2	0·230		0·067	− 0·112	− 0·017
− 1	0·228		0·067	− 0·118	− 0·018
0	0·226	− 0·001	0·066	− 0·125	− 0·019
1	0·225	− 0·001	0·066	− 0·133	− 0·020
2	0·223	− 0·001	0·065	− 0·140	− 0·022
3	0·221	− 0·001	0·065	− 0·148	− 0·023
4	0·220	− 0·001	0·064	− 0·157	− 0·025
5	0·218	− 0·001	0·064	− 0·165	− 0·026
6	0·217	− 0·001	0·063	− 0·174	− 0·028
7	0·215	− 0·001	0·063	− 0·184	− 0·030
8	0·214	− 0·001	0·063	− 0·194	− 0·032
9	0·212	− 0·001	0·062	− 0·204	− 0·034
10	0·211	− 0·001	0·062	− 0·215	− 0·036
11	0·209	− 0·001	0·061	− 0·226	− 0·038
12	0·208	− 0·001	0·061	− 0·238	− 0·040
13	0·206	− 0·001	0·060	− 0·250	− 0·043
14	0·205	− 0·001	0·060	− 0·263	− 0·045
15	0·203	− 0·001	0·060	− 0·277	− 0·048
16	0·202	− 0·001	0·059	− 0·290	− 0·051
17	0·201	− 0·001	0·059	− 0·305	− 0·054
18	0·199	− 0·002	0·058	− 0·320	− 0·057
19	0·198	− 0·002	0·058	− 0·335	− 0·060
20	0·197	− 0·002	0·058	− 0·351	− 0·064
21	0·195	− 0·002	0·057	− 0·368	− 0·067
22	0·194	− 0·002	0·057	− 0·385	− 0·071
23	0·193	− 0·002	0·056	− 0·404	− 0·075
24	0·191	− 0·002	0·056	− 0·422	− 0·079
25	0·190	− 0·002	0·056	− 0·442	− 0·084
26	0·189	− 0·002	0·055	− 0·462	− 0·088
27	0·187	− 0·003	0·055	− 0·482	− 0·093
28	0·186	− 0·003	0·055	− 0·504	− 0·098
29	0·185	− 0·003	0·054	− 0·526	− 0·103
30	0·184	− 0·003	0·054	− 0·549	− 0·108

Vertical Gradient of Atmospheric Pressure

The hydrostatic equation of fluids,

$$\mathrm{d}p = -g\rho\,\mathrm{d}z \tag{33}$$

in which g denotes local gravity, expresses in differential form the approximate relationship between atmospheric pressure p, density ρ and elevation z. The density of dry air, in turn, depends on pressure and absolute temperature T as stated by the perfect gas law:

$$p = \rho_D RT \tag{34}$$

and furthermore, is reduced by a factor of $(1 - 0\cdot378e/p)$ owing to the presence of water vapour exerting the partial pressure e out of the total p.

With $G = 980\cdot665$ cm sec^{-2} for standard gravity (sea-level at $45°$ lat.) and $R = 2\cdot8704 \times 10^6$ erg $g^{-1}°\mathrm{K}^{-1}$ for gas constant of 1 gram of dry air,

$$\frac{\mathrm{d}p}{\mathrm{d}z} = -\left(\frac{g}{G}\right) \times \frac{0\cdot03416\,(p - 0\cdot378e)}{T} \tag{35}$$

where T is in $°\mathrm{K}$, and z is in metres.

Refractive Index Formulae and Derivatives

Refractive index formulae (36), (37a) and (37b) below are essentially those recommended by the International Association of Geodesy (1963). However, the formula of the so-called 'group index of refraction' for modulated light has been left out, since the *refraction* of light—either unmodulated or modulated—depends on the wave velocity alone.

(a) Light: Effective Wavelength λ, in microns

Refractive index of standard air:

$$(n_0 - 1)\,10^7 = 2876\cdot04 + \frac{16\cdot288}{\lambda^2} + \frac{0\cdot136}{\lambda^4} \tag{36}$$

Refractive index at total pressure p (mm Hg), vapour pressure e (mm Hg) and temperature T ($°\mathrm{K}$):

$$n = 1 + \frac{(n_0 - 1)p}{760\alpha T} - \frac{0\cdot55e}{10^7\alpha T} \tag{37a}$$

where α is the expansion coefficient of air ($\alpha = 0.003661$).

Partial derivatives of (37a):

$$\left.\begin{aligned}
\frac{\partial n}{\partial p} &= \frac{n_0 - 1}{760\alpha T} \\[1em]
\frac{\partial n}{\partial T} &= -\frac{(n_0 - 1)p}{760\alpha T^2} + \frac{0.55e}{10^7\alpha T^2} \\[1em]
\frac{\partial n}{\partial e} &= -\frac{0.55}{10^7\alpha T}
\end{aligned}\right\} \quad (38a)$$

(b) Microwaves

Refractive index of air at total pressure p (mm Hg), vapour pressure e (mm Hg) and temperature T (°K):

$$(n-1)\,10^6 = \frac{103.49}{T}(p-e) + \frac{86.26}{T}\left(1 + \frac{5748}{T}\right)e \quad (37b)$$

Partial derivatives of (37b):

$$\left.\begin{aligned}
\frac{\partial n}{\partial p} &= \frac{103.49}{10^6 T} \\[1em]
\frac{\partial n}{\partial T} &= -\frac{103.49p}{10^6 T^2} - \frac{991\,600e}{10^6 T^3} + \frac{17.23e}{10^6 T^2} \\[1em]
\frac{\partial n}{\partial e} &= \frac{495\,800}{10^6 T^2} - \frac{17.23}{10^6 T}
\end{aligned}\right\} \quad (38b)$$

APPENDIX III

States of Vertical Stability in the Atmosphere

(a) Autoconvective Instability

Example:

$p = 1000$ mb

$t = 30°C$

(Lapse rate in excess of $\quad$ $dt = -3.6$ $\quad k_L = -0.01$

$3.4°C$ per 100 m) $\quad\quad$ $U = 10\%$ $\quad\quad k_M = -0.03$

$\quad\quad\quad\quad\quad\quad\quad\quad dU = +2.0$

The decrease of atmospheric pressure with elevation cannot maintain a decrease in air density if the lapse rate of temperature becomes sufficiently steep—the limit for constant density being $3.4°C$ per 100 m, approximately. A steeper lapse rate will produce a state of instability, in which the density of air increases with height and the coefficient of refraction assumes a negative value.

Autoconvective instability may obtain in a shallow layer of air above strongly heated dry ground, such as desert rock or sand, and over a relatively warm water surface in contact with cool air. It is the cause of a certain type of land and sea mirage where the optical illusion is due to the 'inverted' refraction of light.

(b) Unstable Equilibrium

Example:

$p = 1000$ mb

(Superadiabatic lapse rate $t = 30°C$

between 1°C and 3·4°C $dt = -1·5$ $k_L = 0·10$

per 100 m) $U = 15\%$ $k_M = 0·11$

$dU = +1·1$

The condition for unstable equilibrium is a decrease with height of both potential temperature and density of air. It exists where the prevailing lapse rate is steeper than the dry adiabatic but less steep than the autoconvective rate of cooling.

Unstable equilibrium rarely occurs on a large scale, except during extreme solar heating of desert regions where it may reach to a height of several kilometres above the ground. Under such conditions flying is rough because of strong convection currents which produce, often violent, dust whirls. In the more humid regions unstable equilibrium is limited to such a lower portion of the atmosphere that it seldom is a principal cause in the generation of storms or heavy precipitation.

(c) Neutral Equilibrium

Example:

$p = 1000$ mb

(Dry adiabatic lapse rate of $t = 20°C$

1°C per 100 m) $dt = -1·0$ $k_L = 0·14$

$U = 40\%$ $k_M = 0·15$

$dU = +2·0$

The state of vertical isothermalcy of potential temperature is referred to as neutral equilibrium, although this condition is unstable for particles of saturated air.

Neutral equilibrium is merely the border case between the states of unstable and conditionally unstable atmosphere. It is, however, fairly often approximated in the usually unsaturated, frictional layer of the lowest atmosphere where mechanical turbulence may effectively stir the air and, through the process of continued mixing, establish a nearly adiabatic lapse rate.

(d) Conditionally Unstable Equilibrium

Example:

	$p = 1000$ mb	
(Lapse rate steeper than	$t = 15°C$	
saturation adiabatic but	$dt = -0.65$	$k_L = 0.17$
less than 1°C per 100 m)	$U = 50\%$	$k_M = 0.18$
	$dU = +1.5$	
	$(dU = 0)*$	$(k_M = 0.25)*$

Conditional instability occurs wherever the prevailing lapse rate falls between the dry and saturation adiabatic rates. In this equilibrium the atmosphere is unstable for saturated air, but resists the vertical motion of unsaturated air.

Conditionally unstable equilibrium is observed frequently in the lower atmosphere. In the temperate zone it occurs much of the time over the continents in summer and over the oceans in winter, and where the prevailing direction of wind is from colder toward warmer regions.

Cumulus and cumulonimbus clouds, showers and thunderstorms are the typical condensation forms of conditionally unstable atmosphere. They occur when surface heating or cooling from above has advanced a nearly adiabatic lapse rate beyond the saturation level. Along with the steepening of the lapse rate and with increasing turbulence, winds tend to become gusty and flying conditions rough.

(e) Stable Equilibrium

Examples:

	$p = 1000$ mb	
(Lapse rate smaller than	$t = 15°C$	
saturation adiabatic)	$dt = 0$	$k_L = 0.21$
	$U = 50\%$	$k_M = 0.23$
	$dU = -0.6$	
Stable equilibrium exists where	$p = 1000$ mb	
the prevailing lapse rate is less	$t = 15°C$	
steep than the saturation adiabatic	$dt = +1.5$	$k_L = 0.30$
rate. In this equilibrium, which	$U = 70\%$	$k_M = 0.37$
includes all the inversions of tem-	$dU = -7.6$	
perature, the atmosphere resists	$p = 1000$ mb	
the vertical motion of both satur-	$t = 10°C$	
ated and unsaturated air.	$dt = +5.0$	$k_L = 0.54$
	$U = 90\%$	$k_M = 0.67$
	$dU = -31$	

* Decrease of specific humidity with height.

Stable equilibrium is observed in the lowest atmosphere most of the time when the air is warmer than the ground. In the middle and higher latitudes it is generally found over the continents in winter and over the oceans in summer, and where the prevailing direction of wind is from warmer toward colder regions.

Convective clouds cannot develop effectively in such a portion of the atmosphere where stable conditions prevail. If clouds are present there, they are of the fog or stratiform type and, in general, produce little or no precipitation.

Temperature inversions, or layers through which the air temperature increases with height, are frequently found at varying levels in the atmosphere. They represent the highest degree of atmospheric stability suppressing efficiently the vertical motion of air, although an inversion layer as a whole may sometimes rise or subside in the ambient atmosphere. Inversions confined to the air stratum next to the surface are called ground inversions, whereas a high inversion has a positive lapse rate in the layer immediately below.

The base of a high inversion is usually a sharply defined, horizontal or sloping surface which acts as a barrier against the turbulent diffusion of water vapour, dust and smoke. Likewise the upward transport of heat from the ground is restricted to layers below the inversion. If the base of a strong inversion happens to be at a low level, hot and humid weather may prevail at the ground, whereas the air above the base is usually dry and clear.

A frontal inversion is normally observed at low or moderate heights in the transition zone between two air masses of different origin. It is a common type of inversion which differs in structure from those produced by mechanical turbulence or large-scale subsidence, by possessing a relatively high, inverted rate of specific humidity and a considerable slope. Furthermore, as in the case of a warm front, forced ascent of air along the slope of the inversion and the consequent adiabatic cooling process frequently lead to condensation in the form of wide-spread precipitation from layer clouds.

Ground inversions are produced by contact cooling of air at the surface. With the exception of warm air flowing over a cold surface, they are essentially due to the radiational cooling of the ground during the nights, and are often referred to as radiation inversions. The strongest ground inversions, which may persist both day and night, occur in the high latitudes in winter, when the long nights provide excessive periods of radiational cooling. In the middle latitudes during the summer months nocturnal inversions are commonly observed over low-lying land when the sky is clear and the wind is light.

A number of mirage phenomena, such as looming, are associated with

6*

extremely sharp inversions sometimes found in thin layers of air, particularly over calm waters.

References

Brunt, David, 1952. *Physical and Dynamical Meteorology.*
Byers, Horace Robert, 1959. *General Meteorology*, third edition.
Humphreys, W., J. 1940. *Physics of the Air*, third edition.
Höpcke, W., 1964. Über die Bahnkrümmung elektromagnetischer Wellen und ihren Einfluss auf die Streckenmessungen, *Zeitschrift für Vermessungswesen*, No. 6, 183–200.
International Association of Geodesy, Resolution No. 1 of the Thirteenth General Assembly, *Bulletin Géodésique*, No. 70, 390, 1963.
Saastamoinen, J., 1964. Curvature correction in electronic distance measurement, *Bulletin Géodésique*, No. 73, 265–9.
Smithsonian Meteorological Tables, sixth revised edition, 1951.

DISCUSSION

R. C. A. Edge: The most common situation to measure long lines is over the sea. Your method involves radio-sondes which could only be launched at the terminals.

J. Saastamoinen: We have to neglect all changes in reflection in a horizontal plane so it does not matter where we release the sondes. The middle should give a good result. The most important thing is that the measurements with radio-sondes should cover the whole difference in height of the ray path.

R. C. A. Edge: What sort of errors would you introduce by ignoring the path curvature correction?

J. Saastamoinen: From Fig. 2 you will find that for a line of 150 km 0.2 in k would make a difference of 1 m.

G. D. Robinson: I would like to give a word of warning on do-it-yourself sondes. I would advise everybody to consult the local meteorological office before using them. Radio-sondes are not made for this type of work; they do not do anything but smooth out inversions and the results can be very misleading. We can make radio-sondes which will meet your requirement but they are expensive. I would advise a slow ascent for the sonde.

J. Saastamoinen: We do not use a free ascent; instead, the sonde is sent up on a line.

Multiple Wavelength Optical Distance Measurements

J. C. OWENS

Institute for Telecommunication Sciences and Aeronomy,*
Environmental Science Services Administration, Boulder, Colorado

and

P. L. BENDER

Joint Institute for Laboratory Astrophysics
National Bureau of Standards, Boulder, Colorado

One of the main limitations, at present, to the accuracy of geodetic distance measurements by microwave or optical methods is the uncertainty in the average propagation velocity of the radiation due to inhomogeneity of the atmosphere. A direct optical method for finding the desired average by utilizing the dispersion of the refractive index is presented. The technique involves the simultaneous measurement of the electromagnetic distance over the same path using two or more frequencies for which the group refractive indices are different and accurately known. The line integral of the refractive index of the atmosphere over the path may then be determined from the difference in electromagnetic distances, giving the atmospheric correction and permitting the calculation of the true geodetic distance. It should be possible, using this method in conjunction with modern microwave-frequency optical modulation, synchronous detection, and phase measuring techniques, to measure distances of the order of 10 km— roughly an order of magnitude more accurately than by present optical methods. In combination with a simultaneous measurement over the same path as at infra-red or microwave frequency, the method permits an accurate evaluation of the water vapour content of the atmosphere over the path as well.

One of the main limitations, at present, in measuring geodetic base-lines to high accuracy by optical methods is the uncertainty in the average refractive index over the optical path due to inhomogeneity and turbulence in the lower atmosphere (Thompson *et al.* 1960). This uncertainty in refractive index and, hence, in the average velocity of

---

* Formerly the Central Radio Propagation Laboratory of the National Bureau of Standards.

propagation of the light is primarily due to lack of knowledge of the air density, although a small correction for the effects of water vapour is also required. At present, refractive index corrections are determined from measurements of pressure, temperature, and humidity made at one or more points along the path. Except under the most favourable circumstances, these meteorological measurements are inadequate to give the required correction with sufficient accuracy to permit the measurement of true geometrical distance to 1 ppm. Primarily for this reason, current electromagnetic distance measurements are generally limited in accuracy to the order of several parts per million.

A direct method of finding the desired correction by utilizing the dispersion of the atmospheric optical refractive index has recently been proposed (Bender & Owens 1965). The present paper outlines the theory and potentialities of this method, and describes a new instrument, presently under construction, which is suitable for making optical distance measurements using the method. In brief, the method is as follows: the refractive index, n, of the lower atmosphere is dispersive in the visible region, and hence two light signals traversing the same path but having different wavelengths will travel at slightly different velocities. Because $(n-1)$ at a given wavelength is proportional to density, the difference in average refractive index, and hence the difference in transit time, for the two signals, will be proportional to the average density over the path. If the refractive index of the air is independently known from laboratory measurements as a function of wavelength and density, including the appropriate small corrections for variable constituents, then a measurement of the difference in transit time can be used to give the average density over the path. From this quantity, the average value of the refractive index for either wavelength may be calculated, providing the desired correction. In fact, as is shown in the following equations, the density cancels out of the calculation, and it is actually necessary to know the refractive index only as a function of wavelength to find the correction.

The application of this method to interferometric measurements of distances up to a few hundred metres has been discussed previously (Erickson 1962a). In the microwave range, the principle has been applied to the measurement of average water vapour density over a path by using the difference in refractive index on either side of the absorption line at 22·2 GHz (Sullivan 1964). For the purpose of distance measurement, however, the optical range is much more suitable than the microwave, for the available dispersion is much larger and is almost entirely due to oxygen and nitrogen rather than water vapour.

By extending the frequency range considered, however, the same principle may be applied to the measurement of integrated water vapour density over a path. The dispersion between microwave and optical wavelengths is largely due to water vapour, and is much larger than the dispersion due to water in the microwave region alone. Therefore, a simultaneous measurement over the same path using the two optical wavelengths and one microwave wavelength will permit an accurate evaluation of both the average dry air density and the average water vapour density over the path.

The electromagnetic distance between light source and reflector may be written $L+S$, where L is the geometrical path length and S is the contribution due to the atmosphere. Since modulated light is to be used, S must be expressed in terms of the group velocity U. If the group refractive index, n, is defined by

$$n = c/U \qquad (1)$$

in which c is the velocity of light in vacuum, then S is given by

$$S = \int_0^L (n-1)\, \mathrm{d}x \qquad (2)$$

If the red helium–neon laser line at 6328 Å is chosen for one wavelength and the blue mercury line at 3650 Å for the other, the extra optical paths S_B and S_R for the blue and red light, respectively, will differ by about 10 per cent (Edlén 1953, Svensson 1960). This difference in path, $\Delta S = S_B - S_R$, may be written

$$\Delta S = \int_0^L A(n_R - 1)\, \mathrm{d}x \qquad (3)$$

where

$$A = (n_B - n_R)/(n_R - 1) \qquad (4)$$

Because the group refractivity $(n-1)$ is proportional to density, the quantity A, involving a ratio of refractivities, is independent of the atmospheric density along the path. It is true that $(n-1)$ and A depend slightly on the atmospheric composition, the largest changes arising from the presence of water vapour (Erickson 1962b). However, this is a small effect. For dry air, it may be neglected, and A calculated from laboratory data without the use of pressure and temperature measurements taken along the path to be measured. In the more general case of humid air, we may include the effect of variations in composition over the path by replacing A by its average value $\bar{A}$. This quantity may be taken outside the integral sign in equation (3), giving

$$\Delta S = \bar{A} S_R \qquad (5)$$

In general, therefore, we assume that in addition to published values of the refractive index for the two wavelengths, a rough knowledge of the water vapour content is available from independent electromagnetic or meteorological measurements, so that the proper value of $\bar{A}$ can be calculated. A measurement of the difference ΔS therefore permits the determination of the line integral of the refractive index over the path, giving the atmospheric contribution S_R and allowing the calculation of the true geodetic distance L from the directly measured quantity $L + S_R$.

Because the group refractive index for visible light is about 3×10^{-4}, it is clear that the ratio $\Delta S / \bar{A}$ must be known to an accuracy of 1 part in 300 if L is to be found to 1 ppm. For a typical case in which the wavelengths listed above are used and the geometrical path length L is 15 km, the atmospheric contribution S_R is about 400 cm and the difference ΔS is about 40 cm. For the usual type of optical instrument in which the round-trip transit time is measured, the difference in electromagnetic path length for the two wavelengths is $2 \Delta S$, or about 80 cm. To give an accuracy of 1 part in 300, this must be measured to $\pm 2 \cdot 7$ mm. In order to find $\bar{A}$ to the same accuracy, it may be shown from published refractive index data (Barrell and Sears 1940, Svensson 1960, Erickson 1962a) that the average water vapour pressure over the path must be known to about 8 mb, independent of temperature. Of course, the errors in ΔS and in $\bar{A}$ could not be this large simultaneously if the ratio $\Delta S / \bar{A}$ is to be found to 1 part in 300.

The design of an instrument suitable for measuring both $L + S$ and ΔS is indicated in Fig. 1. The superimposed red and blue light beams are simultaneously modulated in amplitude or polarization at a frequency in the u.h.f. or microwave range (Froome and Bradsell 1961, Kaminow 1961). In the present instrument, a modulation frequency of $2 \cdot 9$ GHz is used. For light of either wavelength passing through the modulator, being reflected from the distance mirror, returning through the modulator (which functions as a toothed wheel) and falling upon a detector, the observed intensity averaged over a time long in comparison with the modulation period will be a maximum if the time required for the light to travel from the modulator to the reflector and back is an integral multiple of the modulation period. For the measurement of $L + S$, the ambiguity resulting from the multiplicity is resolved in the usual way by using two or more modulation frequencies. We can find ΔS by measuring the difference in optical path required to give simultaneous intensity maxima for the red and blue light. A pair of dichroic mirrors, a prism reflector, and suitable filters are used to cause the blue light to traverse an additional path compared to the red. Translation of the prism varies this additional path length, allowing

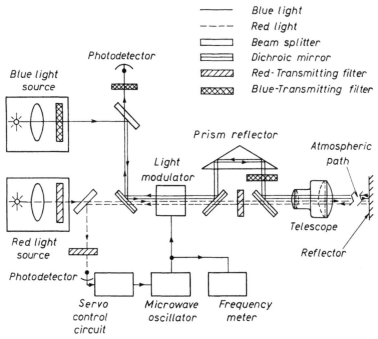

Fig. 1. Schematic diagram of the instrument.

the intensity to be simultaneously maximized. In this condition, the total optical path for the blue light exceeds that for the red by an integral number α of modulation wavelengths λ. The number α may be found from a rough knowledge of the air density. The quantity $2\,\Delta S$ is then found merely by subtracting the contribution of the additional path from the distance $\alpha\lambda$.

A high modulation frequency makes possible a higher precision for a given signal-to-noise ratio compared to existing systems, and also tends to reduce certain systematic errors. It appears possible to measure ΔS to a small fraction of the modulation wavelength even at modulation frequencies of 3 to 10 GHz under good atmospheric conditions. Microwave measurements have indicated that it is not uncommon for the electrical path length through the atmosphere to be stable to a few parts in 10 million over periods of several hours during the night (Thompson and Janes 1964). However, it will probably be desirable to servo-control the modulation frequency so that the observed red light remains a relative minimum during the measurements.

Since we believe that ΔS can be measured to an accuracy of at least

0·3 per cent, the value of L can be found to one part per million or better provided that $L+S$ is sufficiently well known. Therefore, it appears possible to use dual-wavelength optical refractometry to improve the accuracy of optical geodetic distance measurements by about an order of magnitude over present capabilities, and perhaps more.

References

Barrell, H. and Sears, J. E., 1940. The refraction and dispersion of air for the visible spectrum. *Phil. Trans.*, A, **238**, 1–64.

Bender, P. L. and Owens, J. C., 1965. Correction of optical distance measurements for the fluctuating atmospheric index of refraction. *J. Geophys. Res.*, **70**, 2461–2.

Edlén, B., 1953. The dispersion of standard air. *J. Opt. Soc. Am.*, **43**, 339–44.

Erickson, K. E., 1962a. Long-path interferometry through an uncontrolled atmosphere. *J. Opt. Soc. Am.*, **52**, 781–7.

Erickson, K. E., 1962b. Investigation of the invariance of atmospheric dispersion with a long-path refractometer. *J. Opt. Soc. Am.*, **52**, 777–80.

Froome, K. D. & Bradsell, R. H., 1961. Distance measurement by means of a light ray modulated at a microwave frequency. *J. Sci. Instr.*, **38**, 458–62.

Kaminow, I. P., 1961. Microwave modulation of the electro-optic effect in KH_2PO_4. *Phys. Rev. Letters*, **6**, 528–30.

Sullivan, J. F., 1964. Line integral refractometer, SR 112. The Mitre Corp., Bedford, Mass.

Svensson, K. F., 1960. Measurements of the dispersion of air for wavelengths from 2302 to 6907 Å. *Arkiv Fysik*, **16**, 361–84.

Thompson, M. C., Jr. & Janes, H. B., 1964. Radio path length stability of ground-to-ground microwave links. *NBS Tech. Note*, 219.

Thompson, M. C., Jr., Janes, H. B. & Freethey, F. E., 1960. Atmospheric limitations on electronic distance-measuring equipment. *J. Geophys. Res.*, **65**, 389–93.

DISCUSSION

See page 171.

The Use of Atmospheric Dispersion for the Refractive Index Correction of Optical Distance Measurements

M. C. THOMPSON and L. E. WOOD

Institute for Telecommunication Sciences and Aeronomy, *
Environmental Science Services Administration, Boulder, Colorado

The dispersion of the atmosphere to light in the visible spectrum might be large enough to be useful in correcting optical distance measurements. A Model 4D Geodimeter has been used to investigate the magnitude of the dispersion. This magnitude was found to be consistent with the value predicted by the Edlén formula. A programme is under way to improve the precision of the Geodimeter in an attempt to use the dispersion to determine the atmospheric refractive index to 1 ppm for paths of the order of 10 km.

Accurate measurement of long baselines by optical methods requires a correction for the average refractive index of the path. Since the atmosphere is dispersive to light in the visible spectrum, it has been suggested (Bender & Owens 1965) that this dispersion might be useful in determining the refractive index of the path.

When the path is measured using two different wavelengths of light, different optical distances will be obtained owing to the dispersion. From these distances, the average refractive index can be computed.

A commercially available optical distance measuring instrument, a Model 4D Geodimeter, has been used to investigate the magnitude of the atmospheric dispersion. A programme is under way to improve the precision of this instrument in an attempt to use the dispersion to determine the average path refractive index to an accuracy of at least 1 ppm. This work has been supported by the United States Coast and Geodetic Survey.

Refractive Index Determination

Optical distance measurements by Geodimeter-type instruments are performed by measuring the transit time over the path of a modulated

* Formerly designated the Central Radio Propagation Laboratory of the National Bureau of Standards.

optical carrier. This method requires that the group refractive index be used when computing the distance from the transit time. Using the Edlén (Edlén 1953) formula, for phase refractive index, the group refractive index $n_g(\lambda)$ for standard conditions (15°C, 760 mm Hg, dry air, 0·03 per cent CO_2) is computed to be:

$$[n_g(\lambda) - 1] \times 10^6 = \left\{ (64\cdot328 + 29\,498\cdot10) \frac{[146 + (1/\lambda)^2]}{[146 - (1/\lambda)^2]^2} \right.$$

$$\left. + 255\cdot40 \frac{[41 + (1/\lambda)^2]}{[41 - (1/\lambda)^2]^2} \right\} \tag{1}$$

where λ = wavelength in Å.

Table 1 gives values of $(n_g - 1) \times 10^6$ for selected mercury arc wave-

TABLE 1. *Values of* $(n_g - 1) \times 10^6$ *for selected mercury arc wavelengths*

i	Wavelength Å	Colour	$(n_{gi} - 1) \times 10^6$	$(n_{g1} - n_{gi}) \times 10^6$
0	6907	red	282·55	4·40
1	5791	yellow	286·95	0·00
2	5461	green	288·83	− 1·88
3	4358	blue	298·85	−11·90
4	4047	violet	303·46	−16·51
5	3650	u.v.	311·31	−24·36
6	3126	u.v.	328·15	−41·20

Standard conditions: 15°C, 760 mm Hg, dry air, 0·03 per cent CO_2.

lengths. The table also shows the dispersion $(n_{g1} - n_{gi})$ for the various mercury arc wavelengths relative to the yellow (5792 Å) line.

For non-standard conditions, the refractive index of air is given to a first approximation by

$$[n_{air}(\lambda) - 1] = [n_g(\lambda) - 1] \frac{P}{T} \frac{288°\text{K}}{760 \text{ mm Hg}} \tag{2}$$

where

P = total air pressure in mm Hg

T = temperature in °K

If a path is measured at wavelength λ_1, the distance obtained, L_1, is related to the true geometric path L, as

$$L_1 = \bar{n}_1 L \tag{3}$$

where $\bar{n}_1 \equiv n_{air}(\lambda_1) =$ the average refractive index over the path. Likewise, at λ_2, the distance measured is

$$L_2 = \bar{n}_2 L \tag{4}$$

Combining (2), (3) and (4), we obtain

$$(L_1 - L_2) = \frac{(n_{g1} - n_{g2})}{(n_{g1} - 1)} L(\bar{n}_1 - 1) \tag{5}$$

letting $n_{g1} \equiv n_g(\lambda_1)$ and $n_{g2} \equiv n_g(\lambda_2)$. Solving for $\bar{n}_1$,

$$\bar{n}_1 = \frac{\alpha L_1}{\alpha L_1 - (L_1 - L_2)} \tag{6}$$

where

$$\alpha = \frac{(n_{g1} - n_{g2})}{(n_{g1} - 1)}$$

From (5), the sensitivity of $(L_1 - L_2)$ to changes in $\bar{n}_1$ is then

$$\delta(L_1 - L_2) = \alpha L \, \delta\bar{n}_1 \tag{7}$$

For the $(\lambda_1 = 5791$ Å$)$ yellow and $(\lambda_2 = 4047$ Å$)$ violet mercury arc lines $\alpha = 0\cdot058$. Thus for a 10-km path, a resolution of $0\cdot058$ cm is required in the determination of $(L_1 - L_2)$ if $\bar{n}_1$ is to be determined to 1 ppm.

A Comment on the Expected Accuracy of the Two-Wavelength Technique

The two-wavelength technique for the determination of the atmospheric refractive index should be accurate to the extent that (1) and (2) are accurate.

In the determination of the path refractive index by the two-wavelength technique, an error, δ, in the estimation of the dispersion causes an error of δ/α in the estimation of the total refractive index since α is the indicator by which the total refractive index is determined.

The magnitude of n_{g1} obtained from (1) is probably accurate to at least $0\cdot01$ ppm (Svensson 1960) and therefore the computation of $(n_{g1} - n_{g2})$ should be good to $0\cdot02$ ppm. This could result in an error of about $0\cdot3$ ppm in the refractive index determination when using the yellow (5791 Å) and violet (4047 Å) lines.

The use of (2) introduces a more serious error. A more nearly correct form of (2) which includes the effect of water vapour is (Barrell 1951, Erickson 1962)

$$n_{\text{air}}(\lambda) - 1 = [n_g(\lambda) - 1]\left(\frac{P}{T}\,\frac{288°\text{K}}{760\text{ mm Hg}}\right) - \left[17\cdot 0 - 0\cdot 186\left(\frac{1}{\lambda}\right)^2\right]\frac{f}{T}$$

(8)

where P is the total air pressure in mm Hg, T is the temperature in °K, and f is the water vapour pressure in mm Hg. The second term in the equation is a non-dispersive term that reduces the total refractive index and the last term is a dispersive term that contributes positively to the dispersion; thus, the errors introduced by both terms are additive. Table 2 shows the contribution of these terms to the error at various

TABLE 2. *Water vapour contribution to the atmospheric refractive index*

Temperature (°K)	Water vapour pressure f for 100 % relative humidity (mm Hg)	Non-dispersive term of the water vapour (ppm)	Dispersive term of the water vapour for yellow light (5791 Å) (ppm)	Dispersive term of the water vapour for violet light (4047 Å) (ppm)	Error in determination of $\bar{n}_1$ due to water vapour dispersion (ppm)	Total error in determination of $\bar{n}_1$ due to water vapour (ppm)
263	2·15	−0·14	0·014	0·028	0·25	0·39
273	4·58	−0·27	0·028	0·057	0·51	0·78
283	9·21	−0·55	0·054	0·111	1·16	1·71
293	17·54	−1·02	0·099	0·203	1·80	2·82
303	31·82	−1·79	0·174	0·336	2·82	4·61
313	55·32	−3·01	0·293	0·601	5·35	8·36

temperatures and 100 per cent relative humidity for the yellow and violet lines. From this table, it may be seen that for conditions of 40°C and 100 per cent relative humidity, the average water vapour pressure over the path would have to be estimated to within 10 per cent to allow the determination of $\bar{n}_1$ to 1 ppm. For the more favourable conditions of 20°C and 50 per cent relative humidity, a 50 per cent estimate of the average vapour pressure over the path is required to determine $\bar{n}_1$ to 1 ppm.

In principle, of course, a three-wavelength measurement could be performed to determine the water vapour dispersion and thereby to calculate the water vapour contribution to the refractive index. Since the maximum water vapour dispersion is only about 0·3 ppm (yellow and violet lines), this would require the accuracy of measurement of $(L_1 - L_2)$ to be roughly two orders of magnitude greater than is being considered here.

Refractive index gradients perpendicular to the line-of-sight can contribute errors. As these gradients cause bending of the ray paths, the actual distances the rays travel are unequal. Also, since the rays are no longer precisely coincident they may be subjected to slightly different refractive conditions. Work is proceeding in this laboratory to estimate the magnitude of this effect but it is not expected to be a limitation under most conditions, for accuracies of 1 ppm.

Other sources of error are pressure effects on (2), the variability of the atmospheric CO_2 content, etc. Rough calculations indicate that these effects might contribute errors as large as parts in 10^7.

Preliminary Experiments

A Model 4D Geodimeter was used in preliminary measurements to determine the magnitude of the dispersion. In order to use the Model 4D, two minor modifications were required.

The first modification was to replace the internal frequency standard. This standard, which has a stability of about 2 ppm, was replaced by an external standard which has a stability of about 0·01 ppm. This essentially eliminated errors due to oscillator drift.

The second modification was to change the transmitting optical system to pass wavelengths shorter than 4400 Å. This was accomplished by replacing the nitrobenzene in the Kerr cell with orthodichlorobenzene. Nitrobenzene strongly attenuates wavelengths shorter than about 4300 Å, but orthodichlorobenzene has an optical band pass to about 3000 Å. The orthodichlorobenzene requires about twice the driving voltage for the same modulation index as nitrobenzene. It was found, however, that satisfactory performance of the Model 4D could be obtained with the available drive, provided that the 60 Hz bias voltage for the cell was doubled. Additionally, an optical filter was removed from the condenser optical system between the mercury arc and the Kerr cell. This filter absorbs wavelengths shorter than 4400 Å, and is used in the standard instrument to help prevent deterioration of nitrobenzene in the Kerr cell.

Fig. 1 shows the optical spectrum of the output of the Geodimeter before the modification. Fig. 2 shows the output after the modification. In the latter case, wavelengths to 3500 Å are transmitted.

The wavelengths used in the preliminary experiments were selected from the mercury arc spectrum by using optical interference filters. These filters were centred on strong mercury arc lines and were about 100 Å wide at the 10 per cent transmission points.

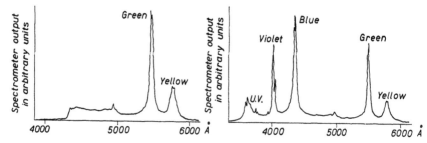

FIG. 1. Spectrum of mercury arc through Geodimeter optics, with no modification of the instrument.

FIG. 2. Spectrum of mercury arc through Geodimeter optics, using orthodichlorobenzene Kerr cell and no condenser filter.

Table 3 shows typical results of measurements in the Boulder, Colorado, area. The measured path length differences are compared with values of the path length difference estimated from meteorological data taken at the path terminals. This estimation was made on the basis of the simple assumption that the path was homogeneous between its terminals.

At least ten determinations of the path length difference were used to calculate one average measured path length difference. It required approximately one hour to obtain these ten determinations. At the beginning and end of this time, the temperature and pressure were recorded at both terminals of the path. The average of these meteorological data was used to determine estimated path length difference. Table 3 shows that the measured path length differences are consistent with the estimated path length differences.

Geodimeter Modification

An attempt is being made to increase the sensitivity of the Model 4D Geodimeter in order to measure $(L_1 - L_2)$ to at least 0·05 cm. The system being developed will provide for the continuous and simultaneous determination of $(L_1 - L_2)$.

The light signal received at the Geodimeter contains components at λ_1 and λ_2 modulated at 30 MHz. These components, which can be selected by optical interference filters, have a difference phase owing to the atmospheric dispersion. It should be possible to measure this 30 MHz phase difference to the required 0·02°, and a phase measuring instrument capable of this sensitivity has been built by this laboratory. However, considerable work is still necessary to use this with optical signals. Careful attention must be given to such problems as transit time

Table 3. *Results of distance measurements on modified model 4D Geodimeter using various mercury-arc wavelengths*

Wavelengths: Y, 5791 Å; G, 5461 Å; B, 4358 Å; V, 4047 Å.

Wavelengths	Path length (km)	Average measured path length difference in ppm	σ	Path length difference estimated from meteorological observation at terminals in ppm
Y–V	7·48	14·3	0·7	13·4
Y–V	7·48	12·6	0·3	13·5
Y–B	15·0	9·1	0·3	10·0
Y–B	7·48	9·0	0·4	9·6
Y–B	7·48	10·4	0·3	9·8
Y–G	7·48	1·3	0·4	1·5
Y–G	7·48	1·2	0·4	1·6
G–V	7·48	12·5	0·5	11·8
G–V	7·48	11·6	0·4	11·9
G–B	15·0	8·5	0·6	8·5
G–B	7·48	9·1	0·3	8·2

Note: In the results of the estimated path length differences, 2 per cent error may be contributed due to uncertainties in the measurements of the meteorological parameter.

variations in the photomultipliers and the effects of amplitude changes caused by the scintillation of the light beam. The results of preliminary work encourages optimism for the completion of a successful system.

References

Barrell, H., 1951, The dispersion of air between 2500 Å and 6500 Å. *J. Opt. Soc. Am.*, **41**, 295–9.

Bender, P. and Owens, J. C., 1965. Correction of optical distance measurements for the fluctuating atmospheric index of refraction. *J. Geophys. Res.*, **70**, 2461–2.

Edlén, B., 1953. The dispersion of standard air. *J. Opt. Soc. Am.*, **43**, 339–44.

Erickson, K. E., 1962. Long-path interferometry through an uncontrolled atmosphere, *J. Opt. Soc. Am.*, **52**, 781–7.

Svensson, K. F., 1960. Measurements of the dispersion of air for wavelengths from 2302 to 6907 Å. *Arkiv Fysik*, **16**, 361–84.

DISCUSSION

This discussion refers to the paper by L. E. Wood and M. C. Thompson as well as that by J. C. Owens and P. L. Bender.

K. D. Froome: Are the authors happy about the shape of the dispersion curve?

J. C. Owens: The absolute value of the refraction coefficient may only be known to an accuracy of about 1 in 10⁷, but the relative values between different colours is much better known. To achieve an accuracy of 1 ppm the dispersion of water vapour is sufficiently well known.

R. C. A. Edge: Was any of this work applied to microwaves?

J. C. Owens: Very small dispersion due to water vapour did not allow this technique to be used.

R. C. A. Edge: Could the equipment be engineered for field use?

J. C. Owens: Plans to do this are in hand and it is expected that a reasonable size and weight will be achieved.

J. J. Gervaise: Have the authors tried lasers of two different wavelengths?

J. C. Owens: Suitable lasers were not freely available. An argon-ion laser has been considered, and it is planned to include one as soon as it can be obtained.

G. D. Robinson: Has anyone had any experience of using a laser over any long distance in the atmosphere?

J. C. Owens: The air is disturbed and dirty and this is troublesome. The Bureau of Standards had been doing some work on the problem. There is no doubt that the laser beams wander and break up, and may have to be re-focused somewhat in order to receive the returning beam reliably. Moreover, the effects of the atmosphere on laser radiation are no more than on conventional incoherent light, and the greatly increased intensity of light from a laser makes its use highly desirable.

Microwave Systems

The Tellurometer Model MRA-101*

J. A. WEBLEY

Tellurometer (Pty) Ltd

The new Model MRA-101 is designed to meet the long-felt
need for a Tellurometer which will be capable of providing
higher accuracy for short ranges, but at a price which will
bring electronic distance measuring within the reach of
commercial surveyors. Among the several important features of the MRA-101 described in this paper is the revolu-
tionary new antenna. This highly efficient antenna virtually
eliminates the shift of the instrument zero calibration with
changes in carrier frequency—thereby making accurate
short range measurements possible. The larger reflector
diameter and narrower beam width further improve accuracy
by reducing multi-path reflection errors and facilitate
operation in built-up areas and among moving traffic. Also,
the use of a pressure moulded glass-fibre case and other
modern design techniques have resulted in a compact,
lightweight robust instrument (the weight of MRA-101 is
almost half that of any previous model).

The paper concludes with a summary of results of carefully controlled accuracy trials.

Tellurometer first introduced electronic distance measurement to the
world early in 1957 with their Model MRA-1, and since then the
Tellurometer system has been proved internationally over several
millions of miles of traverses over all types of terrain and under the most
varied and severe climatic conditions. With the continued successful use
of Tellurometers in long range applications, it became increasingly
apparent that there was a strong need for a commercial (non-military)
instrument which would have sufficient accuracy to enable it to be
used for medium to short ranges—especially in built-up areas. The
model MRA-101 was designed to fill this need at a price which would
bring electronic distance measuring within the reach of commercial
surveyors. The following design requirements were satisfied:

1. Short range accuracy of ± 1·5 cm (see Section 4).
2. Narrower beam width to facilitate measurements in built-up areas
   and generally reduce multi-path reflection errors.
3. Incorporation of the latest circuit design techniques.

* This work was supported by the United States Coast and Geodetic Survey.

4. As survey instruments must essentially be *portable*, weight reduction was of significant importance.
5. Modern construction and layout techniques were employed.
6. Robustness: to meet relevant US 'MIL' Spec. for vibration and shock tests.
7. Ambient operating temperature range − 10° to +50°C.

1. *Instrumentation Improvements*

With the improvement in phase measuring technique, where it becomes possible to measure phase to one thousandth of a complete phase rotation, a shortcoming in antenna design became apparent; this showed as a shift in the electrical centre of the instrument with change in carrier frequency.

In this new model, we are introducing a completely new antenna design, which not only substantially eliminates this source of error, but also results in a cost reduction due to its inherent simplicity of design. The basic propagation principle is based on the Cassegrain optical telescope. In this antenna design, the microwave power is radiated from the open end of a circular waveguide on to a small 'Cassegrain' reflector situated just in front of the waveguide, and then is reflected back into the main parabolic reflector. The mixer crystal is situated across the waveguide in such a position that it does not receive any power from the klystron. The mixer current is then introduced by a small adjustable deflector protruding from the wall of the waveguide. This allows very close control of the phase of the signal which is fed to the mixer, and accounts for the very small amount of electrical zero-shift with carrier frequency that exists in the MRA-101.

Coupled with the new antenna is a 13-in. parabolic reflector which reduces the transmitter and receiver beam-width to 6°. This has resulted in a decrease in multi-path effects, and improves the instrument's accuracy especially on short lines where the error introduced by these effects has the most adverse influence on the overall accuracy.

2. *Circuit Design*

Starting with the new, well proved transistor circuits which are used in our Model MRA-3, the question was put forward: what further improvement in circuit design could be incorporated without adversely affecting the price structure? A significant improvement has been added by the reintroduction of 'phase-lock'. Manual phase-lock was first used in the MRA-1, where the remote operator set up the crystals before each set of readings. The incorporation of *automatic* phase-lock of the remote pattern frequencies has enabled the number of crystals used to be reduced from eleven to six, without detracting from the accuracy of the equipment in any way. In fact, because of the complete elimination of frequency dependence effects in the phase measuring circuitry, accuracy improvements have resulted. The other advantage gained is that we have been able to use a smaller crystal oven of advanced design, which gives very much improved temperature stability with a reduction of battery power requirements.

3. *Layout and Construction*

MRA-101 embodies a simplified circuit layout whereby all the components are accommodated on a single printed circuit board—the only exceptions being the IF amplifier and the Power Supply Unit. This important feature has not only simplified construction, assembly, test and alignment, but has also further improved reliability by reducing the number of soldered connections and plugs and sockets which are always a potential source of failure.

This unitary construction is certainly no retrograde step as far as the service man is concerned either—MRA-3 has shown that the use of high grade components, coupled with conservative design has resulted in a degree of reliability that now renders modular construction unnecessary. Servicing is actually facilitated by the new construction since the complex inter-connecting wiring (with its incumbent wiring looms) has been completely eliminated in MRA-101. Part number markings are silk-screened on to the printed circuit board to identify each component, and adequate test jacks are provided to further facilitate servicing and checking.

The use of a pressure moulded glass-fibre case, has enabled a compact, lightweight, yet robust instrument of almost half the weight of any previous model to be produced. The parabolic reflector is an integral part of the case and is made by metal spraying the surface of the case. A custom-made pressure moulded glass-fibre *transit* case is provided with each instrument. This results in an extremely lightweight, readily portable package, which should withstand the roughest handling encountered in field use.

4. *Field Measurements and Accuracies Obtained*

4.1 *Accuracy*

It has been the custom for manufacturers of this type of equipment to base their accuracy claims on measurements taken under ideal conditions and with instruments which have been specially calibrated. For example, in the 10 Gc/s series of instruments, manufacturers claim similar accuracies of approximately 1 cm ± 3 ppm. This can be achieved with the instruments but implies:

(a) Optimum line conditions, i.e., a line which has minimum reflections and a uniform refractive index along its length.
(b) Special zero calibration of the two instruments being used.
(c) Optimum adjustment and performance of the instruments.

As the majority of users do not operate the instruments under such conditions, it would seem desirable to quote accuracies which are also obtainable under average conditions.

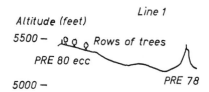

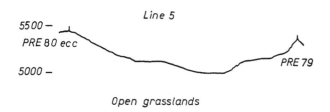

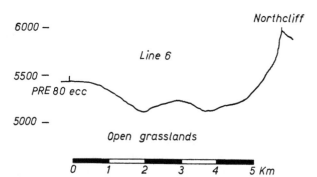

Fig. 1. Profiles of lines: Honeydew test area. *ecc.* = eccentric station.

The need for an indication of the realistic accuracies obtainable by this type of instrument under field conditions has become apparent, and it is our intention to give an indication in future of both the ideal and the practical accuracies applying to instruments of this nature. An estimate of the ideal accuracies can easily be obtained from carefully controlled field tests. To obtain an indication of the accuracies which can be obtained with MRA-101 under field conditions, a series of tests was carried out. The tests included the measurement of all the possible lines on a polygon at the Honeydew Test Site a few miles north-west of Johannesburg (Fig. 1), and a series of measurements taken over town survey marks in Rustenburg (Transvaal) and Cape Town.

The instruments used in the field tests were neither zero calibrated nor specially aligned or adjusted, and could be taken as representing average commercial models.

4.2 *Honeydew Test*

All ten lines in a five-sided polygon were observed (see Fig. 2). The points chosen were suggested by the Department of Land Surveying

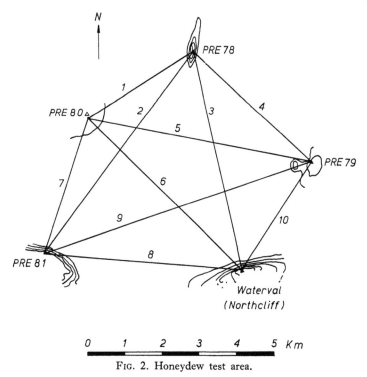

FIG. 2. Honeydew test area.

of the University of Witwatersrand. These proved to be particularly unsuitable lines for Tellurometer measurements. Of the five stations, three had almost sheer drops of several hundred feet between the two stations. The country between the stations was gently undulating grassland with little or no bush and trees.

Ten 'fine' readings were taken in each direction. As expected, the swings observed (Fig. 3) were rather larger than is usual with 10 Gc/s instruments. The minimum swings were ±5 cm on line 1 and the maximum ±60 cm on line 10.

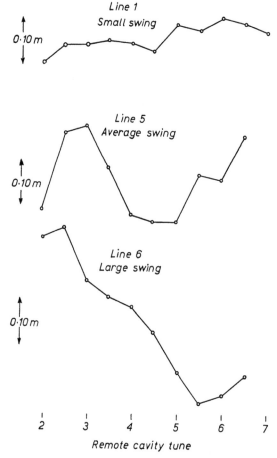

FIG. 3. Typical swings on Honeydew test lines.

Line 1 was measured through two, and possibly three, rows of trees with no adverse effect, although the signal strength was about 50 per cent of what it should have been had the line been clear. Other lines from PRE 80 *ecc* (see Fig. 2) passed between trees, and a slight fluctuation of the null indication was observed in windy conditions, but this did not cause any trouble in ascertaining the null.

In order to be able to obtain the accuracies of the lines without these being affected by other factors, such as accuracy of measured angles, etc., the exercise was treated as a pure trilateration scheme—although this would not be very common in practice. The results of a free net least squares adjustment are tabulated below:

Line	Observed length (m)	Error (m)	Accuracy
1	3423·556	+0·028	1:122,000
2	6812·072	−0·014	479,000
3	5990·468	−0·020	300,000
4	4452·521	+0·025	178,000
5	6349·153	+0·008	793,000
6	5893·070	−0·028	210,000
7	3827·956	+0·034	113,000
8	5506·972	+0·049	112,000
9	7837·767	−0·053	148,000
10	3466·728	+0·043	60,000

Using only two instruments, the observing programme was completed in two days—an average of 1 hour per line—this time being inclusive of warm-up time, measurement time, and travelling time.

4.3   Town Traverse—Rustenburg

Here the object was to obtain measurements in a busy town street, in order to investigate instrument accuracies under these conditions. The lines ran along the pavement of a main street, with various sizes of buildings on the one side and a row of parked cars on the other. Several side streets crossed the lines and the instruments were mounted sufficiently high to clear parked and moving cars. Large vans did, however, obstruct the line occasionally but this did not cause undue delay. The very heavy pedestrian traffic on the lines appeared to have no effect at all.

Three lines were measured and their agreement with the town survey values was as follows:

Line	Observed length (m)	Agreement (m)
RM 48–RM 49	108·196	−0·017
RM 49–RM 50	123·749	+0·011
RM 50–RM 51	92·795	−0·009

7

4.4   *Town Traverse—Cape Town*

Surveying across a multi-lane highway with normal traverse methods is well-nigh impossible due to the difficulty of taping along a stream of fast-moving traffic. Along Rhodes Drive, one of the busiest freeways in Cape Town, a Class A traverse of seven legs which ran along and across the lower carriageway was available, and it was decided to measure these legs with MRA-101. The fast-moving traffic had no noticeable effect on the accuracy.

The results are tabulated below and a sketch is shown in Fig. 4.

Line	Observed distance (m)	Agreement (m)
30 M 13–31 M 13	130·540	+0·023
31 M 13–32 M 13	172·594	+0·039
32 M 13–18 M 12	163·574	+0·018
18 M 12–19 M 12	144·003	+0·009
19 M 12–20 M 12	165·501	+0·022
20 M 12–21 M 12	80·347	+0·017
21 M 12–22 M 12	145·291	+0·020

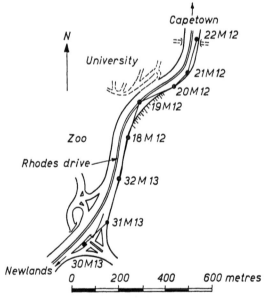

FIG. 4. Town traverse—Cape Town.

From the above limited tests it appears that the following accuracy claim can be made for the MRA-101 Tellurometer:

(*a*) Short line traverses
  (i) Favourable conditions   ± 1·5 cm
  (ii) Average conditions   ± 2·5 cm
(*b*) Medium and long line traverses
  (i) Favourable conditions   ± 1·5 cm   ± 3 ppm
  (ii) Average conditions   ± 2·5 cm   ± 4 ppm

5. *Acknowledgements*

We wish to thank the following organizations for the valuable assistance given during these tests:

Survey Dept. of the University of the Witwatersrand
Survey Dept. of the University of Cape Town
Trigonometrical Survey Office
National Institute for Telecommunications Research of the South African Council for Scientific and Industrial Research.

DISCUSSION

R. C. Cox: Would the cut in the cost of the instrument increase maintenance costs?

J. A. Webley: As this was a very simple instrument, intended for use by the ordinary surveyor, it should be easy and cheap to maintain. The cost of one instrument (without meteorological accessories) should be less than £1,000.

C. C. Brown: Some other tests on this new instrument were made. They were done over some very difficult lines, including some steep hills and some lines over water. The errors, compared with a Geodimeter, were about 0·09 ft for lines up to 10 000 ft. Operators could soon be trained to do a two-way measurement in 20 minutes.

T. Convey: What are the weight and power supplies of this instrument?

J. A. Webley: 25 lb and 36 W at 12 V. There was no internal battery as these small batteries were very expensive.

J. R. Hollwey: Had any tests been done in trenches or in narrow streets between tall buildings?

J. A. Webley: Work had been done along highways with moving traffic, and also along valleys, but not specifically between tall buildings.

W. G. Collins: Could details of the range be indicated?

J. A. Webley: In average conditions the MRA-101 has a maximum range of 50 km; its minimum range was 50 m.

Principles and Performance of a High-Resolution 8-mm-Wavelength Tellurometer

P. J. CABION

National Institute for Telecommunications Research: C.S.I.R.

In this paper, instrumental and propagation problems encountered in distance measurement by the Tellurometer principle are considered. It is concluded that the use of an 8-mm carrier wavelength would significantly reduce ground effects and that the use of higher modulation frequencies would increase the resolution of the system. Equipment has been built and tests undertaken to demonstrate the feasibility and performance of such a system. Preliminary results are given.

Introduction

In 1958, T. L. Wadley[1] suggested the use of an 8-mm wavelength as a carrier for Tellurometers, although at the time the necessary components were not readily available. The first instruments operated at a carrier wavelength of 10 cm, and later at 3 cm. This paper discusses the development of distance measuring equipment operating at 8 mm.

The Tellurometer Principle

Consider two oscillators or clocks M and S separated by an electromagnetic transit time t. If continuous signals transmitted by M and S are received and mixed with signals at S and M respectively, and if the angular frequencies (rad/sec) $Pm - Ps = A$, the phase lag from M to S is tPm and from S to M is tPs. It can be seen from vector diagrams that the low frequencies A have a phase difference $t(Pm + Ps)$ and on telemetering A from S to M, the measurable phase at M becomes $2tPm$ (see Fig. 1). By measuring phase and fixing Pm at various 'pattern' frequencies, t, and hence distance, are uniquely determined.

The pattern signals Pm and Ps are actually carried on microwave carriers in the form of frequency modulation of the carriers to eliminate propagation effects at Pm and Ps. Again it can be shown that the mixing of the carriers gives intermediate frequencies amplitude modulated at

184

A, where A, the detected frequencies, have the same phase as in the previous case (see Fig. 2). The comparison frequency A is normally 1 kc/s for ease of measurement.

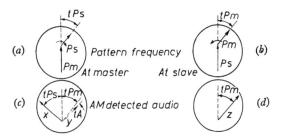

FIG. 1. The Tellurometer principle illustrated by two clocks.

   (a) Phases assumed for Pm and Ps are arbitrary, since rotation of either causes all audio vectors to rotate equally.

   (b) By convention vectors rotate anti-clockwise. A delay causes a backward rotation proportional to delay and to angular frequency.

   (c) Measured phase is angle between x and y, which equals $t(Ps+Pm+A) = 2tPm$.

   (d) Shown for $Pm > Ps$. If $Ps > Pm$, a negative phase $2tPm$ results. The total change of phase measured is $4tPm = 4DPm/C$.

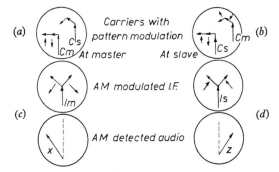

FIG. 2. (a) The phase assumed for carriers and modulation are arbitrary.

   (b) Small arrows indicate the direction of rotation of modulation side frequencies relative to their carrier.

   (c) I.F.'s not generally in phase. Information is in relative positions of I.F. and modulation side frequencies.

   (d) Primary A.M. modulation may be represented by rotating carriers through 90°, leaving I.F.'s unaltered. If $Ps > Pm$, direction of rotation of side tones relative to I.F. is reversed. Measured phase is reversed as in Fig. 1.

The diagram shows the case of low modulation index frequency modulation, but is equally applicable to amplitude modulated carriers. High modulation indices give rise to audio harmonics which must be filtered out and which also represent loss of carrier power. It was pointed out by Dr Wadley that for frequency modulation, the mixing process is independent of the law of the detector. It can, however, be shown that for amplitude modulation into mixers other than of a square law type, even though the carrier harmonic mixing gives rise to a reduction in the intermediate frequency modulation index, it leaves the phases of the audio unchanged.

Limitations to Accuracy

The accuracy of distance measurement by Tellurometers is limited by the accuracy of meteorological measurements, by instrumental errors, and by ground reflections. It will be shown that there is no appreciable change in meteorological effects due to the use of 8 mm. Instrumental accuracy is dependent mainly on the precision of phase measurements and on the pattern modulation frequency. The effect of ground reflections is to cause variations of distance measured with small variations in carrier frequency.

It was shown by Dr Fejer[1] that if the effective amplitude of reflections is reasonably small the effect is to perturb the measured phase angle by an amount

$$\alpha = 2 \tan^{-1} a \cdot \sin \Psi' \cdot \cos \varphi$$

or the measured distance by

$$s = a\Delta \cos \varphi$$

if

$$\Delta \ll \lambda_m$$

where

$$\Delta = \text{excess path to reflection } a$$

$$\varphi = \frac{2\pi\Delta}{\lambda_c} + \text{constant}$$

and

$$\Psi' = \frac{2\pi\Delta}{\lambda_m}$$

Poder has extended Dr Fejer's analysis for single reflections without the assumption of small reflection coefficients. His formula can be slightly simplified to

$$\alpha = \alpha_m + \alpha_s = \tan^{-1} \cfrac{a \sin \psi_m}{a \cos \psi_m + \cfrac{1 + a \cos \varphi_m}{a + \cos \varphi_m}}$$

$$+ \tan^{-1} \cfrac{a \sin \psi_s}{a \cos \psi_s + \cfrac{1 + a \cos \varphi_s}{a + \cos \varphi_s}}$$

The factor $(1 + a \cos \varphi)/(a + \cos \varphi)$ has minimum absolute values of $\pm$ unity when the resultant signal strength is a maximum or a minimum giving maximum and minimum values of

$$\alpha_{max} = + \tan^{-1} \cfrac{a \sin \psi}{1 + a \cos \psi}$$

and

$$\alpha_{min} = - \tan^{-1} \cfrac{a \sin \psi}{1 - a \cos \psi}$$

The swing components α_m and α_s are not exactly in phase owing mainly to the slight difference between carrier frequencies. Dr Fejer also derived these maximum and minimum values.

The Magnitude of Ground Effects

If the beam widths of the antenna units are sufficiently broad to illuminate a surface between M and S, the signal received will be the sum of the direct wave and of a ground reflection. Fig. 3 shows the case of specular reflection with instruments tilted for minimum ground effects.

It will be noted that if a uniphase antenna illumination is used it is theoretically possible to completely nullify the reflected signal at any grazing angle.

In practice, however, the null between the main beam and the side lobes is removed by a variation of phase in the illumination of the paraboloid. In addition it would be difficult to ensure that the null was directed exactly at the point of reflection. Thus in general the effective reflection coefficient is determined by the reflection coefficient of the surface material, by the polar diagrams of the antennae and by the surface shape and roughness as influenced by the grazing angle at the

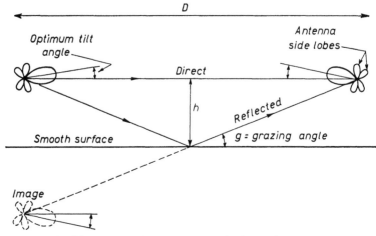

FIG. 3. Diagram showing that, theoretically, reflections can be eliminated by directing the null of either antenna towards the point of reflection, even if the grazing angle is less than the beam width.

surface, i.e.,

$$a = a_m \cdot a_p \cdot a_d \cdot a_s$$

m = material
p = polar diagram
d = shape, e.g., divergence
s = scattering due to roughness
a = effective reflection coefficient

Fresnel's equations[3] show the dependence of the specular reflection coefficient on the dielectric properties of the surface, and on the grazing angle and polarization for a smooth surface. In general these predict values approaching unity for small grazing angles.

At long range over water surfaces where $a_p \cdot a_m \cdot a_s$ are likely to approach unity, reflections are always divergent owing to the spherical shape of the Earth.

Ground reflection coefficients are usually small owing to absorption and scattering, but the path geometry is often convergent, i.e., surface shape can focus energy towards the receiver, and in principle the reflected power could exceed the direct power if there were no protection from the polar diagrams of the antennae, or scattering and absorption from the rough surface of the ground.

It is known that any surface acts as a specular reflector only if it is smooth in relation to carrier wavelength and grazing angle. Lord

Rayleigh suggested a surface be regarded as smooth if

$$\frac{\lambda}{8} > d \sin g$$

where g = grazing angle and d = height of surface perturbations. In addition the area which must be smooth for specular reflection increases rapidly with decrease in grazing angle. On this basis alone a higher carrier frequency should decrease ground effects. By using a higher carrier frequency it is also practical to decrease the antenna beam width which ensures that either the effective reflection coefficient is small or that a path having a high reflection coefficient has a small excess path.

Several attempts have been made to formalize the Rayleigh roughness criterion. A flat surface perturbed by a height change d introduces a phase change $\rho = 4\pi(\sin g \,.\, d)/\lambda$ radians. For a number of such surface perturbations

$$\bar{\rho}^2 = \left(\frac{4\pi \sin g}{\lambda}\right)^2 \times \bar{d}^2$$

From Beckmann's[5] analysis assuming normally distributed slopes on elemental perturbations from a mean flat surface, a scattering reflection coefficient:

$$A_s = e^{-\frac{1}{2} \cdot \bar{\rho}^2}$$

is derived. Hence

$$A_s \doteqdot \mathbf{I} - \frac{\bar{\rho}^2}{2} \quad \text{for small } \rho$$

From Ruse[6] for small perturbations having a correlation length c

$$A_s = \mathbf{I} - \frac{3}{4}\left(\frac{C\pi}{\lambda}\right)^2 \rho^2 \quad \text{when} \quad \frac{c}{\lambda} \ll \mathbf{I}$$

$$A_s = \mathbf{I} - \bar{\rho}^2 \qquad\qquad \text{when} \quad \frac{c}{\lambda} \gg \mathbf{I}$$

Another approach to scattering is to consider a surface periodically perturbed by a height change giving rise to an RMS phase change ρ leading to

$$a_s = J_0{}^2 \rho\sqrt{2}$$
$$= \mathbf{I} - \rho^2 \quad \text{for small } \rho \quad \text{(see Appendix 3)}$$

The energy is then scattered in fixed directions, being more widely

7*

dispersed the shorter the surface disturbances are, thus being scattered more widely vertically than horizontally.

It is feared that these formulae are in error at grazing angles less than the slopes of surface anomalies, owing to local diffraction and reflection effects.

Experiments[4] at long range over land show small reflections at wavelengths less than 10 cm. At short range over flat dry sand, good agreement with Fresnel's equations was obtained. Short grass, however, caused an appreciable reduction in *a*. Measurements over flat sand raked into ridges confirm that the trend predicted by Rayleigh is true, it being reported that if

$$\frac{d \sin g}{\lambda} = \frac{1}{5} \quad \text{then} \quad A_s \doteqdot 0 \cdot 5$$

and if

$$\frac{d \sin g}{\lambda} = \frac{1}{2} \quad \text{then} \quad A_s \doteqdot 0 \cdot 1$$

Interference measurements of reflection at long range over sea have been conducted at 3 and 10 cm using aircraft. Fairly good agreement with Fresnel equations was obtained, though the grazing angles were generally greater than those which could cause trouble with the 8-mm Tellurometer. Experiments at 12·5-mm wavelength across a harbour showed values of a_v from 0·5 to 0·6 at grazing angles of 0·2°. Experiments at 5·81 and 6·35 mm at ranges of 1 to 10 km across the sea at grazing angles from 1·0 to 0·1° showed a_v from 0·25 to 0·8; the sea was reported to be calm, having ripples less than 12 in. and low swell.

Applying formulae of the type $a_s = \exp(-K\rho^2)$ to experiments reported, it appears that values of $K = 0·1$ give reasonable agreement for the sea, though more objective measurements of the surface state would be desirable. Even with low values of K and low grazing angles, it appears that the surface reflection coefficient depends mainly on scattering at 8 mm.

Advantages of an 8-mm Carrier Wavelength

It is clear that the effect of surface roughness is to decrease the reflection coefficient and that the higher the carrier frequency, the greater the effective roughness of any surface.

For a given size of antenna unit and therefore of instrument, the beam width of the antenna is inversely proportional to carrier wavelength. Thus, for example, the 8-mm antenna used has a half power beam width of approximately 1·5° whereas the 10-cm beam width for the same antenna size is 18°.

The use of an 8-mm carrier facilitates the use of higher pattern frequencies, as will be seen in a later section.

The use of a pattern frequency ten times that previously used can be expected to improve the instrumental resolution by the same factor. The use of higher pattern frequencies also decreases certain ground effects.

Expected Ground Swings

Increasing the pattern frequency by ten ensures that the absolute maximum peak-to-peak swing is reduced in the same ratio for any reflection factor.

Using 7·5 Mc/s patterns, the swing is zero if the excess path to a reflection is zero or a multiple of 20 m; for 75 Mc/s pattern the swing is zero for multiples of 2 m in excess path. For example, a line having an excess path of between 10 and 12 m with a reflection factor of 0·03 would produce a swing of ± 10 cm at 7·5 Mc/s. At 75 Mc/s the swing would be zero at 10 and 12 m excess path and would reach a maximum of ± 1 cm at 11 m excess path. The scatter of distance readings owing to multiple weak signals having large excess path should therefore decrease purely on the basis of higher modulation frequency.

When the excess path is less than 2 m and the reflection factor is reasonably small, it is possible to use the approximation

$$\text{swing} = a\Delta \cos \varphi$$

for 7·5 or 75 Mc/s patterns. The peak fractional swing

$$F = \pm \frac{a\Delta}{D} = \frac{ag^2}{2} \quad \text{as} \quad \frac{2h^2}{D} \doteqdot \Delta \quad \text{and} \quad \frac{2h}{D} \doteqdot g$$

For $g < 0·001$ F is less than $\pm 5 \times 10^{-7}$

For $g = 0·025$ (or 1·4°) $a_p = 0·025$

for the antennae used.

$$\therefore \quad F = \pm a_m \cdot a_d \cdot a_s \times 8 \times 10^{-6}$$

For

$$0·025 > g > 0·01 a_p = 0·78(1·3 - 57g)$$

using a straight line approximation for the edge of the polar diagram of the antennae used.

$$\therefore g_{\max} \doteqdot 0·015 \text{ (or } 0·85°) \quad F_{\max} \doteqdot \pm a_m \cdot a_d \cdot a_s \times 3·8 \times 10^{-5}.$$

For $g > 0.025$ it is expected that F will be reduced, possibly marginally just on the first two side lobes, but greatly thereafter: as

$$a_m = 0.6 \qquad a_p \propto \frac{1}{g^2} \qquad a_s = e^{-kg^2} \text{ at } 3.5°$$

and for reasons given previously.

At short range there exists the possibility of a partially developed swing. The minimum cavity frequency shift required to resolve a swing due to excess path Δ cm is $\delta f = 15\,000/\Delta$ cm.

δf is limited to about 1000 Mc/s by the mechanical tuning of available klystrons. Thus an excess path of the order of 10 cm or less could give rise to a small fraction of a swing cycle not apparent to the user. Assuming a 1.5 m height of instrument over a 50 m smooth surface, an unresolved swing of about 3 mm would be expected. If any distance measurement gives an approximately linear swing over the cavity range it should be regarded with caution. Any line less than about 10 km over a smooth surface such as water at grazing angles of less than 1° may be subject to similar errors. The procedure recommended is to mechanically develop a half cycle of swing by changing the excess path by 4 mm. This can be effected by altering the height of instruments and remeasuring as

$$\Delta = \frac{2h^2}{D} \qquad \delta\Delta = 4 \text{ mm} = 4\frac{h\,\delta h}{D} \qquad \text{or} \quad \delta h = 2/g \text{ mm}$$

Therefore if the grazing angle is, say, 1° the vertical shift required is about 12 cm. Where grazing angle corresponds with the first minimum of the antenna beam a similar effect can be had by a slight tilt of the instruments, as the carrier phase reverses across the polar diagram null. Unresolved swing is, however, expected to be random from line to line, and to be unlikely on a moving surface such as the sea.

If the antennae have radomes which are not perfectly transparent over the frequency band used, or the antenna has internal reflections, the local oscillator can in general reach the mixer crystal via different paths. This causes 'antenna swings' which behave in a similar way to 'ground swings' having a reflection coefficient equal to the ratio of direct to indirect voltages. Owing to the geometry of the antenna used, it is expected that such swings will be resolved.

Meteorological Effects

The dielectric constant of the atmosphere at 8 mm differs from the normal microwave dielectric constant because of the neighbouring resonant absorption lines of oxygen and of water vapour, and becasue of increased non-resonant scattering and thermal absorption of rain.

The effect of the oxygen absorption is to increase $n =$ (refractive index $-1) \times 10^6$ by less than 0·5, and that of the water vapour to decrease n by a smaller amount.[2] The effect of rain is to increase n by $\pi/2 \times 0·0854 \times R^{0·84}(N^2-1)/(N^2+2)$ where R is the rainfall rate in mm/hr and a Marshall and Palmer raindrop distribution is assumed (see Appendix 1). Thus for a heavy rain of 16 mm/hr the total increase of n is 1·8 units.

An estimate of the power required to receive measurable signals can be had from noise considerations (see Appendix 2).

It is seen that the minimum power required to measure a 100-km line is approximately 4 mW. If a light rain of 1 mm/hr giving a visibility of 122 m exists and if 30 mW are radiated, the theoretical range is 56 km. For a heavy rain of 16 mm/hr, the range is reduced to 7·7 km.

On theoretical grounds it can be concluded that an 8-mm wavelength is suitable for use in distance measurement.

The Use of Higher Pattern Frequencies

The accuracy of phase measurement appears to be limited to two parts per thousand by cyclic errors in resolvers and by breakthrough in the IF between the 1kc/s AM tone and the 1kc/s tone telemetered to the master.

Increasing the pattern frequency by a factor of ten was at first thought to be impractical using FM because of the limited electronic bandwidths of klystrons, and because of experiments conducted by Dr Wadley at 50 mc/s on 10 cm klystrons. The electronic bandwidth of 8 mm klystrons is usually about two or four times that of 10 cm klystrons, depending slightly on the operating mode. Experiments showed that it was possible to frequency-modulate 8-mm klystrons at a frequency causing side tones outside the static band pass. Spectrum analysis showed that modulation indices considerably higher than desired could be obtained. It was found that if the klystron was far off the static electronic centre, the magnitude of the side frequencies became unequal, as would be expected of a combination of frequency and amplitude modulation. The variation of magnitudes with centring was small if the klystron was operating into a well-matched load. A square law analysis of a system having either inequality of side frequencies or phase shift in the carrier shows that the main effect is to cause single side band or frequency modulation of the intermediate frequency respectively, but that the amplitude detected wave remains unaltered (see Fig. 4).

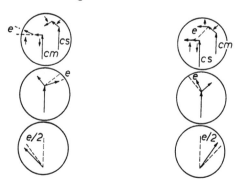

FIG. 4. Eliminating one or both of either upper or lower side tones causes I.F.'s to be S.S.B. with A.M. component as before. Combined A.M. and F.M., represented by a phase shift in one F.M. side tone, gives both I.F.'s an F.M. component. Measured phase is unaltered.

Design of 8-mm Tellurometers

For alignment purposes a small beam width is a disadvantage; however, during rough alignment the automatic gain control effectively broadens the beam width provided there is adequate power available. For final alignment on short lines it is advantageous to have sights on the instrument, and to have a manual gain control of the intermediate frequency amplifier, when stations are too far or for other reasons are not intervisible.

A manual control allows one to set the gain at a suitable constant value. The antenna polar diagram then becomes apparent by observing the signal strength level meter and rotating the instrument.

The effective reflection coefficient is measurable, as the ratio of maximum to minimum signal strength is $(1+a)/(1-a)$. It is necessary to have an adjustment in elevation for narrow beam instruments; the azimuth may be adjusted by rotation in the horizontal bed of an optical plumbing unit.

In earlier Tellurometers the stability of the 1 kc/s measuring frequency was dependent on the difference between two crystal frequencies. A 1×10^{-6} drift of either could cause a 10 c/s drift of measuring time. This was later improved by using 'phase lock', in which the slave oscillator was automatically adjusted to keep the 1 kc/s in phase with a reference oscillator. Increasing the modulation frequency by ten increases the drift similarly as the same 1 kc/s measuring system is employed. A 'frequency lock' servo was developed which uses the high Q reference oscillator LC circuit, the phase discriminator not in use at the slave station and a subsidiary servo to level the output of the tuned circuit. It can capture signals ± 300 c/s off the 1 kc/s and stabilize the slave modulation relative to the master to $\pm 1 \times 10^{-8}$.

In the 10-cm Tellurometer cross polarization at $45°$ to the horizontal is used to isolate the transmitted from the received signal. This was intended to minimize the effect of reflections from smooth surfaces. For grazing angles above the pseudo Brewster angle, surface reflections are cross polarized relative to the receiving plane of polarization. Another advantage is that master and slave instruments are identical.

In the 8-mm Tellurometer the same polarizations are used. Signals are, however, transmitted and received in waveguides. Owing to the higher carrier frequencies and hence narrower beam widths a Cassegrain feed is possible. This has the advantages that the feed is from the back, that the antenna is compact and that the antenna permits symmetry. Minimum blocking geometry has the advantage of highest antenna efficiency but an arrangement is used where the feed blocks less than the area of the sub-dish so as to minimize reflections from the sub-dish. In designing the MRA-101 3-cm instrument, where this problem is more critical, it was found that the reflection from the hyperboloid sub-dish could be prevented by using a tapered point, and this was again used in the 8-mm antenna system.

The small coupling needed between the transmitted wave and the receiving crystal is arranged by reflecting energy from two probes at $\pm 45°$ relative to the transmitted plane of polarization. By spacing these probes a quarter wave apart in the direction of propagation, it is possible to prevent any reflection in the transmitted plane, which could be re-reflected from the klystron.

The transmitted wave and received wave are separated by using two gratings, one of which forms a corner. This allows a good match to be obtained over a wide band to and from the antenna unit. The possibility exists of a cavity resonance between the gratings and the cut-off guides. This can be prevented by slots in the walls of the guides in the same plane as the desired polarization and/or by suitably choosing the distances so that the resonances occur well outside the band of interest.

Measurements are performed as in 3-cm Tellurometers, the additional F or fine pattern being used to obtain the added resolution. The fine pattern is derived from existing 7·5 Mc/s crystal oscillators and this can be achieved by using a times ten step recovering diode multiplier.

The Statistical Nature of Ground Swings

A large number of measurements have been made with 10-cm carrier Tellurometers. Mr I. Watt of the Witwatersrand University Survey Department has allowed me the use of his field books which record measurements on some 150 lines. These lines, ranging from 11 000 mμs

(1 mile) to 530 000 mμs (50 miles) with a median length of 55 000 mμs (5 miles), were measured in the Transvaal over land. Approximately 10 cavity settings were used on each measurement. The peak-to-peak swing s and the best estimate of standard deviation $\rho = \sqrt{[n/(n-1)]}$ RMS were found for each line, where n = number of cavity settings on a line. It was found that if the frequency of occurrence of the various ground swings was plotted on log-probability paper a straight line resulted (see Fig. 5).

In addition it can be seen that the peak-to-peak swing is very nearly 3·4 times the standard deviation for any frequency of occurrence. If swings were random a ratio of $\sqrt{11} = 3·3$ would be expected; for sinusoidal swings a ratio of $2\sqrt{2} = 2·8$. As the normal procedure is to take an arithmetic mean of measured distance it is important that the frequency distribution of swings be symmetrical. 1040 swing readings about the mean of 105 lines were counted and the number of +ve and −ve was found not to be significantly different from what would be expected of a random distribution.

It may be argued that only large swings would cause a non-symmetrical frequency distribution of +ve and −ve swings. Counting the number of lines with standard deviations > 2·0 mμs, it was again found that no significant asymmetry was present.

Normalized autocorrelation for intervals up to three steps of cavity settings, corresponding to about 90 Mc/s carrier shift for the 10-cm instruments, was calculated for each of thirty-eight lines. A test was performed to find whether a larger number of high correlation coefficients occurred than would be expected owing to complete randomness on the size of samples available. There was no significant difference between the number observed and the number expected, though a larger number of high correlation coefficients might be expected if swings were predominantly cyclic. A pooled autocorrelation showed a small significant correlation. However, when lines with swing less than 1 mμs were excluded, there was no significant correlation. The results were:

$$1 + 0·20 \pm 0·07 - 0·13 \pm 0·08 - 0·21 \pm 0·09$$

$$1 + 0·09 \pm 0·10 - 0·06 \pm 0·11 - 0·23 \pm 0·12$$

The 'antenna' swing, i.e., the average of thirty lines with small swing, was = 0·24 mμs and the autocorrelation for this was:

$$1 + 0·41 + 0·0 - 0·27$$

On the basis of measurements made over land, it appears that swings are completely random on any line and a standard error of the mean = $\sigma/\sqrt{n}$ can be assumed where n, the number of readings, is not greater than ten.

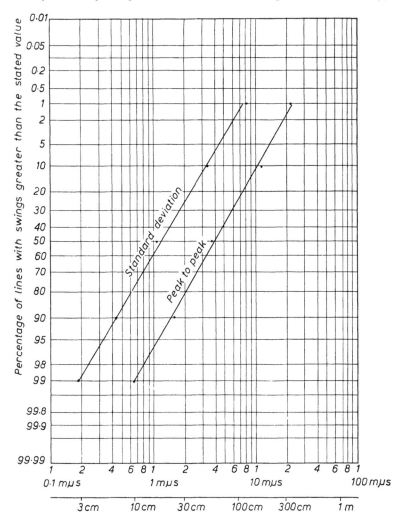

FIG. 5. Frequency distribution of swings on seventy-one lines with 10-cm Tellurometers (approximately ten readings per measurement).

Points were plotted corresponding to swing and distance for 116 lines on log-log paper (see Fig. 6). It is evident that there is no particular correlation between swing and range. However, the fractional expected accuracy due to swing is well correlated with range, i.e. poor accuracy can only occur on short lines. If the frequency of occurrence of various

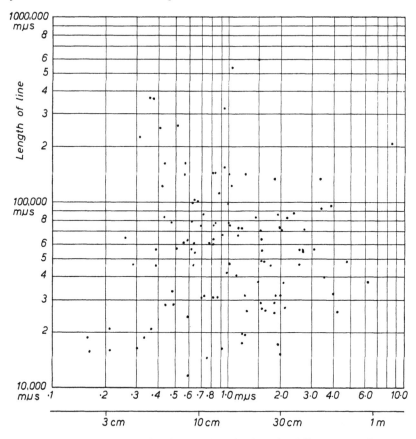

FIG. 6. Scattergram showing length and swing of 116 lines measured
with a 10-cm Tellurometer.

fractional swings is plotted on log-probability paper, a straight line is
again obtained (see Fig. 7).

The median precision is approximately equal to the refractive index
accuracy, while precision three times worse only occurs on lines less than
2·7 miles long.

Measurements with 8-mm Instruments

Initially two MRA-1 10-cm Tellurometers were modified to operate
at 8 mm without fine patterns using 2-kV solid-state supplies. Finally
two MRA-3 3-cm Tellurometers were modified to operate at 8-mm
with fine patterns using improved low-voltage klystrons.

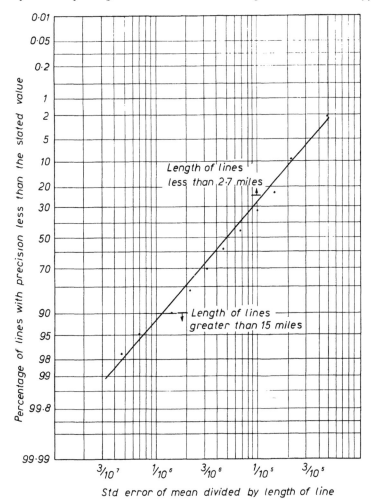

FIG. 7. Analysis of 116 lines from 2 to 50 miles, measured with a 10-cm Tellurometer.

Thirteen lines were measured with the first pair of instruments and about twenty-five lines have at the time of writing been measured with the second set of instruments.

Six lines from 5 to 13 miles measured with the first set of instruments were lines previously measured by Mr I. Watt using 10-cm Tellurometers. These lines are known to have very large ground swings at

10-cm and were surveyed splitting 'bad' lines into two sections. Using 8-mm instruments median swings were reduced by a factor of approximately 9 and the frequency distribution of swings also improved.

Using the second set of instruments on short lines the median swings were approximately forty times less than the 10-cm median. In Fig. 8 it is seen that swings followed the same statistical law and again the spread of swings was smaller.

The six lines previously mentioned had one redundancy. Ten lines measured with 10-cm instruments had three redundancies and included

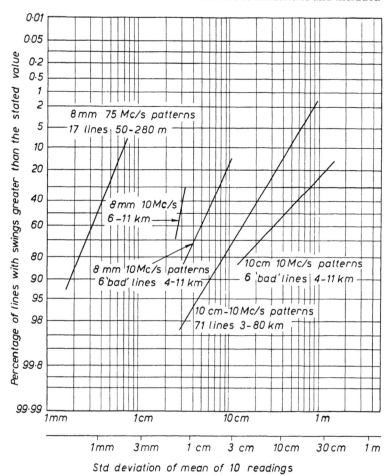

FIG. 8. Frequency distribution of ground swings measured with 10-cm and 8-mm Tellurometers.

the six lines. The two figures were adjusted independently by least squares. The probable error of 10-cm lines was 4·6 in. and the standard agreement between 8-mm and 10-cm lines was 2·6 in. On the basis of errors on the 10-cm lines only an agreement $4·6/\sqrt{6} = 1·9$ in. would be expected.

There are sixteen modes of measurement, namely,

$$\begin{pmatrix} \text{AFC} & \text{AFC} \\ \text{at } m, & \text{at } s \end{pmatrix} \begin{pmatrix} \text{measure} & \text{measure} \\ \text{at } m, & \text{at } s \end{pmatrix} (\text{forward, rev.})(+\text{ve and } -\text{ve}).$$

Modes ($+$ve and $-$ve) can be observed by omitting to reference the instruments and reading phase on reference.

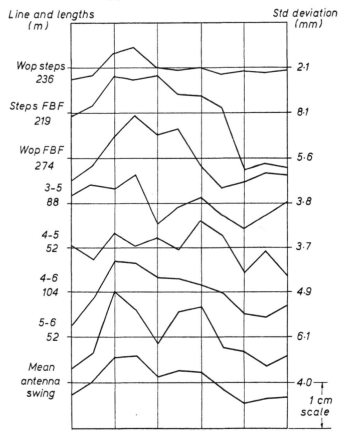

FIG. 9. Actual deviations from mean readings observed on the first seven lines measured with an 8-mm Tellurometer using 75 Mc/s patterns.

Using the second set of instruments on lines from 50 m upward an analysis of variation was performed on measurements. In the interests of simplicity only four modes (at *m*, at *s*) (for., rev.) were observed, with AFC at master and at each of eleven cavity settings over a 650-Mc/s band. This analysis showed that the standard error due to randomness or resolution on a single reading varied from 3 to 6 mm with increasing range or decreasing signal strength. On nine lines from 50 to 300 m the mean at each cavity setting had a standard deviation which varied from 2 to 8 mm. The peak-to-peak swing reached a maximum of 2·1 cm on one line. Fig. 9 and Fig. 10 illustrate the actual readings obtained, and the probable effect of ground and residual antenna swing is obtained by

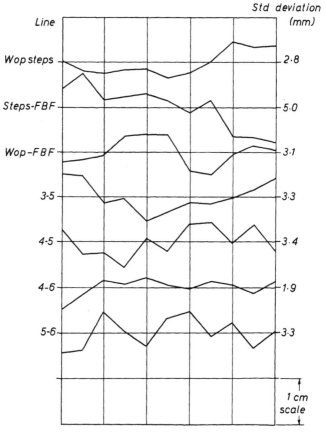

FIG. 10. Deviations from mean readings less average deviation of seven lines shown in Fig. 9.

subtracting the mean deviation at each cavity setting from the individual readings. A previous figure shows the range of swings on most of the lines measured so far.

An investigation into the consistency of zeros or centres of measurement was undertaken. Six 1 ft 6 in. centre-punched pegs were arranged in rectangle around a football field. The distances varied from 50 to 130 m. During the experiment an optical plumbing unit was acquired owing to difficulty in centring with a plumb bob. Eleven lines were measured. Mr Watt then offered to observe angles so as to increase the number of redundancies. Five short vertical markers were made and held over the pegs in three-legged stands. Three arcs were observed at each point and after a least square adjustment the standard error of angle was 1·6 secs of arc. Thus at these short ranges it was possible to determine the relative lengths of lines to better than 1 mm. A pair of goal posts was situated directly along the line between two pegs, and a measurement through this obstruction showed an error of 4 cm and was therefore not included in the zero determination. The standard variation of zeros was 4·8 mm with 9 degrees of freedom using the optically plumbed and plumb-bob lines without bias, and using 1 degree of freedom for scaling the figure.

The probable accuracy of the length of the lines is thus about ± 1·5 mm. The nature of the swing curves obtained leads one to expect higher accuracy. Pegs were then concreted into the ground and angles are being re-observed. Improvements are being made to the resolvers and the lines will be re-measured using greater care in centring over the pegs.

Conclusion

The median resolution obtainable appears to be about forty times better than with 10-cm instrument. For many purposes sufficient accuracy may be obtained by measurement at a single or a few cavity settings.

APPENDIX I

The Dielectric Constant of Rainfall
By M. C. HODSON

The polarizability

$$P = (\epsilon - \epsilon_0)/\text{No. particles/Unit vol.}$$

and

$$P = \frac{\pi\epsilon_0}{2} \frac{(N^2 - 1)}{(N^2 + 2)} d^3$$

where

$$d = \text{drop diameter}$$

$$\therefore \quad \epsilon_r - 1 = \sum_i \frac{\pi}{2} \frac{(N^2 - 1)}{(N^2 + 2)} d^3$$

for a distribution of drops. Assuming a Marshall and Palmer drop size distribution, number of drops per cubic metre in an interval of drop diameters measured in metres is:

$$\text{no.} = 8 \times 10^6 e^{-\tau d}$$

where

$$\tau = 4100 \cdot R^{-0.21}$$
$$R = \text{rainfall rate in mm/hr}$$

and

$$d = \text{diameter of drops in metres}$$

$$\epsilon_r - 1 = \pi/2 \frac{(N^2 - 1)}{(N^2 + 2)} 8 \times 10^6 \int_0^\infty e^{-\tau d} \cdot d^3 \cdot dd$$

$$= \pi/2 \frac{(N^2 - 1)}{(N^2 + 2)} 8 \times 10^6 \times \frac{6}{(-\tau)^4}$$

$$\therefore \quad \delta n = \frac{(N^2 - 1)}{(N^2 + 2)} \times 0.0854 R^{0.84}$$

where N is the refractive index of water and $(N^2 - 1)/(N^2 + 2)$ is therefore approximately unity.

APPENDIX 2

An Estimate of Power Requirements

Klystron and mixer noise are approximately equal.

Klystron noise at 33 Mc/s off the carrier is approximately -160 dB at 1 mW mixing level and mixer noise figure is approximately 14 dB.

Thus the total noise is -157 dBm/C/S(*sic*).

The IF and image band are approximately 1·5 Mc/s giving an IF noise level of -95 dBm.

The desired IF signal level is -88 dBm for a 6 dB signal to noise ratio.

Practical 17-in. diameter antennae have a measured gain of 38 dB at 8 mm and the inverse square path loss on a 100-km line is therefore $164 - 76 = 88$ dB.

The normal attenuation of the atmosphere at 8 mm is 0·06 dB/km, thus the minimum power required to measure a 100 km line is 4 mW. Light rain, 1 mm/hr, adds 0·25 dB/km attenuation. Heavy rain, 16 mm/hr, adds 4·2 dB/km.

APPENDIX 3

Scattering by Reflection and Diffraction

Consider a surface of length L having sinusoidal height variations in one direction.

Let m = number of cycles of phase change in length L
 ρ = peak phase change in direction of specular propagation
 g = angle of incident radiation relative to reflecting surface
 g' = angle of reflected radiation relative to reflecting surface
 $l = L \sin g$, the effective re-radiating aperture
 $\dfrac{Vl}{2}$ = the distance from the centre of the effective aperture.

Put

$$U = \frac{\pi L \sin g}{\lambda} \quad \text{and} \quad \theta = \pi \, \mathrm{m}V$$

$$\beta = \frac{\pi L \sin (g' - \pi)}{\lambda}$$

The far field E is due to radiation from the effective aperture having a field distribution with periodically varying phase

$$E = \frac{l}{2}\int_{-1}^{+1} \exp\left(i\rho \sin \pi \, \mathrm{m}V\right) \cdot \exp\left[i(U+\beta)V\right] \cdot \mathrm{d}V$$

$$= \frac{l}{2}\int_{-1}^{+1} [J_0(\rho) + J_1(\rho)(e^{i\theta} - e^{-i\theta}) + J_2(\rho)(e^{2i\theta} - e^{-2i\theta}) \ldots]$$

$$\times e^{i(U+\beta)V} \cdot \mathrm{d}V$$

$$= \frac{l}{2}\int_{-1}^{+1} [J_0(\rho) + J_1(\rho)e^{i\theta} - J_1(\rho)e^{-i\theta}$$

$$+ J_2(\rho)e^{2i\theta} - J_2(\rho)e^{-2i\theta} + \ldots]e^{i(U+\beta)V} \cdot \mathrm{d}V$$

$$= J_0(\rho) \cdot \frac{\sin(U+\beta)}{(U+\beta)} + J_1(\rho) \cdot \frac{\sin(U+\beta+\pi m)}{(U+\beta+\pi m)}$$

$$- J_1(\rho) \cdot \frac{\sin(U+\beta-\pi m)}{(U+\beta-\pi m)}$$

Sin$(U+\beta)/(U+\beta)$ is unity in the direction of specular reflection when L is large, and the reflected beam shrinks to a reflected ray having an amplitude $J_0(\rho)$ where ρ is a measure of roughness of the surface. If the surface has sinusoidal variations in two dimensions, the amplitude of the reflected ray becomes

$$J_0^2(\rho) = J_0^2(\sqrt(2)\rho_{\text{RMS}})$$

$$= \left(1 - \frac{\rho^2}{4} \ldots\right)^2 \quad \text{for} \quad \rho \ll 1$$

$$= 1 - \frac{\rho^2}{2} \doteq 1 - \rho_{\text{RMS}}^2.$$

The angle from the specular to the first side ray is approximately

$$\frac{m\lambda}{L} = \frac{\lambda_{\text{radiation}}}{\lambda_{\text{surface wavelength}}}$$

References

[1] Wadley, T. L., 1958. Electronic principles of Tellurometer. *Trans. S.Afr. Inst. Elect. Eng.*

[2] Van Vleck, J. H., M.I.T. Radiation Laboratory Series, Vol. 13, Chapter 8.

[3] Stratton, J. A., 1941. *Electromagnetic Theory.* (McGraw-Hill).

[4] Kerr, D. E., Fishback, W. T. and Goldstein, H., M.I.T. Radiation Laboratory Series, Vol. 5, Chapter 5.

[5] Beckman, A. Spizzichino, 1963. *Scattering of E.M. Waves from Rough Surfaces.* Pergamon Press.

[6] Jasik H., *Antenna Engineering Handbook*, pp. 2–40.

[7] Poder, K. Reflections. *Proceedings of the 1962 Tellurometer Symposium.*

DISCUSSION

G. D. Robinson: How does this 8-mm instrument work in rain or in cloud? Equipment using 8-mm waves is used to locate clouds, and so the back scattering must be large.

P. Cabion: We have not done any actual measurements through clouds or heavy rain.

J. C. De Munck: With this high-frequency equipment the frequency spread of the cavity may be so small that only a part of the swing curve will be shown.

P. Cabion: The frequency can be changed through 650 Mc/s, and this should show up enough of the swing curve.

F. J. Hewitt: Mr Cabion has concentrated on describing a short-range high-resolution instrument, but it is also capable of accurate long-range (20-mile) single-frequency reading.

P. Cabion: While I always recommend making use of the fine patterns, a single reading should give precisions of 1 cm at ranges of 50 km.

R. C. A. Edge: What about working through cloud?

P. Cabion: The method would be to line up roughly with AVC, and then change to manual gain for final alignment. The instrument would not line up on a side lobe, but might be deflected by a large reflecting surface such as the side of a mountain.

J. Kelsey: How about weights and power supplies?

J. A. Webley: The equipment would weigh 25 lb and take 40 W. The price would be high—at least £2000. On the other hand it would be capable of reading to an accuracy of 3 mm if the full frequency spread were used. This figure applied to errors due to reflections, zero errors, etc., and did not include uncertainties due to refractive index.

K. Poder: Has this instrument been tried over smooth water?

P. Cabion: We have not tried it over water but have successfully used it over a smooth road.

K. Poder: It may be enough to have waves on water of only 1 or 2 cm to give a good scatter.

Tellurometer Zero Error

S. A. YASKOWICH

Geodetic Survey of Canada

Test measurements for determination of zero error were made with a pair of MRA-3 Tellurometers. The results show the presence of a cyclic zero error of small amplitude and magnitude. These results are compared with the cyclic zero error observed on the earlier model Tellurometers. Laboratory tests were also made in which Tellurometer signals were directed through a metal pipe containing layers of microwave absorbing material. Several short distances were measured with all models of Tellurometers and the observed lengths were compared with tape distances. Ground swing was large but repeatability was good and standard errors of mean observed lengths were only slightly larger than for ground base measurements. There was no apparent correlation between observed zero errors and those computed from data observed on outdoor bases. The laboratory base may, however, be used for testing the operation of Tellurometers.

Introduction

The zero error of the Tellurometer has been the subject of much investigation and many reports. Canadian work previous to 1963 was reported at the Berkeley meeting of the International Union of Geodesy and Geophysics by J. E. Lilly, in a paper entitled 'Tellurometer Cyclic Zero Error'. The paper dealt with work done on Tellurometer models 1 and 2. Since then we have made zero error tests with a pair of MRA-3 instruments, the results of which are discussed in this paper.

In addition, we have attempted to devise a laboratory procedure for evaluating Tellurometer zero error. The method involves making short range measurements with the Tellurometer signals directed through a metal pipe of circular cross section having layers of microwave absorbing material in it to reduce signal levels. A discussion of test measurements made and the results obtained is given in the second part of this paper.

Ground Bases

Measurements with MRA-3 Tellurometers for zero error determination were made on the same two bases that were used for the earlier model instruments. These are designated base A, 394 to 408 m and base B, 192 to 208 m. Each base consists of eight line lengths varying in 2-m steps.

The bases are laid out over generally level grassland. Base A is on rocky earth covered with grass approximately $\frac{1}{2}$ m high at time of measurements and has some shrubbery near the middle portion of it. Base B is on smooth earth with short grass cover.

MRA-3 Measurements

Our MRA-3 instruments are equipped with null meters and circular dial readout. The master A modulation frequency is 10 Mc/s so that the readings are in millimicrosecond units.

Each measurement consists of 17 fine A readings observed at ten-division intervals in the cavity tune range 30 to 190. The mean of the seventeen fine readings, combined with coarse readings, was accepted as one length measurement. Meteorological data were obtained at only one station. Measurements on each line were made in pairs, with one instrument as master for one measurement and the other instrument as master for the second. Pairs of measurements on each line were taken on different days and a total of six measurements were made on each line. The mean of six measurements was accepted as the observed length for each line.

Summary of Results

In Table 1 is a summary of results obtained with all instrument pairs on the two ground bases. Lengths observed with MRA-1 instruments were revised so that only measurements that deviated from the mean of a group by more than 10 cm were rejected. Of the seven MRA-1 observations rejected, six were observed with the earlier model instruments. All measurements made with MRA-2 and MRA-3 instruments were within 10 cm of the group means, and no rejections were made.

Instruments are referred to by the model number, followed by the serial numbers in brackets, master preceding remote in the case of a single measurement. Thus 3(309/314) refers to MRA-3 instruments 309 and 314, with 309 as master if reference is to a single measurement. The two sets of MRA-1 instruments, 1(117/153) and 1(283/397), are representative of early and late production models respectively.

Ground swing on a measurement is defined as the difference between maximum and minimum A reading. Measurements with the three pairs of instruments made on both bases had larger ground swings on base A than on base B. This is most probably due to the difference in topography of the bases. Note that ground swing for MRA-1 and 2

TABLE 1. *Summary of field tests*

Instruments	Base	Accepted measures per line	Measures rejected	Mean ground swing (cm) Av.	Max.	Average range of measures (cm)	Std. error of mean (cm) Av.	Max.	Zero error $E = b_1 + b_2 \sin 7.2 A$ b_1(cm)	b_2(cm)
1(117/153)	A	7–9	6	38	73	12·6	1·5	2·1	+4·0	+2·7
1(117/153)	B	4	—	27	36	4·4	1·0	2·0	+2·4	+4·1
									+3·2	**+3·4**
1(283/397)	A	8–10	1	33	60	11·4	1·3	1·8	+6·8	+3·6
1(283/397)	B	5–6	—	21	30	7·1	1·3	2·2	+3·9	+4·0
									+5·4	**+3·8**
2(573/708)	A	6	—	44	73	5·8	0·8	1·4	+2·0	+4·8
3(309/314)	A	6	—	13	26	6·4	1·0	1·4	−3·2	+2·0
3(309/314)	B	6	—	10	19	5·7	0·8	1·3	−1·2	+2·0
									−2·2	**+2·0**

Note: Average zero errors shown in bold type.

measurements is two to three times larger than for MRA-3 measurements. The average ground swing for the latter is 12 cm.

Ground Swing and Standard Errors

Examples of ground swing curves for individual measurements made with all four sets of instruments are shown in Fig. 1. They were observed on the 402-m line of base A. The observation represented

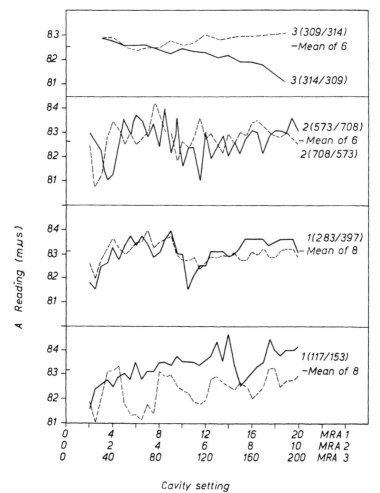

Fig. 1. Ground swing, 402-metre line, Base A

by the broken line for set 1(117/153) was rejected because the measurement was 12 cm shorter than the group mean. The curve shown for instruments 3(314/309) represents a ground swing of 1·75 mμs or 26 cm which is the largest observed with the MRA-3 set.

The slope on this curve is typical of that observed on several lines with instrument 314 as master. It does not represent true ground swing but rather a klystron characteristic in which is involved a critical alignment of the tracking between klystron frequency and power output in the cavity tune control. Because of this, the manufacturer has recommended that observations be limited to the cavity tune range 40 to 160. Considering only observations in this range the ground swing would be 13 cm for this measurement.

The standard error of the mean of each line measured was computed. The averages and maximum values of mean standard errors for each base are given in Table 1. These errors are smallest for MRA-2 and MRA-3 measurements. In the case of the latter, a relatively large difference between pairs of measurements on some lines (maximum 5·5 cm), contributed to the magnitude of the standard errors. This factor is also reflected in the average range of measurement. With the MRA-2 instruments we also interchanged the master and remote sets but found no appreciable differences in measurements.

Zero Error

The 2-m increments on both bases were carefully re-taped and slight revisions to base length were made on some lines. Errors, Tellurometer minus tape, are shown in Fig. 2.

For purposes of comparison, the data were reduced by least squares using the formula, zero error, $E = b_1 + b_2 \sin 7\cdot2\, A$. The computed terms b_1 and b_2 are given (in cm) in Table 1. It is interesting, and perhaps significant, that on base B, where the smaller ground swings occurred, the b_1 zero error terms were smaller for all instrument pairs. It should also be noted that for the MRA-3 instruments only, the b_1 term is negative.

During 1964, eighteen geodetic lines, 2·6 to 15·7 km in length, were measured with model 4 Geodimeters as well as with Tellurometers 3(309/314). The discrepancies, Tellurometer minus Geodimeter, without zero correction were in the range −12·5 cm to +6·2 cm and, with a mean zero correction of +2·2 cm applied, −9·9 cm to +7·9 cm. The standard errors were 5·6 cm and 4·5 cm respectively. Six of the eighteen lines were measured in the city of Hamilton, where topography and high temperatures resulted in larger than normal ground swings and suspected index of refraction errors. The mean Tellurometer minus Geodimeter, without zero correction, for the six lines was

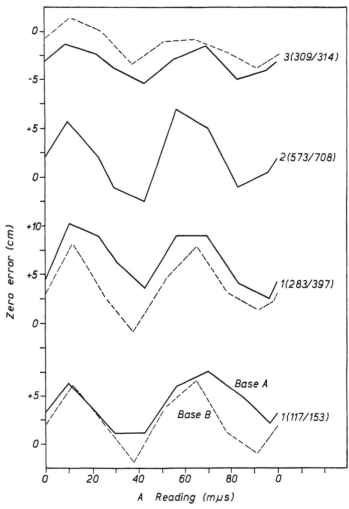

FIG. 2. Observed zero error.

−7 cm. Standard errors computed for the other twelve lines were 3·6 cm with no zero corrections applied, and 3·8 cm with mean zero corrections applied.

The results on these latter twelve lines are felt to be more representative of general MRA-3 measurements and confirm our earlier contentions that applying corrections for zero error on lines of geodetic length does not significantly improve Tellurometer measurement accuracy.

8

Last spring we made three pairs of measurements on lines of our roof-top base. The measurements with Tellurometer 309 as master were all within 4 cm of the base lengths, but the measurements with set 314 as master were in error on an average of − 15 cm. The swing curves for the latter measurements showed a rapid decrease in A reading with increasing cavity setting. This extreme slope of the swing curve was partly caused by the klystron characteristic and tuning alignment in set 314 as mentioned earlier, and partly because of a misadjustment in the read-out mechanism. The instrument was repaired but time did not permit further measurements on the roof-top base.

Laboratory Zero Error Tests

Tests were conducted to see if it is possible to determine zero error in a laboratory. The technique used involves measuring very short distances by directing the Tellurometer signals through a cylindrical metal pipe in which are placed baffles, or layers of microwave attenuating material. The measured length is then compared with the tape distance between instrument plumb points.

Cylindrical, galvanized iron pipes, 4·6 cm in diameter and 76 cm long were used. One end was slightly flanged to permit pipes to be joined together to make a continuous cylinder and to enable variation of the base length. The attenuating material used has the trade name Ecosorb and is obtained in sheets 5 cm thick. It is easily cut to fit inside the pipe at right angles to its axis.

The majority of measurements, including all preliminary tests, were made with Tellurometers 2(573/708). The preliminary trials were made to determine the number of baffles required to reduce the signal (AVC) to a normal level, and the spacing within the pipe to give most constant crystal current readings. It was found that four or five baffles were sufficient to reduce AVC readings to less than sixty. All subsequent measurements were made using either four or five baffles. The position of the two baffles nearest the instrument was quite critical with respect to stability of crystal current readings. By trial and error, positions for the baffles were determined which resulted in a minimum variation of crystal current over the entire range of carrier frequency. A typical laboratory set-up is sketched in Fig. 3.

The initial plan was to measure a series of lengths with the MRA-2 instruments using odd numbers of pipes yielding observed lengths corresponding to equally spaced A readings over its whole range. Measurements with three and five pipes went reasonably well but difficulties were encountered with the seven-pipe set-up. At this range, very large variations in crystal current were noted and the circle and

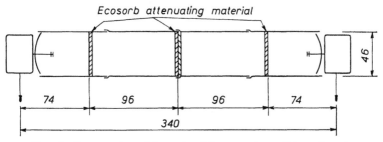

FIG. 3. Cross-section of four-pipe laboratory base. Dimensions in centimetres.

break were very poor at some cavity settings. The ground swing was very large, 2 to 3 m, with maximum and minimum readings occurring at cavity tune settings where extremes of crystal current were noted. It was decided to limit observations to lengths corresponding to 2, 3, 4 and 5 pipes.

The four lengths were each measured several times. The most stable operation occurred on the four-pipe base and it was decided to make measurements on this base with all test instruments. In addition, four measurements were made with instruments $1(117/153)$ on the two-pipe base. A summary of results of all measurements is given in Table 2.

The built-in antenna on the MRA-3 instruments made it necessary to place the instruments close to the pipe to avoid stray signals and this

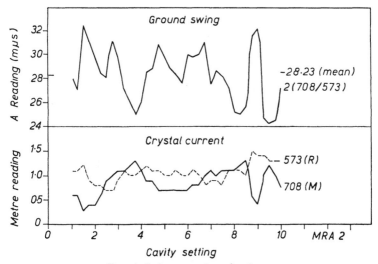

FIG. 4. Laboratory five-pipe base.

TABLE 2. *Summary of laboratory tests*

Instruments	Base line	Tape length (m)	No. of measures	A reading (mµs)	Mean ground swing (cm)	Mean observed (m)	Mean range of measures (cm)	Std. error of mean (cm)	Telluro-meter minus tape (cm)	Mean zero error from grd. bases (cm)
1(117/153)	2 pipes	1·980	4	13·63	159	2·042	15·6	3·5	+ 6·2	+6·7
1(117/153)	4 pipes	3·400	8	23·10	168	3·462	8·2	1·0	+ 6·2	+4·0
1(283/397)	4 pipes	3·400	8	23·60	154	3·537	10·0	1·3	+13·7	+6·1
2(573/708)	2 pipes	1·980	9	13·45	136	2·015	15·6	2·2	+ 3·5	+6·8
2(573/708)	3 pipes	2·724	6	17·94	124	2·688	12·2	1·7	− 3·6	+5·7
2(573/708)	4 pipes	3·400	8	23·35	90	3·499	11·4	1·2	+ 9·9	+3·0
2(573/708)	5 pipes	4·120	9	28·09	118	4·209	12·4	1·2	+ 8·9	+0·2
3(309/314)	4 pipes	3·300	15	22·10	99	3·312	8·6	0·6	+ 1·2	−1·5

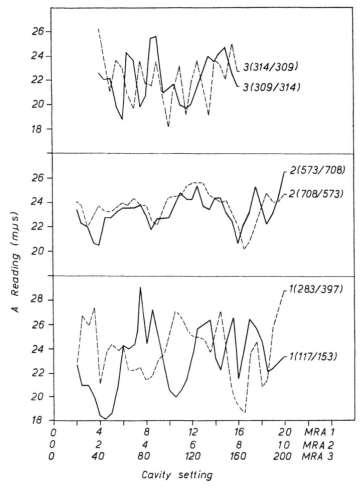

FIG. 5. Ground swing, laboratory four-pipe base.

resulted in a slightly shorter base length. Using four baffles, AGC meter readings for the MRA-3's were approximately thirty-five. Thirteen fine readings per measurement were taken over the cavity tune range 40 to 160 for seven measurements and twenty-five fine readings over the same range for eight measurements. The measurements were combined because the results were consistent.

The mean ground swing for measurements with all instruments is very large, 90 to 168 cm. Examples of ground swing curves observed on the four-pipe base are shown in Fig. 4. Despite the large ground swings, the mean range of measurements and standard errors of the mean measurements are only slightly larger than those for the ground base line measurements.

Fig. 5 shows the ground swing and variations in crystal current for one measurement with the MRA-2 sets on the five-pipe base. The measurement is representative of cases where large crystal current fluctuations were encountered. In general, peaks of crystal current, high or low, correspond to maxima and minima on the ground swing curve.

Included in Table 2 are mean discrepancies, Tellurometer minus tape, and the corresponding zero errors computed from the mean results determined on the two ground bases. There is no apparent correlation between the two sets of figures.

It was found that the laboratory base is useful for testing the operation of Tellurometers. Poor receiver alignment and other malfunctions in the measurement circuits are readily apparent when measurements are attempted.

Conclusions

The zero error determined on short baselines for a pair of MRA-3 Tellurometers shows a cyclic variation with A reading similar to that observed with MRA-1 and MRA-2 instruments. However, limited comparisons with Geodimeter measurements indicate that application of zero corrections to measurements on geodetic lines does not improve the overall accuracy.

Attempts at determining zero error by measuring an enclosed path in a laboratory failed to give results consistent with outdoor measurements. Large ground swings and fluctuations in crystal current create uncertainty in the results. The technique can, however, be used to test the operation of Tellurometers.

Abnormal lines, such as those on our roof-top base and laboratory base, can be measured with an accuracy of a few centimetres with the MRA-3 instruments, which is better than the accuracy obtainable with the earlier model instruments. However, large errors can result from equipment malfunctions or misadjustments, but these can be detected from the Tellurometer data and corrective measures taken.

DISCUSSION

R. C. A. Edge: I suggest that tubes were used in order to cut down ground swing, but they seem to have had the opposite effect.

S. A. Yaskowich: Yes, even movements of someone's hand outside the pipes could affect the readings.

F. J. Hewitt: Are the baffles put at an angle in the tubes to reduce the reflections?

S. A. Yaskowich: They are at right angles.

F. J. Hewitt: The absorbent material might reflect as well as absorb, unless it were a perfect match to the transmissions.

I. Brook: I am surprised that results were not improved when Mr Yaskowich applied zero error corrections. I have tried two different types of Tellurometer, together with a Geodimeter. When I applied zero error corrections the spread of the Tellurometers was greatly reduced, though their mean was no nearer to the Geodimeter readings.

S. A. Yaskowich: My data are limited, and I am glad to hear that others had improved their figures by zero error corrections.

J. A. Webley: What were the base lines used in the trials?

S. A. Yaskowich: Bases measured by catenary were used, with marks at 2-m intervals at one end of each base.

R. C. Gardiner-Hill: In the analysis of the results, was any investigation made of harmonics higher than the $b_2 \sin 7 \cdot 2A$ term?

S. A. Yaskowich: No.

K. Poder: I wonder if anyone familiar with microwaves can explain the cyclic errors. They seem to be common with all microwave instruments. Admittedly they happen when an MRA-3 is badly adjusted, but they also occur when everything seems to be correct.

S. A. Yaskowich: I have used forward and reverse readings.

H. D. Hölscher: One possible explanation is contamination of the comparison frequency signal at the master by the returned signal from the remote. There are two possibilities: (a) if the reversal does not give an exact 180° change of phase, some residual contamination error will remain, (b) if the contamination occurs after the forward-reverse switching point, this error would not be removed and a resultant cyclic error will be evident.

K. Poder: With 10 per cent contamination, first-order effects are removed by reversal, but 1 per cent of second-order effects remain, and this would be enough to account for the cyclic errors.

J. A. Webley: The phase detector may have an error of $1\frac{1}{2}$ parts per 1000. However, there are three sources of error: the recorder, contamination, and errors in the antenna and feed due to unwanted small reflections.

Electro-Optical Systems

The AGA Geodimeter Model 6

R. SCHÖLDSTRÖM

AGA
Aktienbolag, Lidingo 1, *Sweden*

The reflectors used for Geodimeter measurements consist of one or more retrodirective prisms. Such prisms reflect light centric to the light source. Therefore, if the optical system of the receiver is placed next to the optical system of the transmitter, the receiver will be unevenly illuminated. Over long distances, atmospheric turbulence will contribute to an equalization of the light, but over shorter distances deviation wedges must be placed in front of the prisms. These wedges are, however, optimal for one specific distance only. Wedges of different strength are therefore required to cover ranges below 3–4 km.

It has been possible to eliminate this problem by incorporating a coaxial arrangement of the two optical systems. The rotating symmetrical light distribution due to the reflector is thus best utilized by the receiver optics which are concentric to the transmitter optics. The use of deviating or dispersing arrangements is thereby avoided completely.

The optical system now permits elevations from $+90°$ to $-55°$; the electronics are transistorized to reduce weight and power; and the daylight range increased.

Finally this paper includes a discussion of different factors contributing to errors in Geodimeter measurements. With respect to this discussion three methods are presented for achieving higher than normal accuracy under certain conditions. Results from practical tests are shown.

In designing the Model 6 AGA Geodimeter—successor to the Model 4 —three chief objectives were drawn up:

1. Reduce weight and power consumption.
2. Increase the daylight range.
3. Simplify pointing to the reflector.

It soon became apparent that such major improvements could be realized only through a radical change in design, particularly of the optical system. A reduction in weight and power consumption could be achieved comparatively easily through the use of transistors in the electronic part of the instrument.

In earlier Geodimeter models, the optical systems for transmitting and receiving the modulated light signals were practically identical.

When measuring in darkness, the illumination of the reflector is proportional to the effective aperture of the transmitter optics and to the light intensity, while the receiver's efficiency is proportional to the size of the aperture of receiving optics and to the sensitivity of the photocell. The darkness current of the photocell is disregarded in this discussion. Consequently, the useful electrical signal i measured at the photocell's output for a specific distance can be expressed as

$$i \sim A_t . A_r . I . S$$

where A_t and A_r are the effective optical surfaces of the transmitter and receiver lenses respectively, I the light intensity and S the sensitivity of the photocell.

At the cost of increased weight and power consumption, I can be increased considerably by using super-high-pressure vapour lamps. The sensitivity of the photocell, S, is so high for modern photocells that no significant improvement can be expected. For practical reasons, it is preferable to have $A_t = A_r$ in a Geodimeter when measuring at night. In daylight measurements, the same formula is also valid for signal transmission. However, in daylight measurements, another limiting factor is the undesired stray light which enters the photocell by various channels and generates an electric noise signal which quenches the desired signal. There are three different kinds of disturbing light which must be contended with:

1. Light, such as background light, which enters the receiver optics and after reflection by the tube walls, etc. reaches the photocell.
2. Light which originates from around the reflector but which is able to enter the aperture owing to imperfections in the receiver optics or to turbulence in the atmosphere.
3. Light coming from sources inside the Geodimeter housing or from a light background behind the instrument which after reflection in the reflector is picked up by the receiver optics.

Light as in (1) can be reduced by introducing a reflection absorbing device in the receiver optics. Of course a reduction can be achieved by reducing the receiver optics aperture and increasing its focal length. The aim was thus to design an optical system with high contrast.

Light as in (2) can be limited by improving the image definition. Here again a smaller aperture ratio would be a step in the right direction. A variable focal stop which can be adjusted to limit the active field of view to that of the reflector is an important part of the optical system.

Light as in (3). By using reflectors with the smallest possible dispersion, it is possible to limit the reception of noise from the Geodimeter surroundings. This solution is not suitable for optical systems such as those used in the Model 4. On the contrary, the placement of the transmitter and receiver tubes next to one another makes it necessary to introduce a deviating element (glass wedges) to make daylight measurements of distances below 1 km at all possible.

When examining the possibility of realizing the requirements outlined above, we find that noise light as in (3), for example, can only be reduced to any great degree by arranging both optical systems coaxially. With such an arrangement, all distances can be measured using the ideal corner prisms and the problem of adapting different wedges for different distances is completely eliminated. The Model 6 has consequently been designed with coaxially arranged transmitter and receiver optical systems. Fig. 1 shows its construction and the principal optical elements.

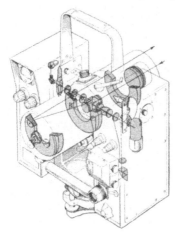

FIG. 1. The Geodimeter.

The Optical System

As can be seen in Fig. 1, the modulated light source and the photocell are placed on opposite sides of a tiltable optical tube. In fact, they are on the pivoting axis of the tube. At the cross section between this axis and the tube's axis of symmetry is a plane mirror placed at an angle of 45° to both axes. Thus, irrespective of the tube's angular elevation, the optical axes of both systems will coincide.

The main element in the transmitter system is a spherical mirror and in the receiver a three-lens optics with a high degree of correction.

The centrally placed receiver optics has a free aperture of about 40 mm and the system's focal length has been extended by means of a negative-lens system to about 600 mm. It has a high resolution and the picture sharpness is in practice limited only by turbulence in the atmosphere. It is possible to focus from infinity to 15 m by means of an inner focusing arrangement. This arrangement has the advantage of having a constant light path and it is thus possible to eliminate systematic errors which would arise when focusing is achieved by moving the mirror. A variable aperture has been placed in the focal point in the receiver optics and it can be adjusted from outside for apertures of about 0·1 to 0·5 mm in diameter. The aperture opening is, in its turn, projected through an astigmatic lens system to the photo cathode in the form of a line. In this manner, it is possible to keep the error caused by variations in the transit time of the electrons emanating from different parts of the photo cathode within reasonable limits.

To facilitate pointing to the reflector, a telescope sight with twofold magnification can be used and for fine pointing it is possible to deflect the received light beam on to a reticle which is visible in an eyepiece. The magnification is then 25 times and picture quality is uniform over the entire field of vision, which has a diameter of about 1·5°.

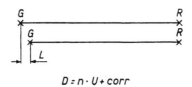

$$D = n \cdot U + corr$$

Fig. 2. Errors due to changes in delay line calibration.

The modulated light beam leaving the Kerr cell travels via an inner focusing system (which can be adjusted with the same control as the receiver focusing system) to the previously mentioned central mirror and a negative lens system and further on to the spherical mirror from which the light is sent out as a parallel beam. The receiver optics screens off the central part of the beam and its active area has an outer diameter of about 100 mm and an inner diameter of about 50 mm. The area is thus considerably larger than that of the receiver optics. The image definition is designed for projecting a luminous area 0·2 × 1·0 mm² (= the Kerr cell's electrode gap) placed in the optical axis. It is thus possible to meet the requirements for high light intensity at the expense of the image field. This should not be a disadvantage since the transmitter optics is used only for projecting and not—as in earlier Geodimeter models—for finding the reflector by visual means.

As in the previous models, the measuring principle used for the Model 6 requires the measurement of each unknown distance to be supplemented with a measurement of a known distance. This known distance consists of an inner light path running through the tube along which the beam is led, bypassing the central mirror via a system of mirrors. The inner light path has been designed so that a sharp image of the Kerr cell electrode gap is obtained at the focal point of the receiver system. This image is also used when adjusting the light source.

The Model 6 can be used with two different light sources. The standard one is an ordinary tungsten-filament incandescent lamp and the second a super-high-pressure mercury vapour lamp, called here the Hg-lamp. The incandescent lamp is mounted in an easily removable holder. Adjustment of the lamp position is accomplished by means of four screws accessible from outside the instrument. Checking of the lamp position is carried out by visual observation using the calibrating light path.

To permit quick change from standard to mercury lamp, the latter has been mounted together with its ignition unit in a special holder which can be attached to the Geodimeter. The standard lamp holder is then removed and replaced by a projection system which is screwed into the same opening. This projection system projects the mercury lamp arc on to a point where the incandescent lamp filament would normally have been located. When changing to the Hg-lamp it is possible to check its position in the eyepiece in exactly the same manner as when changing standard lamps.

The Electronic System

The operating principle for the Model 6 is much the same as the principle upon which earlier models were based. Since we can assume that these are already well known, the electronic construction is described here only in brief.

By transistorizing the electronic circuits, it has been possible to reduce power consumption, weight and volume in comparison with the Model 4. The entire electronic system including the power supply (for battery operation) is housed in the two side compartments. The left compartment houses the transmitter components and the power supply while the receiver components are housed in the right compartment.

The three transmitter frequencies are generated by thermostat-controlled crystals ($29\,970 \cdot 000$ kHz, $30\,044 \cdot 920$ kHz and $31\,468 \cdot 500$ kHz). The transmitter output signal is fed both to the Kerr cell and to the photocell. The signal to the latter is conducted via an electric delay line placed in the receiver side.

The delay line has been fitted with a three-figure digital read-out device. As was the case with the earlier models, four phase adjustments are made in each measurement series but the read-out device's capacity is designed so that the sum of the four phase values is approximately equal to the adjusted delay in millimetres. Interpolation is not necessary and only the nearest whole number is recorded. Conversion to length measure is carried out with the aid of a calibration table.

A new design has been used for the null detector and its sensitivity has been increased. Furthermore it is possible to compensate the decreased sensitivity of the photocell caused by low battery voltage by increasing the receiver amplification.

Power Supplies, etc.

The operating voltage for the Model 6 Geodimeter is 12 volts. Depending upon the number of lamps lit, whether the Kerr cell heater is switched on, etc., current consumption varies between 1·5 and 3·0 amps. Hermetically-sealed lead-cell accumulators allow an operating time of about two hours. There is space for this type of battery in the instrument transport case.

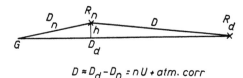

$$D \approx D_d - D_n = nU + atm. \ corr$$

Fig. 3. Error in the Geodimeter constant.

If the Hg-lamp is used, battery operation is no longer possible and power must be supplied by a gasoline-driven generator. A compact generator of this type has been designed especially for the Model 6 and developed by AB Hägglund & Söner in Sweden. This generator provides a stabilized 12-V d.c. voltage for the instrument and a 50-V d.c. voltage for the mercury lamp ($I_{max} = 6$ A).

Another useful feature of the Model 6 is the incorporation of a horizontal scale graduated in 400 ^{g} on one side and 360° on the other that can be used to facilitate finding the reflectors.

Since the majority of the instrument's assignments will be for polar measurements and traverses, it has been fitted with a centring device which can fit the T2's optical plummet, the Kern centring tripod, etc.

During transport, the instrument is stored in a glass-fibre-reinforced plastic case. This case has been designed so that it can be carried by one or two men or attached to a light harness for backpack transportation. The latter method is especially useful when moving over long distances in rough terrain.

In comparison with the Model 4 the daylight range is two to three times greater while the darkness range and accuracy are the same, despite the fact that the power consumption has been reduced to less than half (when using the standard lamp). The total weight of the entire equipment including the Hg-lamp attachment is also less than half.

Furthermore, the following features have been incorporated:

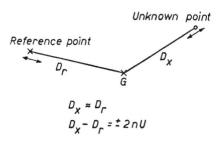

$$D_x \approx D_r$$
$$D_x - D_r = \pm 2nU$$

FIG. 4. Determination of small distance changes over long time intervals.

With a levelled tripod it is possible to measure from $Z = 0°$ to $Z = 140°$. This is possible through the tiltable optics. The change-over from standard to Hg-lamp takes only a minute or so and requires no tools. The aiming of the instrument has been improved in several respects, thanks particularly to the tiltable coaxial optics. There is much less danger of reading errors due to human factor owing to digital readout of the delay line readings.

Technical Data

Range (depending upon visibility):

Daylight	3 km	Standard	6 km	Hg-lamp
Darkness	15 km	lamp	25 km	

Accuracy: 10 mm + 2 mm/km (mean square error)

Weights:

Instrument	16 kg
Transport case, incl. battery and harness	12 kg
Power generator (incl. backpack)	16 kg
Hg-lamp attachment	1·5 kg

Power requirements:

Standard lamp operation	20–35 W, 12 V from battery or power generator
Hg-lamp operation	300 W, 12 V and 50 W from power generator

Error Discussion

In the following we will discuss the different factors influencing the accuracy of a Geodimeter measurement and also show how it is possible to increase the accuracy by use of a special measuring technique.

The total uncertainty in a Geodimeter measurement depends mainly on the following partial uncertainties:

1. Setting
2. Calibration
3. Geodimeter constant
4. Eccentricity of Geodimeter and reflector
5. Frequency
6. Meteorological data

1. *Setting Error*

This is defined as the mean square error for a series of four phase readings obtained under average conditions, i.e. stable atmospheric conditions, indicated by a good image quality. The mean square error has been found to be about 0·3 cm. Now for a measurement

$$\tfrac{1}{3}\sum L = \frac{\sum(R - C)}{3}$$

and since this error has a random character, the mean square error would be

$$0\cdot 3\sqrt{\frac{2}{3}}$$

or about 0·2 cm.

If the signals are very weak, this error will increase. It is, however, difficult to define an upper limit, since theoretically one may use signals just above the noise level, in which case very large delay changes are required for a noticeable movement of the null indicator pointer.

2. *Delay Line Error*

The delay line calibration is made by measuring known distances at intervals of 0·1 m over a length of 3 m, say between 20 and 23 m. The delay settings are then plotted on graph paper as a function of the distance. A smooth curve is drawn to give the best fit. Values from this curve are fed to a computer for final adjustment. The adjusted values can be found in the calibration tables supplied with each instrument.

The mean square error of a tabulated delay value is of the order 0·3 cm. Since each L-value is computed as a difference $R - C$, it follows that the delay error is zero when $R = C$. For large $R - C$ differences the delay error may increase to $0·3\sqrt{2}$ cm $\approx$ 0·4 cm. However, because of ageing effects of the electrical components associated with the delay line, the actual delay may differ from the tabulated one. These effects are indicated by a systematic difference between the three L-values obtained. If this difference, which for a newly calibrated Geodimeter seldom exceeds 1·5 cm, increases to 4 or 5 cm, evidently the error of the mean is larger. However, experience has shown that, as a rule, this error does not increase so much as the above-mentioned increase would indicate. This fact is due to the circumstance that the delay errors generally have different signs for F_1 and F_2 on one side and F_3 on the other one. In the following estimation the delay error has been set to 0·7 cm.

Analysis of a large number of measurements has shown that this assumption is justified.

3. *Geodimeter Constant*

The Geodimeter constant is computed at the factory by the precise measurement of a known distance, selected so that $R = C$ for one of the frequencies. This measurement is made with a mean square error of about $\pm$0·3 cm.

4. *Eccentricities of Geodimeter and Reflector*

Using optical plummets or corresponding centring devices the mean square error will be about 0·1 cm for each set-up.

5. *Frequency*

The frequencies are adjusted at the factory to within $\pm 1 \times 10^{-6}$ from the nominal value. The ageing of the quartz crystals will result in frequency error of the order 1×10^{-6} per year, gradually decreasing with time.

6. *Meteorological Data*

Under average meteorological conditions the difference between the temperature and pressure readings taken and the actual conditions integrated along the light path will result in an error of about $\pm 1 \times 10^{-6}$. Under extreme conditions (large height difference in combination with temperature inversion), however, this may increase to perhaps $\pm 3 \times 10^{-6}$.

The effects of the individual errors on the end result can be found in the following table:

	Mean square error (cm)	Proportional error $(10^{-6} \times D)$
1. Setting	± 0.2	
2. Delay	± 0.7	
3. Geodimeter constant	± 0.3	
4a. Eccentricity, Geodimeter	± 0.1	
4b. Eccentricity, Reflector	± 0.1	
5a. Frequency setting	—	1
5b. Frequency ageing	—	1
6. Meteorological data	—	1
Resulting mean square error	± 0.8	± 2 approx.

The methods described in the following will eliminate or reduce the contribution of the individual errors as follows:

Method 1: Error 2 is eliminated: $dD \approx 0.4 \, \text{cm} \pm 2 \times 10^{-6} \times D$

Method 2: Errors 2, 3 are eliminated: $dD \approx 0.2 \, \text{cm} \pm 2 \times 10^{-6} \times D$

Method 3: Only setting errors remain: $dD = (0.2\sqrt{2})/\sqrt{n}$, where n is the number of observations.

Methods for Increased Accuracy

In some cases it may be necessary to increase the measuring accuracy above the normal level. In the three methods outlined here the first one eliminates the errors caused by changes in the delay time calibration; the second one also eliminates errors in the Geodimeter constant, while the third method is intended specially for the determination of small distance changes over long time intervals, since it eliminates the delay line errors, the error of the Geodimeter constant and reduces or eliminates the proportional errors (frequency error and errors due to meteorological uncertainties).

1. This method is characterized by a displacement of the Geodimeter (or reflector) equal to one of the L-values obtained by the normal measuring procedure. Assume that the uncorrected L-values are $L_1 = 0.435$ m, $L_2 = 1.372$ m and $L_3 = 1.065$ m. If the Geodimeter is moved 0.435 m towards the target and measurements for frequency U_1 are made, it is obvious that the R_1 and C_1 values will be almost equal. If a series of measurements C_1, R_1, R_1 and C_1 is now made the difference of the mean R_1 and the mean C_1 can be computed. If this difference is multiplied by the scale constant as determined from the calibration table, any errors in the delay line calculation will be eliminated.

2. The limits of accuracy of the previous method are determined by the accuracy of the Geodimeter constant, the frequency and the meteorological data. Of these, only the first is independent of distance, while the remaining ones introduce proportional errors. The following method allows an elimination of the uncertainty in the Geodimeter constant. The distance to be measured must however be located in such a way that it is possible to place the Geodimeter at least 100 m from the nearest point and in line or almost in line with the points. Furthermore the two reflectors should be selected so that their apparent sizes as seen from the Geodimeter are about equal.

At first a determination of the distance between the two points is made in the normal way. Then the Geodimeter is replaced by a reflector which is moved so that the distance between the two reflectors will be a whole number of unit-lengths for one of the three frequencies, i.e. the nearest reflector should be moved $(L - 0\cdot130 - 0\cdot030)$ m towards the distant reflector. When alternating measurements are now made to the two reflectors R_n' and R_d', the difference between the mean R_n and the mean R_d (expressed in metres) will be the length by which the nearest reflector should be moved away from the distant one in order to make the distance between the two reflectors a whole number of unit lengths.

The reflectors must be arranged in such a way that while measuring towards one of the reflectors, the other does not have any influence on the signal. This can usually be arranged by a slight lateral displacement of the Geodimeter. The corrected distance D can be computed from the following formula:

$$D = D_d - D_n + \frac{h^2}{2} \cdot \frac{D_d}{D_n(D_d - D_n)}$$

where D_n is the distance to the nearest reflector, D_d the distance to the distant reflector and h is the shortest distance from the nearest reflector to the sightline connecting the Geodimeter and the distant reflector.

To give the order of magnitude of this correction, for $D_n = 100$ m, $D_d = 1100$ m and $h = 1$ m, the correction is $+5\cdot5$ mm.

In this case an error of $0\cdot05$ m in the h-value will change the correction by $0\cdot5$ mm.

3. This method is particularly well suited to the measurement of long time changes of the position of a point, e.g. in earthquake research, determination of slow movements in hydropower dams, etc. One necessary condition for its successful use is, however, that it is possible to find at least one reference point with high stability.

The distance to the unknown point is first measured in the usual way.

Then the reference point is located. It must be placed so that, for one of the frequencies, the R readings are nearly equal and of the same sign. It is not absolutely necessary that the two distances be equal, but they should be nearly equal in order to eliminate the necessity of any changes of the focusing of the Geodimeter and also to make the reflectors appear equally large as seen from the Geodimeter, e.g. 100 and 120 m, 300 and 345 m (using U_1, etc.). One should also make a kind of slide arrangement for the reference reflector so that its position can be adjusted to make the R readings to the two reflectors almost identical. The difference between the R measurements translated into metres gives the needed adjustment of the reference reflector position.

If the distances are equal, the proportional errors, the calibration error and the error in Geodimeter constant have evidently been eliminated and only setting errors and eccentricity errors of reflector and Geodimeter remain. Since the setting errors are of a random character, their magnitude can be diminished by an increased number of measurements. Tests have shown that it is possible to arrive at an accuracy in the determination of the relative position of the two reflectors of about ± 1 mm for distances up to several hundred metres, good visibility of course assumed.

DISCUSSION

R. Schöldström showed a film about the Geodimeter. He then pointed out how the nature of the reflection from a corner cube reflector made it very desirable to adopt coaxially-mounted transmitter and receiver systems and how this had made it possible to use improved corner cube reflectors for all ranges of operation without, as in Model 4, using wedges in front of reflectors. He described the basic layout of Model 6 and emphasized that assessment of individual and total errors given in the paper were generous and that proportional error of $2 \times 10^{-6}D$ could be relied on.

First Experiences
of the Model 6 Geodimeter
in the Field

I. R. BROOK

Geographical Survey Office of Sweden

During tests in early 1965 some 50 lines, varying from
1–17 km, were measured. It was found possible to measure up
to 6 km in daylight and up to 8 km in half light. The use of
new, coated prisms gave a markedly improved reflected
signal and are particularly recommended for lines over
15 km long.

From limited field trials it appeared that the Model 6 was
far superior to Model 4, pointing being improved by the
incorporation of a telescope and horizontal circle and recording facilitated by a digital read-out.

Several recommendations are made for minor improvements to the system.

During the late spring and early summer of 1965 the Geographical
Survey Office of Sweden carried out a series of field measurements
with an Aga Model 6 Geodimeter. The Model 6 Geodimeter was presented for the first time at the Lisbon Photogrammetric Congress
during the autumn of 1964 and was in use in Sweden at the beginning of
1965. The early Model 6 Geodimeters were equipped with tungsten-
filament lamps, the mercury-vapour lamp unit and generator not being
available. The Geographical Survey Office was interested in the possibilities of using the new model for measurements in the Swedish second-
order triangulation network and town and city networks where advantage could be taken of the claimed increased daylight range and improved
portability. We were, therefore, particularly interested in the Geodimeter with the mercury-vapour lamp unit. Unfortunately, and owing to
circumstances beyond their control, Aga were unable to deliver the
mercury-vapour lamp unit and motor-driven generator before the
beginning of April; as the instrument was required in the field, no
time was available for laboratory tests other than calibration of the
modulation frequencies and a summary check of the crystal oven. It is,

therefore, the purpose of this paper to present our first, practical experiences of the Model 6 Geodimeter in field use.

In an article in *Svensk Lantmäteritidskrift* 5–6 1964, Mr Schöldström of Aga wrote that the constructors of the Model 6 instrument considered that a new instrument should have the following four principal advantages over the Model 4 instrument, namely:

(i) Lower weight and thereby improved portability
(ii) Reduced power consumption
(iii) Increased daylight range
(iv) Be easier to use—this applied particularly to location of the prisms with the help of a small sighting telescope and built-in searchlight.

The Model 4 Geodimeter is a heavy and bulky instrument and appears to have been designed more as a scientific instrument than a practical instrument suitable for field use. This is particularly true of Model 4 instrument with the mercury-vapour lamp modification for which a motor-driven generator and a heavy and bulky power unit is required. In areas of forest and broken terrain where loads must be carried by labourers over long distances the Model 4 instrument compares unfavourably with for example, the MRA-2 Tellurometer. An extra labourer is normally required if the Tellurometer is to be replaced by the Geodimeter. The Model 6 Geodimeter, on the other hand, complete with mercury-vapour lamp unit and generator weighs only 50 kg, which is approximately half the weight of the Model 4 instrument, and it can be comfortably carried by two persons. The Geodimeter is carried in a practical, robust, waterproof, plastic case which is reinforced with fibreglass. The case can quickly and easily be mounted on a specially constructed haversack frame. The frame appears to be robust and is anatomically correctly shaped. The instrument rests in the carrying case on a number of stable, rubber cushions and our experience is that it is well protected from the shaking and jolting which is inevitable with pack transport. The motor-driven generator and the mercury-vapour lamp unit are carried on a specially constructed haversack frame. The case in which the mercury-vapour lamp unit is housed under transport fits into the lower part of the frame under the generator. The carrying case, which is made of steel, is partly rubber lined to protect the lamp unit from damage during transport. The tungsten-filament unit when not in use is housed in a small tubular compartment in the same case. We were afraid that the mercury-vapour lamp unit and particularly the quartz envelope would be susceptible to damage during pack transport as the carrying case is placed in a somewhat vulnerable position under the generator. Our fears were soon proved to be completely unjustified:

the lamp case fell from a 25-m steel tower without damage to the mercury-vapour unit. Not even the filament of the tungsten lamp was
broken.

The generator which is included in the standard equipment is driven
by a two-stroke engine which has been specially constructed for use
with the Model 6 instrument and can be used as power source for both
tungsten and mercury-vapour lamps. This unit together with haversack frame weighs a little over 15 kg and is a practical construction well
adapted for field use. The motor is easily started and is surprisingly
economical: a full tank which is somewhat over one litre is sufficient for
a running time of approximately seventy-five minutes. A practical detail is the breather-screw on the filler-cap which can be closed to prevent
fuel spill under pack transport. Low fuel consumption and low weight
distinguish this unit from the motor-driven generator which Aga in
Sweden recommend for use with the Model 4 instrument. A transistorized power unit which is built into the Geodimeter replaces the
heavy power unit which is a part of the Model 4 standard equipment.
The power unit which is required for the Model 4 instrument as modified for use with the mercury-vapour lamp is a particularly unpractical
unit from the field surveyor's point of view.

The Model 6 Geodimeter with the tungsten-filament lamp can also
be operated using special gas-tight accumulators which are transported
in a compartment in the instrument carrying case. The capacity of these
accumulators is sufficient for approximately six full measurements. The
power requirement for the Model 6 Geodimeter with tungsten-filament
lamp is 35 W, 12 V which is approximately half of that of a Model 4
instrument. The weight of the carrying case plus accumulators is 12 kg.

The only other item of equipment required is a tripod.

Field Measurements

During the first field period approximately 50 distances were observed
and these varied in length from under 1000 m to almost 17 000 m. Since
then a side over 20 000 m has been measured.

According to Aga, the range of the Model 6 Geodimeter is as follows:

	Tungsten lamp	Mercury-vapour lamp
In daylight	2–3 km	5–6 km
In darkness	10–15 km	20–25 km

All measurements with the exception of two 6000-m sides and between twenty and thirty traverse legs, none of which was over 1500 m
in length, were measured with the Geodimeter equipped with the mercury-vapour unit. The larger part of the measurements were made

during the month of June at which time of the year the nights are never truly dark in the part of Sweden where the field work was carried out. Throughout the period, which was notable for several weeks of fine weather with visibility up to 30 km and light north-easterly winds, measurements were possible on all but two nights.

The larger part of the distances measured were between 5000 m and 12 000 m and formed part of an interlocking triangulation network. It was, therefore, normally possible to observe several distances from each set-up and the observing programme was drawn up so that the first observation of the shortest side could begin approximately two hours before sunset and sides of increasing length were measured as darkness fell. After the first few evenings' observations it became clear that measurements of sides of up to 6000 m presented no difficulties whatsoever before sunset and that distances of up to 8000 m could be measured in conditions of half-light. For distances of over 8000 m, darkness was necessary before observations could begin.

For all measurements between 5000–7000 m a minimum of nine prisms was the rule; for distances between 7000–10 000 m, between twelve and fifteen prisms were used, the number depending upon the location of the reflector station. A larger number of prisms was, for example, used when the measurement was in the direction of the very light northern horizon. Aga, in the Geodimeter handbook, suggest the use of nine prisms for distances of the order of 5000 m in daylight and one prism for measurements at night; for distances between 5000 m and 10 000 m Aga recommend the use of two prisms (night measurements). Aga's recommendations are, I feel, extremely optimistic and one would be ill-advised to go to the expense of visiting a triangulation station to set up one or two prisms when the likelihood of one being able to complete an observation in anything but the most ideal weather conditions would be very small. I would suggest a minimum of three prisms for night observations of distances between 3000 and 6000 m and a minimum of six prisms for observations of distances between 6000 and 10 000 m. I refer here to the older types of prisms which are marked with three white dots and which were available for use with the Model 4 Geodimeter and not to the newer, much improved type, which Aga later supplied to the Geographical Survey Office. Aga's figure of eighteen prisms for distances of the order of 20 000 m agrees with our estimate of the number of prisms that is required.

The longest line measured during the first field period was of the order of 17 000 m and was between a station on the Swedish mainland and a station on the island of Öland. The first determination of the distance was made with the Geodimeter on the mainland in a 17-m tower and the prisms in a 6-m tower on the island. Eighteen prisms of the older type were used and weather conditions were ideal. The measurement was made at the darkest time of the night. In spite of all attempts to point the Geodimeter as carefully as possible and although the maximum aperture was used in combination with a fully open grey-wedge position it was not possible to obtain more than a weak flutter of the pointer on the control instrument and the movement of the pointer of the null indicator meter was too sluggish to enable a measurement to be made. The amplitude setting was between 0·5 and 1·0, as recommended in the handbook. By increasing the amplitude setting to *ca.* 2·0, it was possible to obtain a low dip reading on the control instrument and movement of the pointer of the null indicator was less sluggish. The pointer was extremely unstable when a check calibration reading was made. The instrument was then pointed using the standard daylight observing procedure, i.e. pointing the instrument and adjusting with the aid of the slow-motion screws so as to obtain the fastest possible movement of the null indicator pointer when the phase selector switch was switched between phase 1 and phase 2, and ignoring the control instrument reading. Despite the relatively sluggish movement of the null indicator needle during the reflector measurements and over-sensitivity during calibration measurements, the internal agreement was good. With a Model 4 Geodimeter with mercury-vapour lamp the measurement would not have presented any difficulties and normal instrument readings would have been obtained.

Previous measurements of sides up to 10 000 m had presented no difficulties although several more prisms than were recommended in the handbook had been used. For distances over 10 000 m the 'dip' had been weak but not so weak as to make observation difficult. From experience with the Model 4 instrument we were convinced that the source of our difficulties lay in the quality of the prisms which were not suitable for measurements over long distances with the Model 6 instrument. Mr Schöldström of Aga was contacted and he arranged for ten prisms of a new and improved type to be sent to us in the field. Mr Schöldström will, doubtless, be prepared to describe in detail how the new prisms differ from the older type; but in brief it can be said that they are surface treated which results in a markedly improved reflected signal.

It was decided to remeasure the line between Öland and the mainland using a combination of new and old prisms. The total number of prisms used was eighteen as the intention was to investigate whether, under similar weather conditions, a similar number of prisms, half of which were of the improved type, would result in an improved reflected signal. Ideally eighteen prisms of the new type should have been used, but at that time only ten new prisms were available. When the measurement was made weather conditions were, as with the first measurement, ideal but the night was much lighter than when the first measurement was made. The reflected signal appeared to be stronger and the dip was improved. Nevertheless, with a Model 4 Geodimeter, the 'dip' would have been much greater than was now the case with the Model 6 instrument. The null indicator needle showed a normal degree of sensitivity and reacted for the smallest movement of the digital reading scale adjustment screw. The internal agreement between the three determinations and the over-all agreement between the two independent measurements were good.

It would appear that a very low 'dip' reading is to be considered as normal over longer distances.

Later measurements made using the newer prisms confirmed these first impressions. The new prisms definitely appear to give a sharper reflected signal and are to be recommended for measurement of distances over 15 000 m.

A Geodimeter traverse with legs of the order of 2000–3000 m was measured between small islands which lie between the mainland and Öland. The measurements were made in full daylight and strong sunlight which caused strong reflections from the calm sea. Despite these conditions which must be considered as extreme the observations could be carried out quickly and without difficulty. Similar measurement would have been impossible with a Model 4 instrument.

Our impressions of the new Geodimeter which are based on a relatively short field period are largely positive for there is no doubt that as a short-to-medium range instrument the Model 6 Geodimeter is far superior to the Model 4 instrument.

The new instrument is a more practical field instrument than the Model 4 instrument. The provision of a real sighting telescope is a great improvement on the somewhat impracticable arrangement on the Model 4 instrument, and this, together with the engraved horizontal circle, makes pointing of the instrument easy to carry out. The digital reading system is superior to the previously used reading arrangement, and its use greatly reduces the possibility of erroneous readings being made.

In conclusion we would like to make the following comments on the general design of the instrument:

1. The locking screws and slow-motion screws are badly placed and are difficult to manipulate, particularly when the Geodimeter is set up in a tower. Similar difficulties will, I imagine, arise when the instrument is set up on a pillar.

2. The unit on which the Model 4 is mounted on the tripod is a stronger and more stable construction than the Wild T2 tribrach which

is supplied with the Model 6 instrument. The instrument has a tendency to shake with even moderate wind forces. This is not altogether un-expected as the Geodimeter is much heavier than the T2 theodolite and much larger. Mr Schöldström has stated that, in his opinion, the new Geodimeter will be mainly used for traversing and radial measurement, and it is for this reason, one assumes, that the forced centring tribrach arrangement has been adopted. I wonder, however, if the larger survey organizations will be so restrictive in their use of the instrument; at the same time, it would be interesting to hear to what extent such organizations intend to take advantage of forced centring in, for ex-ample, Geodimeter traverses.

With its high portability and greater crystal stability and using re-fined observational procedures, the Model 6 may prove to be an ideal instrument for high-order measurements. For such measurements a stable instrument set-up is a primary requirement.

3. The placing of the power socket under the instrument combined with a long plug makes the instrument difficult to use in towers and, I imagine, on pillars without an adaptor between the tower ring and the tribrach.Without an adaptor the cable can be easily trapped and damaged.

4. The mercury-vapour unit must be mounted on the Geodimeter before each set-up and removed after observations have been completed. Despite the fact that the three adjuster screws are spring-loaded, mounting and dismounting of the unit does make readjustment of the lamp necessary before measurement can be begun at each new station. This is, of course, not the case where the distances to be measured are short. Is not a more rigid mounting of the lamp unit possible and, indeed, preferable?

5. Focusing of the mercury-vapour unit is by no means difficult, but as the easiest method of focusing the unit is to place a piece of tracing paper in front of the transmitter/receiver optics and adjust the position of the lamp with the two adjuster screws until the illumination is even over the whole field of view, focusing is best carried out indoors. Would it not be possible to supply a frosted-glass disk which could be fitted over the optics when refocusing was necessary?

6. At present the searchlight is fitted with a tungsten-filament lamp. I feel that an arrangement similar to that in the Model 4 instrument in which the mercury-vapour lamp is also used as a searchlight would be preferable to the present arrangement.

7. The connecting cable as supplied with the generator is short and with measurements from towers the generator must be placed on a special platform under the observer's platform.

8. The motor-driven generator appears to be reliable, but, as the unit vibrates heavily, the four resistances which are rigidly mounted on the

generator housing have a tendency to vibrate loose (the mounting having fractured).

DISCUSSION

L. Asplund: I accept some responsibility for the paper and stress that we were searching for faults and that we were very favourably impressed with the instrument.

R. C. A. Edge: Would it not be advantageous to equip the instrument with an accurate horizontal circle?

R. Schöldström: Almost invariably the users of the instrument are already employing theodolites. The instruments are most often used to measure traverses when they only occupy alternate stations; therefore, it is considered best to separate the two operations of distance and angle measurement. In any case a circle of comparative accuracy with the distance accuracy would add greatly to the cost.

I. R. Brook: The horizontal circle was designed primarily to assist in pointing and is very useful for this. I suggest that an accurate vertical circle might be more useful than a horizontal one.

J. R. Hollwey: Why is it not possible to eliminate the delay line tables and why is a direct read-out in metres not incorporated in Model 6?

R. Schöldström: Manufacturing difficulties are such that it is not possible to construct identical delay lines in each instrument and individual calibration is therefore necessary.

J. C. de Munck: I understand that the mercury lamp cannot be used to assist in the pointing of the telescope. Could this not be made possible?

R. Schöldström: Yes, a modification for this purpose is in hand.

A. R. Robbins: Is there likely to be any trouble caused by the breakdown of the nitrobenzene in the Kerr cell by ultra-violet light when operating in the tropics?

R. Schöldström: In the Model 6, the shielding of the cell has been improved and a filter fitted to reduce the ultra-violet light reaching the cell from the mercury-vapour lamp. I do not think this is a likely cause of trouble.

The Accuracy of a 50-km Over-Sea Geodimeter Distance and a Study of Temperature Correction by means of a Temperature Gradient Formula

E. BERGSTRAND

Geographical Survey Office of Sweden

Presented by S. SUNDQVIST

The measurement of a 50-km distance over sea was performed with the Geodimeter NASM-2. The light source was a Hg-lamp. In good visibility there was plenty of light. The vertical temperature-gradient was obtained from the curvature of the ray. A simple formula was deduced using the dip of the horizon for obtaining a value on the temperature-gradient. Including all reasonable errors the accuracy was 1 ppm.

The task was a check of the co-ordinate distance from the mainland to Gotland, the large island in the Baltic sea, previously determined by ordinary triangulation under unfavourable conditions. In this study we were particularly interested in the main part of the distance of 50 km between Öland and Stora Karlsö, as shown in Fig. 1. Except for the reflector station at Ängjärnsudden, all stations in the figure are of first

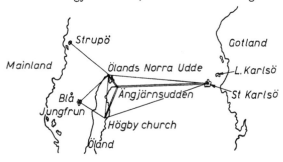

|| Directly measured distances

FIG. 1. Arrangement of measured lines.

order. By aid of a 30-m steel mast the forty-two reflector prisms were placed 37 m above sea level. At the station Karlsö the Geodimeter-stand was placed on the ground 57 m above sea level. According to ordinary formulae the lowest part of the sight then should pass 5 m above the sea surface, leaving a small margin for unfavourable refraction.

It appeared, however, that at the measurements in August the summer-warm sea caused such a great negative deviation from normal refractive condition that, after sunset, the reflector was seen above the horizon only once during a whole month. This occurred in a night with a north-westerly half-gale (8 Beaufort). The visibility was excellent and good distance records were obtained on that one night. The waves on the sea, reaching a max. height of 3 m, at no time interrupted the line of sight. From this it was concluded that the line of sight in its lowest part averaged at least a height of 3 m above sea level. On the other hand a week later under very similar circumstances, the reflector was observed just before the Sun set and then gradually disappeared under the horizon. Thus it may be concluded that even during the measurement at night the height above sea level probably could not have been large and it was estimated to be 5 m, or corresponding to normal refraction.

In this way one got an estimate of the curvature of the line of sight and thereby of the temperature gradient above the sea-surface. Temperature was observed at the Geodimeter station as well as at Karlsö lighthouse, situated directly under the line of sight, and at Ölands Norra Udde lighthouse, all of them at different altitudes. After application of the temperature-differences, computed from the curvature of the line of sight, the standard error of the computed average temperature of the line was estimated to $\pm 0.3°C$. This figure also includes a small systematic gradient towards the somewhat warmer mainland. By increasing the figure to $\pm 0.5°C$ we certainly also allowed for possible local gradients, thermometer errors, etc. An estimate of standard errors is listed in the following table:

The value of c	0.6 ppm	$(\pm 0.2$ km/sec)
Distrib. of obs.	0.4 ,,	(From the three frequencies)
Temperature	0.5 ,,	$(\pm 0.5°C)$
Geodimeter	0.3 ,,	(Internal instr. error ± 1.5 cm)
Colour	0.2 ,,	$(5490 \pm 30$ Å by colorimeter)
Pressure	0.2 ,,	$(\pm 0.5$ mm Hg)
Frequency	0.2 ,,	(Noted change in 4 years)
Centring	0.2 ,,	$(\pm 7$ mm at each end of the distance)
Humidity	0.1 ,,	$(90 \pm 10\%$ above the sea)
$\sqrt{(\Sigma\Delta^2)}$	1.0 ,,	$(\pm 5$ cm)

It is obviously difficult to make an accurate estimate of the various sources of errors. Therefore the limits in the table, except for the first two, are amply chosen and sometimes correspond more to maximum errors.

For comparison with co-ordinates, the reflector station at Ängjärnsudden must be linked to the first-order net. For that purpose a Geodimeter NASM-4 was used on the first-order stations Ölands Norra Udde and Högby church and directions were obtained also from Strupö and Blå Jungfrun. The difference in Ängjärnsudden coordinates as determined from north and south was 7 cm, indicating a scale error in first-order network of about 1 : 300 000. The difference between Geodimeter and co-ordinate distances Ängjärnsudden–Karlsö was 53 cm. Regarding the above-mentioned scale error, the real measurement difference between the old triangulation and the new determination amounted to 70 cm.

In the above over-sea measurement there arose the possibility of determining the temperature gradient. Such information could be useful in many cases. Suppose equal altitudes h_1 for both ends of the distance D km. In vacuum, a levelling sight from one end would pass $D^2/(2 \times 6363)$ km above the other station (6363 km is the Earth's radius.) In the ordinary case in the atmosphere, the sight passes lower by an amount of $D^2/(2 \times R)$ km, where R is the curvature radius of the line of sight. By aid of the theodolite we measure the remaining vertical angular distance down to the height h_1 according to

$$\frac{D^2}{2 \times 6363} - \frac{D^2}{2 \times R} = \frac{\Delta_{h_1} \times 2\pi\, D_{km}}{40\,000}$$

where Δ_{h_1} is the measured angle in centigrade minutes (Fig. 2). We get

$$\frac{1}{R} = \frac{1}{6363} - \frac{\pi\,\Delta_{h_1}}{10\,000.D_{km}}$$

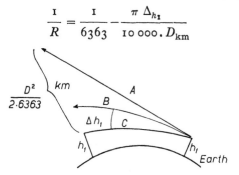

FIG. 2. Relation between lines of sight. A = horizontal line of sight in vacuum. B = horizontal line of sight in atmosphere. C = line of sight to target in atmosphere.

9

Now, according to Newcomb, we have the inverted curvature radius of the line of sight:

$$\frac{1}{R} = c \cdot \rho \cdot \left(\frac{1}{h_p} + \frac{1}{\tau} \cdot \frac{d\tau}{dh} \right) \sin z$$

where R is the curvature radius as above, c = constant, ρ = atmospheric density and $2c\rho = (\mu^2 - 1)$, where μ = refractive index; h_p = 'pressure-altitude', that is the equivalent height of the atmosphere for constant pressure along the vertical (h_p is around 8 km); τ is the abs. temperature (°K) and h indicates height in km. For the present purpose, between 0° and +20°C and 720–770 mm Hg one may put $c \cdot \rho = 0 \cdot 00028$, $h_p = 8 \cdot 3$ km, $\tau = 283$ and $\sin z = 1$.

Equating the $1/R$-equations we obtain:

$$\frac{d\tau}{dh} = 123 - 314 \cdot \frac{\Delta_{h_1}}{D_{km}} \, °C/km$$

where we would call Δ_{h_1} the dip under the horizon of an equal height (expressed in centigrade minutes).

In case the heights are not equal one must reduce the angle by

$$6363 \cdot \frac{(h_2 - h_1)}{D_{km}} \text{ minutes}$$

and replace Δ_{h_1} by

$$\Delta_m + \frac{(h_2 - h_1)}{D_{km}} \cdot 6363$$

where Δ_m is the measured angle from horizontal direction to the target, the height of which is h_2 km.

Starting with the temperature observed at the end(s) of the distance and applying the obtained value of $d\tau/dh$ we may compute the temperature along the line of sight. Windy weather will give best results or generally when the influence of the near-ground atmospheric layer is comparatively small. Usually it is so.

DISCUSSION

L. Asplund: This is the first description of the measurement of a line of over 50 km with Geodimeter. I think 60 to 70 km would be possible. I should emphasize the importance of the study of temperature correction by determining the curvature of the ray.

Geodimeter 2A Measurements in Base Extension Nets

H. GROSSE

Institut für Angewandte Geodäsie, Frankfurt/Main

Recommendation No. 5, composed at the Symposium for the Readjustment of the European Triangulation Network held in Lisbon, and submitted to the Twelfth General Assembly of the IUGG held in Helsinki, requested the member nations to determine the greatest possible number of triangulation sides by means of electronic distance-meters. In implementing this recommendation the Institut für Angewandte Geodäsie (Second Division of the German Geodetic Research Institute) measured, by means of an NASM-2A Geodimeter purchased in 1958, the base lines and the base-extension nets of the recently determined base lines of Munich (1958) and Heerbrugg (1959) partially, as well as the base lines of Meppen (1960) and Göttingen (1961) completely. The results illustrate the accuracy obtained in the electro-optical measurement of distances. They make it doubtful whether in future the classical base line measurements with invar wires and angular measurements in the respective base-extension figures, will still be reasonable because of their considerable wastage of time and labour.

1. *General*

In accordance with the Resolutions adopted in Rome in 1954 by the International Union of Geodesy and Geophysics, four base lines had been measured with invar wires in the years from 1958 to 1961 in West Germany and at Lake Constance on Swiss territory: namely Munich (1958), Heerbrugg (Switzerland) (1959), Meppen (1960) and Göttingen (1961). The pertinent base-extension networks had been determined anew by means of angular measurements.

The Resolutions No. 3 and 4, adopted by the Permanent Commission for the European Triangulation of the International Association of Geodesy at a meeting held in Munich from 22nd to 26th May, 1956,[1] and the Resolution No. 5, adopted by the Symposium for the New Adjustment of the European First-Order Triangulations held in Lisbon from 19th to 24th April, 1960,[2] invited studies on the accuracy of the first-order triangulation networks and studies on scale as well as experimenting with the electronic distance meters, which have recently appeared on the market. In accordance with these Resolutions, the four baselines mentioned above as well as the newly-determined base-extension networks were remeasured by means of an NASM-2A Geodimeter, procured late in 1958. During these measurements the accuracy of this instrument should be controlled, and further possibilities of application should be found in controlling the scale of the first-order triangulation network. The results of these measurements will be published in the publication series B of the Deutsche Geodätische Kommission (German Geodetic Commission) in summer 1966.

2. Results of Base line Measurements with Invar Wires and 2A Geodimeter

In the following the results of measurements are given which are comparable, since they refer to the same centres and to the same heights.

2.1   *The base line of Munich* (Fig. 1), situated east of the city of Munich in the Ebersberger Forest, is 8·2 km in length and has been measured by means of 18 invar wires.[3,4] Before and after measurement these wires were compared with the Standard Metres in the 'Bureau International des Poids et Mesures' at Sèvres near Paris, and in the 'Physikalisch Technische Bundesanstalt' in Brunswick.[5] An additional comparison had been made before, during and after the measurement of the base line on the standard line of Munich in the Ebersberger Forest, which is in close neighbourhood of the base line. The angles of the base extension net have been measured by means of the T-3 and T-4 Wild instruments.[6]

The length of the base line of Munich (Ebersberger Forest) amounts to (measured with 18 invar wires) 8231·847 m.[3,4]

The results of the 2A–128 Geodimeter measurements made in October 1959 are as follows:

> 8231·873 m (21st October) with frequencies 1, 2 and 3[7]
> 8231·868 m (30th October) with frequencies 1, 2 and 3
> 8231·870 m mean value out of six measurements.

The base line of Munich has also been re-measured by means of the Geodimeter 2 of the Austrian Bundesamt für Eich- und Vermessungswesen (Federal Board of Standardization and Surveying), resulting in a value of 8231·879 m.[10]

Both these Geodimeter values coincide quite well, but they differ by ⌒ 26 mm from the value obtained from the measurement by means of invar wires. As the Geodimeter measurements refer to the surface beacons (first-order pillar), which have been set in the meantime, the

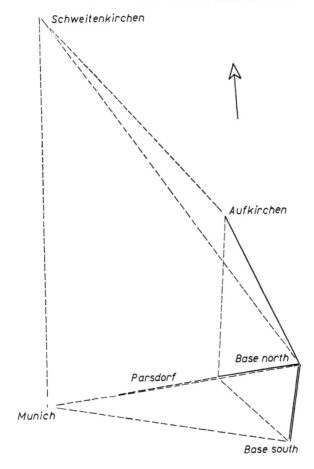

Scale 1: 400 000

FIG. 1. Munich base-line and base-extension net 1958.

difference may be caused by the centring with respect to the under-ground marking point. This may especially concern the southern base terminal, this terminal being situated on a slope.

Moreover, as to the Geodimeter used by a German team, it must be admitted that this has been the first ever Geodimeter measurement, and that a certain lack of experience may also have caused the consider-able difference.

2.2. *The base line of Heerbrugg* (Fig. 2), is situated south of Lake Constance near the boundary between Switzerland and Austria. It has been measured by a joint team from Switzerland, Austria and Germany in 1959 by means of 12 invar wires. The calibration of these 12 invar wires had been executed in the same way as in the year before in

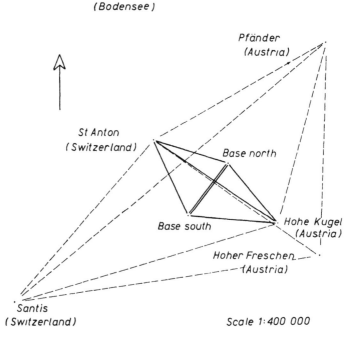

FIG. 2. Heerbrugg base-line and base-extension net 1959.

Munich. The angles in the base extension network had also been measured in common by observers of Switzerland and Austria by means of a T-3 Wild instrument. The results are published within the series of publications of the Bodensee-Konferenz. The result of the measurement with invar wires is:

7253·514 m, reduced to levelling datum.

The Geodimeter measurements in the same region were made in June 1960.

The base line has been measured once with three frequencies:[7]

$$7253 \cdot 503 \text{ m } (f_1)$$
$$7253 \cdot 512_5 \text{m } (f_2)$$
$$7253 \cdot 524 \text{ m } (f_3)$$

Mean value:

$$7253 \cdot 513 \text{ m, reduced to levelling datum.}$$

This base line has also been measured with the Geodimeter 2 by an Austrian squad; the results are not known to us.

2.3   The *base line of Meppen* (Fig. 3) is situated south of Emden,

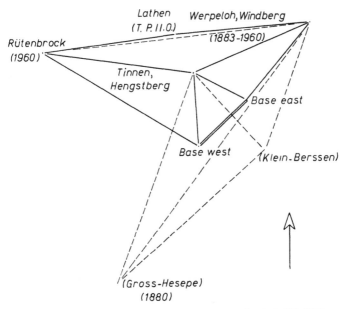

FIG. 3. Meppen base-line and base-extension net 1883/1960.

near the Dutch boundary. For this reason, it is of special importance also for the Dutch triangulation network. Therefore this base line has been measured in collaboration with the Dutch Geodetic Commission in 1960 by means of 10 invar wires. It was possible to find the underground permanent marks of the year 1883 and to use them as base line terminals, so that a direct comparison with the earlier measurement of Schreiber's was made possible. The base extension network was observed by means of T-3 Wild and Tpr-Askania instruments (visually and with electric eye).

As a basis of comparison there were available: laboratory comparisons of Paris and Brunswick; standard line comparison at the standard line of Munich; standard line comparison at the standard line of Loenermark.

As in accordance with the Resolutions of the IAG only the values of comparison with the standard lines were to be used for the calculation, the value of the length of the base line amounts to:

7039·457 m, reduced to levelling datum
(with comparison value of NS Munich)
7039·453 m, reduced to levelling datum
(with comparison value of NS Loenermark).

For the purposes of comparison the value of 1883 is given:

7039·481 m, reduced to levelling datum.

The Geodimeter measurements were made in April 1961 and in April/May 1962. In 1961 the signals still existed on the supplementary first-order points, so that it was possible also to calculate the base line twice by sections.

24 April, 1961	7039·469 m, reduced to levelling datum
25/27 April, 1961	7039·444 m, reduced to levelling datum (composed)
25/27 April, 1961	7039·458 m, reduced to levelling datum (composed)
Mean value 1961:	7039·457 m
6 April, 1962	7039·461 m, reduced to levelling datum
8 April, 1962	7039·456 m, reduced to levelling datum
3 May, 1962	7039·448 m, reduced to levelling datum
3 May, 1962	7039·454 m, reduced to levelling datum
4 May, 1962	7039·460 m, reduced to levelling datum
Mean value 1962:	7039·456 m
Total mean value 1961–62:	7039·456 m, reduced to levelling datum

Thus the value is between the two values of lengths determined by comparison in Munich and at Loenermark. This confirms the difference found between the measurements with invar wires and the measurement of 1883.

2.4 *On the base line of Göttingen* (Fig. 4) the subsurface marks of 1880 had been found and used as terminals. The length of this base line had been measured in 1961 by means of 10 invar wires, which had been calibrated in Paris, Brunswick and on the standard line of Munich. The result was 5192·901 m, reduced to levelling datum. The 1880 value was 5192·929 m, reduced to levelling datum. The base

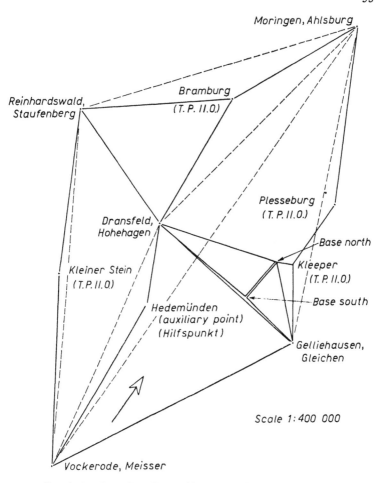

FIG. 4. Göttingen base-line and base-extension net 1880/1961.

extension net had been observed by means of a T-3 Wild instrument.

The Geodimeter measurements were made immediately after the base line measurements, because the signal 'northern base terminal' had to be dismantled by the end of 1961. The results are:

18 October, 1961	5192·899 m, reduced to levelling datum
22 October, 1961	5192·900 m, reduced to levelling datum
25 October, 1961	5192·904 m, reduced to levelling datum
Total mean:	5192·901 m, reduced to levelling datum

9*

Here, in the same way, the difference between the 1880 measurement and the measurements with invar wires is confirmed by the Geodimeter measurement.

3. *Re-measurement of Old Base Lines with the 2A Geodimeter*

Some base lines in the West German portion of the European triangulation network could not be re-measured with invar wires or tapes, as they are now situated in built-up areas or because important traffic lines now cross the base line. The systematical deviations of the old base lines from the new observations near Meppen and Göttingen make a control desirable which can be performed with the 2A Geodimeter.

3.1   *The base line of Braak* (near Hamburg) (Fig. 5) was first measured in 1820 by Denmark, and later in 1871 with the Bessel apparatus. The first value obtained in 1820 was not introduced into the newer adjustments. But also the base line value of 1871 had not been used generally, because afterwards a considerable correction was applied to it[8] (a small steel ball had a wrong calibration value).

1820:	5875·274 m, reduced to levelling datum
1871:	5875·322 m, reduced to levelling datum
	5875·297 m (corrected).[9]

In the course of the new triangulation of the north-west coast of Germany, signals were also built on the base line of Braak, so that Geodimeter measurements could take place from 12 May till 19 May, 1965. On 12 May, 1965 it was hazy and calm, from 13 till 16 May, 1965 it was impossible to make measurements, and from 17 to 19 May, 1965 the visibility was very clear with a wind force amounting to 5–6. In order to eliminate the internal instrumental errors we measured not only to the central point but also to three satellite stations which required in each case a different light conductor. The results are:

Date	Geodimeter	South II	South I	North II
12.5.1965	5875·316 m	—	5875·315 m	5875·320 m
17.5.1965	5875·318 m	5875·310 m ·318	—	5875·308 m
18.5.1965	5875·318 m	5875·310 m ·312	—	5875·306 m ·305
19.5.1965	5875·311 m	5875·310 m ·314	—	5875·312 m ·313

Mean value: 5875·313 m ± 3·0 mm

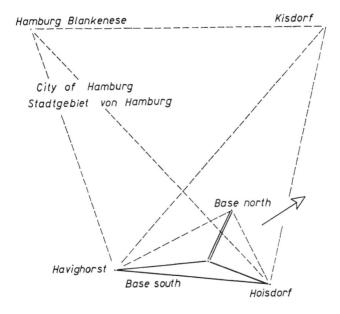

Scale 1:400 000

Fig. 5. Braak base-line and base-extension net 1871.

Each of these measurements consists of three complete pairs. The double values result from the fact that at both the satellite stations (South I and North II) we have measured with light conductors 1 and 10, with 2 and 11 respectively. For these above-listed values we must still carry out some calibration measurements; thus the values are still preliminary. But now the question arises, whether the later correction of the value of 1871 was correct for all scale control resulted in a somewhat smaller new value. The doubts with respect to the value of 1871 were justified, and a clarification is possible with the aid of the Geodimeter.

3.2 *The length of the base line of Døstrup* situated in Denmark was determined with invar wires in 1933 by Denmark. It cannot be re-measured today, because the base line is now crossed by a highway, a railway, a bridge and a very high roadway. In course of the North German coast triangulation in 1964 the connection with Denmark has also been established. For this reason it was possible also to re-measure with the 2A Geodimeter the first-order triangulation side of Rømø–Vongshøj, which is expanded from the base line of Døstrup.

Because no signals were constructed, we had to measure on the ground (on the astronomical pillars). This proved to be a very great disadvantage, because in this way the measurement was performed in the layers immediately above ground; 8 km of the 25-km-long distance crossed the shallow sea, another 10 km crossed a marshy land. In four weeks it was possible to measure the side on only three evenings.

30 September 1964	25 084·057 m, reduced to levelling datum
30 September 1964	25 084·084 m
4 October 1964	25 084·007 m
18 October 1964	25 084·013 m
Mean	25 084·040 m, reduced to levelling datum

As a comparison value obtained from the invar wire measurement taken from ancient observation documents: 25 084·035 m. An official confirmation of this value has not yet been received. By Tellurometer measurements, the Institut für Angewandte Geodäsie obtained a value of 25 084·02 m.

4. *Re-measurement of Base Extension Networks with the 2A Geodimeter*

Beside the base lines we also re-measured the appropriate base extension net. The comparisons with the networks which were observed with the old classical method have not yet been completed, as the calculation is not yet carried out in all cases, and the results are not published.

4.1 *In the base extension network near Munich* (Fig. 1) only the sides, northern base terminal to Parsdorf and northern base terminal to Aufkirchen were re-measured. The observations near Munich were the first with the Geodimeter; a considerably greater number of first-order sides will be observed in this area during 1966.

4.2 *In the base extension net near Heerbrugg (Switzerland)* only the first graduation of extension was measured (Fig. 2), because at that time the Geodimeter was equipped with the standard lamp (white). The measurements executed in June 1960 gave the following values of comparisons:[11]

Side	Geodimeter (m)	Trig. (m)	Diff. (mm)
Northern base terminal—St Anton	8 731·667	·670	+ 3
Southern base terminal—St Anton	9 122·297	·292	− 5
St Anton–Hohe Kugel	16 158·316	·360	+44
Northern base terminal–Hohe Kugel	8 215·805	·802	− 3
Southern base terminal–Hohe Kugel	9 553·507	·528	+21

At the sides extending to the 'Hohe Kugel' the meteorological data were registered by means of captive balloons, but the values have not yet been introduced into the calculation. Perhaps the differences of trilateration to this point will become somewhat smaller after this correction.

4.3   *In the new base extension net 'Meppen 1960'* (Fig. 3) interesting comparisons are resulting after completion of the calculation of the angular observations, because the network had been observed with the Wild T-3 and with the Askania Tpr instruments (visually and with the electric eye). A comparison with the provisional results of angular adjustment gives the following table:

Side	Geodimeter (m)	Wild T-3 (m)	Ask. vis. (m)	Ask. el. eye (m)
Hengstberg–eastern base terminal	6 340·627	·628	·648	·648
Hengstberg–western base terminal	7 653·382	·405	·405	·425
Hengstberg– Windberg	13 643·720	·718	·668	·728
Hengstberg– Rütenbrock	16 680·688	·644	·554	·734

This table is interesting, not least as to the comparison with the Geodimeter values, which here partly arise from considerable fluctuations with the different methods of angular observation.

4.4   *In the base extension network near Göttingen* (Fig. 4) the triangulation sides measured have not yet been calculated completely. The trigonometric station of Dransfeld, Hohehagen is located near a quarry, so that the solid stone tower was displaced by about 20 cm during the observation time. All measurements carried out near Göttingen by the different offices must be reduced to a fixed datum by means of auxiliary observations. This is also the reason why the base extension net was expanded to the first-order station of Staufenberg.

5. *Re-measurement of First-order Triangulation Sides with the 2A Geodimeter*

At those places of the network where a control of the scale seems to be urgently necessary, we can re-measure a portion of the first-order net with the 2A Geodimeter without great difficulty and thus determine the scale. Only the construction of signals is required.

5.1　This case of a control of a net has been carried out once in a portion of first-order network near Lüneberg. (Fig. 6) In the German first-order network two triangles were measured with the 2A Geodimeter. The measured figure is situated at the junction of two net portions which had been adjusted geometrically so that there appear large discrepancies. It is not possible here to make comparisons with other measuring methods of equal accuracy. It is known only that there is a discrepancy in the network. The measurements are quoted here in this paper only because they demonstrate the desirable method of a scale control in triangulation networks. The results are published in [12].

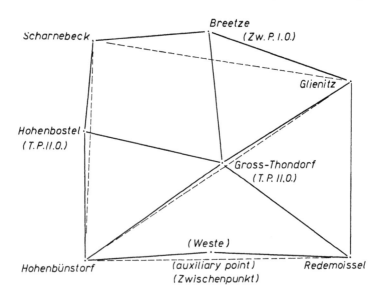

Scale 1 : 400 000

FIG. 6. Portion of the first-order net near Lüneburg.

6. *General Recommendations*

6.1　All Geodimeter measurements since 1959 have been executed with the original Aga equipment. For the prisms we constructed in our own workshop a special housing taking up to 37 prisms. A second housing of the same type facilitates two measurements to several points during the same night when the conditions of weather are favourable. For the measurements up to a distance of 45 km planned for 1966 in South Germany, a prisms' housing for 75 prisms is being constructed.

6.1.1 The *frequency* f_1 of our Geodimeter is unusually constant and surpassed that of all other instruments of electronic distance measurement. From 1958 till March 1965 the frequency had a value of 10·000 000 Mc/s, the variation during the measurements, which lasted for hours, not exceeding 0·2 c/s. In March 1965 the Kerr cell had to be exchanged; at this operation the frequency adjustment knob was moved, so that we have now a frequency of 9·999 995 Mc/s. Although a power supply of 220 V is necessary, we found that the frequency is also stable at 170 V. When we work with the gasoline generator, we do not expect any influence on the frequency, or, consequently, on the scale of the measurements either.

6.1.2 Experiments on the *additive constant K and the light conductor constant L* have been made systematically since 1961. They show that we must count upon great variations (∼ 20 mm) when any changes are made in the instrument; moreover damages in connection with transport are possible. The experience of the Institut für Angewandte Geodäsie has shown that the constants of the 2A Geodimeter have to be calibrated more frequently than the frequency. Just as with invar wire measurements, the constants and the frequency have to be calibrated before and after a period of field measurements. Frequencies were gauged by the frequency decade (SCHOMANDL), the constants by a standard base line or a similar arrangement. Short distances are less recommendable. A publication on the calibration of the 2A Geodimeter is in preparation.

6.2 When in April 1963 the Geodimeter models 4B were delivered with the *super-pressure mercury-lamp*, the Institut sent its 2A Geodimeter to Stockholm in the same month in order to have this instrument changed, so that we could also make use of this very intensive radiant lamp. With the meteorological conditions prevailing in Germany and with the mist or haze in the industrial regions, we were only able to measure 18 or 20 km with the standard lamp at very good visibility. After installation of the mercury-lamp we directly measured 31 and 35 km near Göttingen, and we hope to measure 42 and 45 km in 1966 in the mountains. The distances between the single first-order triangulation points are in Germany nearly 30 km, so that in former times auxiliary points were necessary, where the prism units had to be placed and which had to be determined geodetically. By means of the new lamp it is possible to measure these triangulation sides directly.

Shorter sides can also be measured with less favourable weather conditions. The installation of the super-pressure mercury-lamp is the most important improvement of the instrument in recent years. It must be noted that the correction for colour is $-1 \cdot 2 \times 10^{-6}$ for all instruments and phototubes, because now the wavelength of the mercury-lamp is authoritative and no longer the maximum of the colour sensitivity of the photocell.

6.3 The *arrangement of the networks of measured distances* (trilateration) at any rate depended on the position of the first-order sides, as well as on the old classic figuration of the net. Within the scope of this figuration all available distances should be measured, even when thereby a point of second order must be included in the measurements. Therefore the base extension network near Göttingen (see Fig. 4) is not favourable in its arrangement, because one attached great importance only to the re-measurement of the classic figure. In many cases traverses have brought an improvement of the figure (see Fig. 4). In practice this was not possible because of the indispensable building of signals. Professor Höpcke uses for his Tellurometer measurements a ladder of steel tubes, which can be built up to a height of 35 m by a building squad within one day. By tests this ladder has proved to be extremely suitable as a reflector station, because owing to the lateral bracing only a trifling turn movement may occur but no movements in the directions of both the co-ordinates are possible.

6.4 As to the question of a *comparison of costs*, the following remarks shall be made:

One Geodimeter-measuring team consists of one observer, one recorder as well as one observer at the reflector station and a driver. Instruments and personnel are transported with a special measuring car (1·75 tons). Without salaries one month of field work costs 4000 DM. If we add the calibration cost of 1000 DM, one month of field work will cost 5000 DM. Within this time it should be possible to measure a base network like Meppen (see Fig. 3) or a portion of the first-order net like Lüneburg (see Fig. 6). Larger base extension nets like that near Göttingen require (including waiting-time) nearly 9000 DM.

Relatively independent of the dimensions of the base extension net, we had to spend 70 000 DM for measurements with invar wires as well as for angular measurements within extension network (both sums do not include signal construction).

6.5 In conclusion we can state that it is possible to control old base lines which, for different reasons, cannot be re-measured with invar wires or tapes, by means of the 2A Geodimeter. The accuracy of this instrument is sufficient (careful measurement with well-determined constants) to find out the differences of scale existing with the old base lines which were measured before 1900. When the Geodimeter measurements are expanded on the base extension net, the first results available show that the accuracy of measurement is equal to the angular measurement in the extension net. The saving of time and personnel, as well as the resulting reduction of costs, are in favour of the Geodimeter measurement. The use of the Geodimeter is especially valuable at those points of the first-order triangulation network where, because of the difficult conditions of the ground, no base line and no extension net can be planned, but where nevertheless a checking of the scale within the whole net is necessary. Owing to the modern means of communication and transportation there is no limitation with regard to the use of the Geodimeter. The re-measurement of the base line near Bonn (1892) and Oberbergheim (1877), as well as that of the pertinent extension nets is urgently desired. Two other controls of scale near Coburg and near Heidelberg would be sufficient to bring the whole West German portion of the European primary triangulation network to a uniform and well-controlled scale.

References

[1] Kneissl, M., 1957. Die Arbeitstagung der Permanenten Kommission für die Europäische Triangulation in der Internationalen Assoziation für Geodäsie vom 22–26.5.1956 in München. *Veröffentl. d. Deutschen Geodätischen Kommission*, Series B, No. 42/I.

[2] Kneissl, M. & Kirschmer, G., 1960. Bericht über das Symposium für die Neuausgleichung der Europäischen Triangulationen vom 19–24.4.1960 in Lissabon, *Veröffentl. d. Deutschen Geodätischen Kommission*, Series B, No. 68.

[3] Gerke, K., 1958. Basis München *Veröffentl. d. Deutschen Geodätischen Kommission*, Series B, No. 56/I and 56/II.

[4] Sigl, R. & Herzog, H. Europäisches Dreiecksnetz (Retrig)—Basis München im Ebersberger Forst, Bayrische Akademie der Wissenschaften, *Mathem. –Naturwiss. Klasse*—Neue Folge, No. 103.

[5] Hoffrogge, Ch. & Rumert, H., 1961. Meteranschluss von Invardrähten für Basismessungen. *Veröffentl. d. Deutschen Geodätischen Kommission*, Series B, No. 71.

[6] Weigand, A. Winkel- und Richtungsmessungen im Basisvergrösserungsnetz München–Ebersberg. *Veröffentl. d. Deutschen Geodätischen Kommission*, Series B, No. 97.

[7] Gerke, K. Die Geodimetermessungen 1959–60 des Instituts für Angewandte Geodäsie. *Veröffentl. d. Deutschen Geodätischen Kommission*, Series B, No. 58.

[8] *Trigonometrische Abteilung der Landesaufnahme*: Die königlich preussische Landestriangulation (Hauptdreiecke), Part 6, 116, Berlin, 1894.

[9] Strasser, G., 1950. Die Grundlinienausgleichung im Zentraleuropäischen Netz. *Veröffentl. d. Instituts für Erdmessung*, No. 5, Bamberg.

[10] Kneissl, M. & Kobold, F. Bericht über das Symposium über die Neuausgleichung der Europäischen Hauptnetztriangulationen vom 9–12.10. 1962 in München, Association Internationale de Géodésie, Section I, Triangulations, Commission Permanente Internationale des Triangulations Européennes, Publication No. 2.

[11] Gerke, K. Über die Grundlinienmessungen und die elektronisch gemessenen Dreiecksseiten I.O. in Westdeutschland, *Veröffentl. d. Deutschen Geodätischen Kommission*, Series B, No. 99.

[12] Dirk, H. Massstabskontrollmessungen mit dem Geodimeter NASM-2A in zwei Dreiecken I.O. des westdeutschen Hauptdreiecksnetzes, *Veröffentl. d. Deutschen Geodätischen Kommission*, Series B, No. 106.

DISCUSSION

R. C. A. Edge: Is it always justifiable to assume that the positions of the old station marks have not changed? If movement has taken place, and then new measurements of base lines are used to readjust the old triangulation, it is likely to make things worse.

H. Grosse: I have no information about whether the stations have moved or not, but in all cases the firmly emplaced underground marks have been used and in each case a complete network has been measured, and not just the base line.

R. C. A. Edge: Has there been any change in the angles of these networks? If the figures seem to have changed shape, it may be because of real crustal movement.

H. Grosse: It is difficult to give specific examples, since the analysis of the work is not yet complete, but the angles have changed very little.

L. Asplund: The Commission for the Readjustment of the European Triangulation recommended that in all cases at least a portion of the network should be observed and not just the base line. Has any check been made on sides spaced through the triangulation between the base lines? Dr Grosse should easily be able to measure sides of 45 km or more.

H. Grosse: Only the base networks have been measured.

Distance Measurement by Means of a Modulated Light Beam yet Independent of the Speed of Light*

K. D. FROOME and R. H. BRADSELL

National Physical Laboratory, Teddington, U.K.

There are no frequency standards in 'Mekometer II'; instead the basic length standards are UHF cavity resonators filled with dry air but allowed to acquire atmospheric temperature and pressure. Two cavities are used to determine the modulation wavelength of the light beam, and the speed of light does not enter into the basic measuring equation which simply expresses the required distance as a function of the cavity dimensions (INVAR) and the (substantially constant) ratio of two refractive indices. Polarization modulation is used, the modulation wavelength being variable between 2 ft and 2 × 20/19 ft (corresponding to frequencies near 500 Mc/s). By a combination of phase-settings and frequency changing, the desired distance is obtained directly in feet and there is no atmospheric refractive index correction to apply. The resolution at the moment is about ±0·003 ft (1 mm) and the accuracy is about 3 ppm with a range of about 5000 ft (1·5 km). It is hoped greatly to increase the resolution later this year.

Introduction

'Mekometer II' has been designed as a prototype instrument to investigate the measurement of lengths up to about 5000 ft (1·5 km) in the simplest possible manner and the machine described herein is to be produced by Hilger & Watts Ltd. The potential users for such an instrument would be builders, engineers, architects and those surveyors interested in town and property measurement.

For these purposes it is highly desirable that the required distance should be obtained directly without any calculations, calibrations, or atmospheric refractive index observations. Also, the accuracy should be better than 1 in 10^5 in order to satisfy engineering requirements, and the sensitivity should be at least 1 mm and preferably better.

Mekometer II fulfils all these requirements by (*a*) using a light-ray

*Reproduced by consent of the Director, National Physical Laboratory.

modulated as a UHF frequency of nearly 500 Mc/s, and (b) using specially-designed cavity resonators to determine the modulation wavelength independent of a knowledge of the velocity of light and thus of refractive index.

The Basic Method

For any form of cavity resonant to electromagnetic vibrations, the frequency (F) may be written:

$$F = \frac{c_0}{\mu_r \, f(L)} \tag{1}$$

where c_0 is the velocity of electromagnetic waves in vacuo, μ_r is the refractive index of the medium filling the cavity, $f(L)$ is a linear function of the cavity dimensions.

For a light ray travelling through the atmosphere, modulated at a frequency F, the modulation wavelength (λ) is given by

$$\lambda = \frac{c_0}{\mu_g F} \tag{2}$$

where μ_g is the averaged value of 'group' refractive index of air appropriate for the modulated light path used ($\mu_g = \mu - \lambda \, d\mu/d\lambda$). Thus if a cavity resonator is used to control F, the modulation wavelength becomes

$$\lambda = f(L)\mu_r/\mu_g \tag{3}$$

Thus λ is expressed as a function of cavity dimensions multiplied by the ratio of two refractive indices; μ_r may represent a gas (e.g. dry air) filling the cavity and if this is allowed to acquire atmospheric pressure and temperature, the ratio μ_r/μ_g is substantially constant over a wide range of conditions.

How constant is the relation μ_r/μ_g? Assuming refractive index to be proportional to density we can expand the ratio to:

$$\mu_r/\mu_g = 1 + D(K_r - K_g) \tag{4}$$

where D is the density of air (for convenience taken as equal to unity at standard conditions), K_r is the constant relating air density in the standard cavities to corresponding refractive index. K_g is that relating to air along the light path, and it ($K_g \approx 300 \times 10^{-6}$ at s.t.p.) will of course be a function of the effective wavelength of the light-ray. (With the xenon flash tube used in the present equipment this wavelength is

0·45 μm.) If the standard cavities are set for a 'standard atmosphere',

$$\mu_r/\mu_g = (\mu_r/\mu_g)_s + (K_g - K_r)(D_s - D) \qquad (5)$$

where $D_s = 1$ and the suffix, s, denotes these standard values. For light of wavelength 0·45 μm and dry air filling the standard cavities,

$$K_g - K_r = 23\cdot0 \times 10^{-6}$$

For infra-red radiation at 1·0 μm, $K_g - K_r = 3\cdot3 \times 10^{-6}$. We are assuming here an average value of atmospheric water vapour pressure of 10 mm Hg. Variations of this have so small an influence on μ_g that there is never any need to measure relative humidity along the air path.

At high altitudes $D_s - D$ may become a considerable proportion of D_s, namely about 0·5 at 20 000 ft. Thus a small altitude correction can be applied to allow for the deviation from perfect compensation arising from these small changes in μ_r/μ_g with considerable changes in atmospheric density. The effect of temperature variations on $(K_g - K_r)$ $(D_s - D)$ is very small and can usually be neglected. Alternatively, this small residual temperature effect (parts in 10^8 per degree Centigrade, depending on the wavelength of the light) can be eliminated by appropriate adjustment of the temperature characteristics of the standard cavities.

Table 1 gives the correction in parts per million (ppm) to be *added* to the distance given by the Mekometer for carrier radiation of the two wavelengths mentioned, at different altitudes.

TABLE 1. *The compensating altitude correction (ppm)*

Altitude of transmitter (ft)	Effective wavelength of light beam (μm)	
	0·45	1·00
0	0	0
10 000	6·0	0·9
20 000	10·9	1·6

This altitude correction can be eliminated, or avoided, as follows: Firstly μ_r can be made equal to μ_g by the addition of a small percentage of gas of high refractive index at radio frequencies (e.g. carbon dioxide); secondly, the standard cavities can have a small tuning trimmer fitted which is set manually or automatically when the transmitter is used at extreme altitudes.

So far we have implied that the average value of μ_g along the light path is the same as at the transmitter. This assumption is completely justified for horizontal sites, but a very simple additional form of altitude correction is needed if the far reflector is above or below the transmitter.

In Table 2, it is assumed that the reflector is at a higher altitude than the transmitter, the correction (ppm) then being *added* to the distance read off from the Mekometer. The correction changes sign if the target is below the transmitter. The required altitude difference is of course obtained from the Mekometer reading and a rough measurement of angle of elevation.

TABLE 2. *The fundamental altitude correction*
(In ppm on measured distance relative to the transmitter. When the target is above the transmitter the quantities given are to be added to the indicated distance; when below, subtracted.)

Altitude of transmitter (ft)	Elevation of target above transmitter (ft)			
	0	1000	2000	3000
0	0	4·4	8·8	13·2
10 000	0	3·4	6·8	10·2
20 000	0	2·7	5·4	8·1

The basic altitude correction can be eliminated experimentally by making two measurements of the required distance, interchanging transmitter and target between them. It does not matter if the atmospheric conditions change in the interval between the observations.

The total altitude correction is the sum of Tables 1 and 2.

Thus we see that the modulation wavelength (which is the basis of the distance measurement) is independent of a knowledge of the speed of light; in fact, this distance is measured in terms of the dimensions of the standard cavity resonators.

Thus

$$d = \frac{\lambda}{2}(N+\delta+p) \qquad (6)$$

where d is the distance from modulator to far reflector, λ is the modulation wavelength, N is an integer, δ is the 'fractional excess' of d above an integer number of half-waves and p is a phase factor depending on the nature of the demodulation (i.e. detection) process.

In practice we can vary the modulation wavelength between the limits λ_1 and λ_2 determined by reference to standard cavities. Then

$$d = \frac{\lambda_1}{2}(N_1 + \delta_1 + p) \tag{7}$$

and

$$d = \frac{\lambda_2}{2}(N_2 + \delta_2 + p) \tag{8}$$

If we count the cyclings (n) of the phase detector output in changing the modulation wavelength from λ_1 to λ_2, then

$$N_1 = \frac{\lambda_2}{\lambda_2 - \lambda_1}[n + (\delta_1 - \delta_2)] - \delta_1 - p \tag{9}$$

$$(\text{since } n = N_1 - N_2)$$

Now we know that N_1 has to be an integer so that equation (9) is used to determine *the order of interference* by rounding to the nearest whole number. The ratio

$$\frac{\lambda_2}{\lambda_2 - \lambda_1} = \frac{L_2}{L_2 - L_1} = R$$

can be fixed at some convenient value or varied with n fixed. L_1, L_2 are the relevant dimensions of the standard cavities.

The best method of measuring phase and thus δ_1 and δ_2 is by means of a variable light path within the instrument and if δ_1 is first made zero by adjusting this variable path before counting is started, then,

$$N_1 = nR - R(\delta_2 + p/R) \tag{10}$$

In one method of operation we made $\lambda_1 = 2$ ft and $\lambda_2 = 2 \times 20/19$ ft, so that $R = 20$. With the type of modulator and demodulator herein described, it is best to set the variable light path to a detector minimum and in this case $p = \frac{1}{2}$, so that the order of interference is obtained by rounding the following equation to the nearest whole number.

$$N_1 = 20n - 20(\delta_2 + 1/40) \tag{11}$$

Since $20n$ is always an integer this value can be obtained directly from an appropriately calibrated mechanical digital counter giving units of 20 ft. The range 0–19 ft is covered by means of a dial divided into 20 sections and driven from the variable light path so that one revolution corresponds to 20/19 ft.

The measurement is made as follows:

1. Set the modulating oscillator to produce a modulation wavelength of 2 ft as determined by reference to the first standard cavity. (Cavity A in Fig. 1.)
2. Set the variable light path to a detector minimum. Read off in fractions of a foot the actual position of the variable light path. Set the 0–19 'rounding' dial mid-way between '0' and '1'.
3. Press the 'start count' button which steadily increases the modulation wavelength until that given by Cavity B is reached.
4. Using the second standard wavelength ($2 \times 20/19$ ft) as given by Cavity B, set the variable light path to a detector minimum. [This determines δ_2 in equation (11).] Read the 0–19 'rounding' dial to the nearest foot and subtract the reading from the '$20n$' counter reading. This gives the order of interference in feet from modulating crystal to far reflector and back again. The 'initial setting' has already determined the fraction of a foot.

This method of counting interference minima in a predetermined modulation wavelength change has been found very satisfactory at ranges up to 1500 ft (0·5 km) but beyond that conditions of bad heat shimmer can cause lost counts. Thus, for longer ranges the following method of obtaining order of interference is to be preferred.

In this case R [equation (10)] can be made variable whilst n is always made equal to unity, i.e. the observer, having made the Cavity A settings, varies the modulation frequency until the order of interference changes by one (thus $\delta_2 = 0$). In practice this is accomplished by a small plunger inserted through the side of the modulating cavity and actuated by a cam, the movement of which is noted by a pointer attached to it. This gives a reading of 500 to 5000 ft in units of 500 ft for a 'count of one'. At the farthest extent of its motion it makes $R = 500$ highly accurately. Then, by means of an additional scale engraved on the 0–19 ft disk, units of 20 ft up to 500 ft can be obtained by one additional phase (variable light path) setting.

As with the first NPL machine, which used a microwave modulation frequency, the light beam is polarization-modulated, although the method is basically applicable to any form of modulation. The polarization modulation is by means of a crystal of ammonium dihydrogen phosphate (ADP) which exhibits the linear electro-optic effect (Pockels effect) and can thus be used as a combined modulator and demodulator.[1,2] Plane polarized light passing along the z-axis of the crystal becomes elliptically polarized at the modulating frequency, the ellipticity being either enhanced or cancelled after reflection from the far reflector and re-traverse of the crystal. The instantaneous value of the returned light intensity falling on the photo-detector after passage

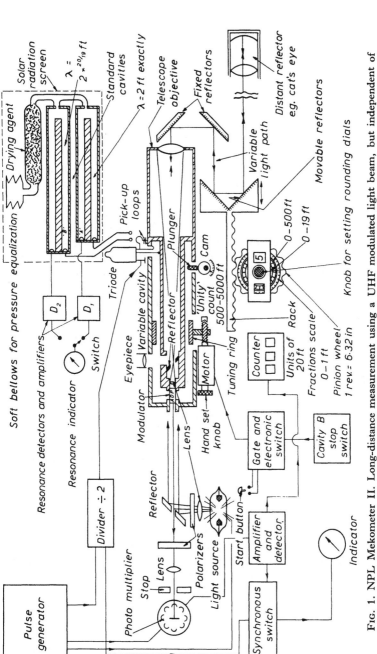

Fig. 1. NPL Mekometer II. Long-distance measurement using a UHF modulated light beam, but independent of the speed of light.

through a second plane polarizer (crossed relative to the first) is a
function of the ellipticity and is given by[1,2]

$$I = \sin^2\tfrac{1}{2}\theta \qquad (12)$$

where

$$\theta = \frac{2\pi V}{V_m} \sin\left(\omega t + \frac{2\pi d}{\lambda}\right) \cos\frac{2\pi d}{\lambda} + e \qquad (13)$$

where $V \sin \omega t$ $(= V \sin 2\pi Ft)$ is the modulating voltage applied to the
crystal (V is the peak value), λ is the modulation wavelength in air,
d is the distance from crystal to far reflector, e is additional ellipticity
supposed to be introduced along the light path, such as might occur
with the use of a trihedral reflector at the far station, or off-axis passage
of the light through the crystal. V_m is the value of the d.c. voltage
(16 kV for ADP) required on the crystal to produce maximum electro-
optic effect corresponding to $I = 1$ (i.e. to produce plane-polarized
light at right angles to the initial plane).

The photo-detector output is the time average of equation (12),
namely,

$$I = \frac{1}{t} \int_0^t \sin^2\left[\frac{\pi V}{V_m} \sin\left(\omega t + \frac{2\pi d}{\lambda}\right) \cos\frac{2\pi d}{\lambda} + \frac{e}{2}\right] dt \qquad (14)$$

After some manipulation this reduces to

$$I = \frac{1}{2} - \frac{\cos e}{2} J_0\left(2\pi \frac{V}{V_m} \cos\frac{2\pi d}{\lambda}\right) \qquad (15)$$

where J_0 indicates a Bessel function of zero order.

Thus we see that the photo-detector output is a minimum when

$$d = \frac{N\lambda}{2} + \frac{\lambda}{4} \qquad (16)$$

where N is an integer, namely the order of interference. We also note
the important fact that the ADP crystal used in this manner is an
error-free demodulator of its own modulation. The effect of any change
in ellipticity along the light path is simply to degrade the minimum,
not to shift it. At low modulation levels ($V \ll V_m$), I varies sinusoidally
with increasing d, but at high levels of modulation ($V \to V_m$) the
minima become sharpened and the maxima become broadened or
even doubled. Thus it is desirable to work with interference minima
using the highest possible modulation levels in order to produce the
sharpest minima.

The standard cavities are best set by making the instrument read correctly on an accurately known distance, in which case no knowledge of the velocity of light is required. Alternatively, they may be adjusted by setting their resonant frequency to the appropriate value when the temperature and pressure of the gas inside them is accurately monitored.

The Apparatus

The heart of the Mekometer II is the light modulator which imposes elliptical polarization modulation on the transmitted light beam and demodulates it on its return from the distant target reflector.

Fig. 1 shows the modulator, which is a coaxial cavity resonator excited into strong UHF oscillation by a pulsed triode valve. Under pulsed conditions the relatively low anode impedance allows matching to be achieved with the anode connected to the inner line near the short circuit end of the cavity. A low Q tuned loop dips into the magnetic field inside the cavity and feeds r.f. energy to the valve cathode in such a phase as to enforce oscillation.

Across the high voltage end of the coaxial resonator is situated the ADP crystal which is the light modulator.

A motor-driven tuning ring within the modulator allows the modulation wavelength of the cavity to be varied over a range of 5 per cent. A further hand-operated capacitive plunger in the cavity wall imposes an exact 0·2 per cent wavelength shift when fully inserted.

The absolute wavelength generated by the modulator is monitored by two 'standard' cavity resonators, designated A and B in the figure. These cavities have wavelengths of 2 ft and $2 \times (20/19)$ ft exactly. Atmospheric temperature and pressure are communicated to the cavities and the air inside is dried.

Light from the pulsed xenon flash tube is rendered parallel and plane polarized. The beam-splitting reflector, with the aperture at its centre, projects an annular ring of light through the ADP crystal where the UHF electric field modulates it. A microscope objective forms a diminished real image of the source which in turn acts as a point source at the focus of the 2·5-in. telescope objective.

The projected light beam, which has a milliradian spread, is directed towards the distant target reflector by way of a folded variable light path. This light path allows the distance between modulator and reflector to be adjusted.

Returning light from the target reflector follows the same path and retraverses the crystal where the modulation is reinforced or diminished depending on the instantaneous phase of the electric field in the crystal relative to the phase of the modulation already imposed on the light.

The beam is now transmitted through the aperture in the beam-splitting reflector and passes through a second, but crossed, polarizer. The intensity of the light now depends on the degree of elliptical polarization on the light. This intensity is measured by the photomultiplier and displayed on a meter.

For a given modulation wavelength the variable light path can be adjusted until the meter indicates a minimum intensity. The distance the light path is moved is a direct measure of the excess of the distance over an odd number of modulation half-waves.

The 'standard' cavity resonators warrant more detailed description as the accuracy of the instrument depends upon them. From the mathematical treatment it can be seen that the measuring equation has the dimensions of length. In fact distance is measured in terms of length standards which, to a near approximation, are the length of the inner conductors of the coaxial 'standard' cavities.

Fig. 2. Mekometer II in use on NPL 300-m base-line.

Advantage is taken of the fact that the air-spaced capacitor formed between the end of the inner rod and the end plate of the outer sheath increases the wavelength produced above that which would have been expected from the length of the inner line of the quarter-wave coaxial cavity. Now the inner lines are made of copper-plated invar (low expansion alloy) and the outer sheaths of copper. The relative expansion between these two metals has the effect of reducing the end capacity as temperature increases. If the end capacity has the right absolute value, the capacity change will exactly compensate for the thermal expansion of the invar rod. The process is analogous to that of the compensated clock pendulum. Thus constant wavelength can be maintained as if the inner rod of the coaxial resonator had zero thermal expansion.

The air in the 'standard' cavities is allowed to pick up atmospheric temperature easily by the situation of the cavities in the instrument. Atmospheric pressure is communicated by means of a soft plastic reservoir whose volume changes with the external pressure. The internal air is maintained dry by a connected container of silica gel.

The target reflector shown is of the 'cats-eye' type, that is, simply a concave reflector at the focus of a lens. However, the cube corner can also be used, although it does itself introduce a small degree of elliptical polarization. Both these reflectors are noncritical as to their angular alignment relative to the transmitted light beam.

A standard 'cats-eye' target reflector is employed with the Mekometer II which has a 2·5-in. aperture $f4$ lens. At the longer ranges a cluster of three such target reflectors is used.

Mekometer II is provided with an eyepiece which, together with the main objective, form a telescope with a sufficiently wide field of view to locate the target reflector. When this has been done the returned light signal can be viewed by means of a drop-in reflector, before it reaches the second polarizer so that the returned light spot can be centred in the aperture through the beam-splitting reflector before a measurement is commenced.

The Electronic Circuits

The power requirements of the Mekometer II are supplied by a 12-V, 15-A.h silver-zinc accumulator. The basic 12 V is converted to the various necessary potentials by a transistorized d.c./d.c. converter operating at 400 c/s. A second converter supplies the high voltage needed for the photomultiplier and has a potentiometer-controlled output, variable between 500 and 1600 V, which acts as a detector gain control. The total power consumption is about 12 W, allowing at least 100 measurements per battery charge.

The circuit arrangement is shown in Fig. 3. Apart from the UHF planar triode modulator, semiconductors are used throughout.

A 100-c/s multivibrator controls the basic repetition rate of the device; it triggers a silicon controlled rectifier, feeding a miniature ignition coil,

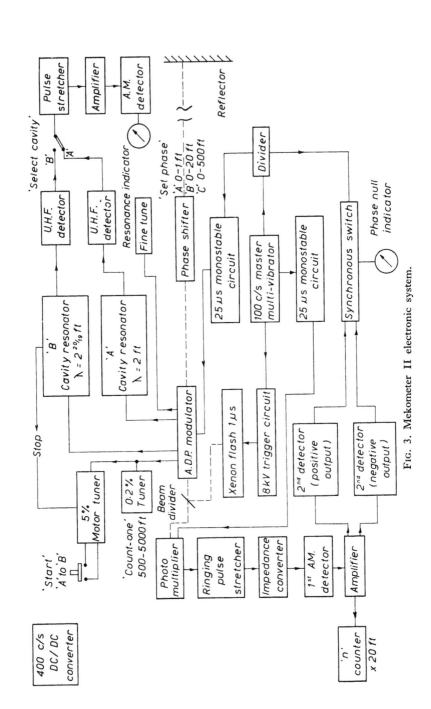

FIG. 3. Mekometer II electronic system.

to produce the 8 kV pulse necessary to trigger the xenon flash-tube light source. This tube produces 1 μs flashes at an energy of 0·04 joule. The power consumption of the flash circuit is minimized by charging the 0·05-μF storage capacitor through a 200 H ringing choke instead of the usual resistor with its associated power loss.

The 100-c/s multivibrator also triggers a 25-μs monostable circuit which pulses the fifth dynode of the photomultiplier to its normal operating potential. This very greatly increases the photomultiplier sensitivity for the duration of the 25-μs pulse.

After division by two, by a bistable circuit, the multivibrator pulses trigger a further 25-μs monostable circuit which grid pulses the modulator valve. The valve produces 25-μs-duration pulses of UHF energy of about 100 W peak at a repetition rate of 50 per sec. The actual light modulation is effected by the ADP crystal situated at a position of high electric field within the modulator cavity resonator.

The 1-μs light pulse detected by the photomultiplier, while it is gated 'on', sets the 30 kc/s tuned circuit in its anode rigging, thus producing a train of damped oscillations and effectively stretching the pulse. The high dynamic impedance of the tuned circuit is matched to the detector by an impedance converter. The time constant of the detector is such as to produce a saw-tooth waveform which is the envelope of the damped oscillations. This is amplified and fed to a transformer, with a centre-tapped secondary winding, which produces two antiphase outputs. These outputs are each rectified. A synchronous switch, activated by the bistable divider at 50 c/s, selects each rectified output in turn and connects it to the null indicating meter. Since the light signal is modulated on alternate flashes only, the synchronous system serves to eliminate unwanted signals detected by the photomultiplier.

The signal applied to the meter is also d.c. amplified and fed to a Schmitt trigger which in its turn operates a 20 ms monostable circuit when the amplifier output reaches a predetermined d.c. level. The 20-ms pulse switches an electromagnetic counter one digit, which in this case represents 20 ft.

The modulation wavelength of the ADP modulator is monitored by two 'standard' cavity resonators. The cavities are separated by a 5 per cent wavelength change. Changing the wavelength of the modulator between cavities A and B is achieved by a sliding capacitive ring within the modulator cavity, which can be moved by an electric motor and lead screw. The motor is automatically stopped at the cavity *B* position.

The standard cavities, which have an unloaded *Q* factor of about 2000, are very loosely coupled to the modulator. Some of the UHF input is mixed with the phase-shifted signal that has been transmitted through the cavity so that, on resonance, a phase-balanced null is achieved.

The output pulse from the respective cavity is detected, stretched and amplified. A second detector rectifies the consequent signal and the resultant direct current is displayed on the resonance indicating meter.

An 0·2 per cent wavelength shift, which is also required by the measuring process, is achieved, without a third standard cavity, by means of an accurate, mechanical stop-controlled capacitive plunger in the modulation cavity wall.

Mekometer II consists of two units; the first contains the accumulator and power converters, and the second the instrument proper plus its electronic systems. The semiconductor circuits are contained on seven plug-in printed circuit boards. A flexible cable is used to connect the power unit to the instrument.

Several months' experience with the Mekometer II suggests that the electronic systems employed are stable and trouble-free over a wide temperature range.

Results

A base line of approximately 1000 ft (300 m) has been established at the NPL for the purpose of testing the Mekometer II. This base was measured (with the generous co-operation of the Ordnance Survey) by means of the three NPL 50-m invar tapes during the summer of 1964. The result, believed to be accurate to ± 2 ppm is: $d = 984 \cdot 065$ ft.

On 8 March, 1965 with an air temperature of 8°C and a barometric pressure of 770 mm Hg, the Mekometer result was: $d = 984 \cdot 060$ ft. On 23 March, 1965 with an air temperature of 12·5°C and a pressure of 744 mm Hg, the Mekometer gave exactly the same result, although the atmospheric refractive index had fallen by about 14 ppm.

Future Developments

It is hoped to improve greatly the sensitivity of setting on an inter-ference minimum and on the standard cavity resonances by introducing a small frequency modulation of the modulating cavity. This can be accomplished either mechanically, by means of a vibrating plunger, or electronically by means of a variable reactance placed inside the modulating cavity. By these means the modulating frequency will be changed in a cyclical manner between alternate light pulses. Thus both the interference minima and the standard cavity resonance will appear as meter 'zeros' and should give a greatly enhanced precision of setting.

This should approach 0·1 mm for distances up to one or two hundred metres, and 0·3 mm thereafter.

Acknowledgement

The Mekometer described has been developed as part of the research programme of the National Physical Laboratory and has been described above by permission of the Acting Director.

References

[1] Froome, K. D., & Bradsell, R. H., 1961. *J. Sci. Instrum.*, **38**, 458.
[2] Bradsell, R. H., 1964. *Trans. Soc. Instrum. Tech.*, **16**, 30.

The following British patent applications are involved: 919368; 44480/63, 4564/64, 7473/64, 7687/64, 7858/64 and the cognated version of these.

DISCUSSION

For the discussion on this paper see the end of the next paper, by D. V. Connell.

NPL–Hilger & Watts Mekometer

D. V. CONNELL

Presented by

J. A. ARMSTRONG

Hilger & Watts Ltd, London

After the successful use of elliptical polarization modulation of a light beam at 9375 Mc/s by the National Physical Laboratory (NPL) in 1961 to measure distances up to 50 m to an accuracy of about 1 part in 10^6, the principle has been applied to the development of a surveying instrument having a range up to 1–2 km. This is being done in close collaboration with the NPL. For surveying distances a modulation frequency of 600 Mc/s was found to be more suitable, and a system employing two reference cavities (substantially independent of the velocity of light) was developed. The Mekometer is expected to give an accuracy in the order of 3–10 parts in a million and to take only a few minutes from the first pointing of the telescope towards the target reflector. No corrections for humidity and pressure will be required, and it is hoped that automatic temperature compensation will, for general work, make unnecessary the application of temperature corrections. The instrument is expected to operate through a full range of climatic conditions.

The NPL–Hilger Mekometer will measure distances by using a modulated light beam. The system is based on the 500-megacycle Mekometer II described by Dr K. D. Froome and R. H. Bradsell[1] and in NPL Patents.[2] The NPL Mekometer II model, reading in feet and decimals of a foot, is best suited to nearly level lines of sight, from survey pillars. To make it of more general application to the surveyor, provision has been made for inclinations up to 30°–40°, and for the location of the instrument and target by tripod over, or perhaps under, a survey point. It was, therefore, necessary to incorporate into the NPL design many of the features of the familiar theodolite, i.e. a tilting optical system with means of measuring the inclination, rotation about a vertical axis, optical plumbing and provision for levelling and interchanging with other survey instruments. It was also decided to make the first model with a metric readout and therefore a primary modulation wavelength of half a metre was adopted.

The instrument and target described are experimental models, and it may be expected that after field trials the design will undergo a few further changes. It is also hoped that it will be possible to reduce the size of the instrument.

The instrument (Fig. 1) consists of a swinging telescope mounted between uprights containing the microwave units and the remainder of the optical system. The right-hand upright houses the modulator cavity and the reference cavities which control the modulation wavelength, and some associated electronic components. The left-hand upright contains a vertical circle and two-side reading system providing measurement of the slope of the line of sight to 1 second of arc. The circle-reading eyepiece is alongside the telescope eyepiece, and the various controls and meters are conveniently placed for the operator.

Built into the instrument is an optical plummet (11) for centring it over a ground mark. The power supply for the instrument is contained

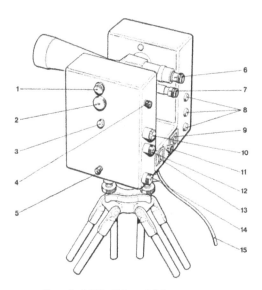

FIG. 1. NPL–Hilger Mekometer.

1 Circle illuminating mirror	9 Meter
2 Vertical-circle micrometer	10 Vertical tangent screw
3 Altitude-bubble viewer	11 Optical-plummet eyepiece
4 Telescope clamp	12 Telescope tangent screw
5 Azimuth clamp	13 Meter
6 Sighting telescope	14 Azimuth tangent screw
7 Circle-reading eyepiece	15 Power supply
8 Controls	

in a metal case placed on the ground and connected by cable to the upright of the Mekometer.

The 12-volt supply comes from silver–zinc accumulators. The instrument is attached to a levelling base by a circular fitting which enables it to be precisely interchanged with the target or other surveying instruments.

The vertical-circle reading system is shown in detail in Fig. 2 together with the optical components of the telescope and the optical plummet. The circle system is largely that from the Microptic No. 2 Theodolite. Readings from graduations 180° apart are automatically averaged to eliminate centring error and the micrometer provides readings direct to single seconds. The telescope and Mekometer system are shown more clearly in a later figure.

Fig. 3 shows the 'cats-eye' target, the optical unit being mounted on trunnion bearings so that it can be tilted ±45° and rotated and pointed

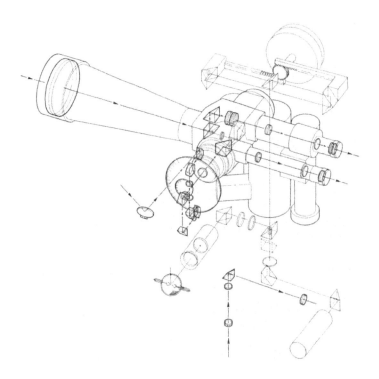

FIG. 2. NPL–Hilger Mekometer, vertical-circle system.

in azimuth. This also has a built-in optical plummet for centring and the instrument is mounted on the levelling base by an adaptor similar to that fitted on the Mekometer.

Fig. 4 shows a diagrammatic layout of the complete Mekometer optical and sighting system, showing the position of the xenon lamp (28), modulation cavity (22) and ADP crystal (19), variable path unit (18) and telescope objective (25). A field-splitting prism (24) allows visual observation through the sighting telescope which can be focused by the lens (16) and has a graticule (17).

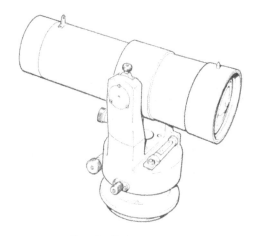

Fig. 3. Cat's-eye target.

The field splitting between measuring and sighting systems is brought about by a small aperture in the reflective layer of the prism (24), and results in a small central blind spot in the visual field of view. The modulated light therefore, is projected by the telescope objective lens as a collimated beam, unseen by an observer looking through the telescope. The optical system controls the divergence of the beam to about a minute or two of arc.

When pointed towards the 'cat's-eye' reflector at the other end of the line, the 'cat's-eye' will be seen until the pointing is exactly made, when it disappears inside the visual blind spot. When the 'cat's-eye' is within the blind spot it can be seen illuminated through the target-viewing eyepiece (32) when the appropriate control knob is operated.

The light has then retraced the outgoing path, i.e. through the telescope, the variable path and modulation cavity to the beam splitter (30) which allows the light to reach the photomultiplier (23) or the target-viewing eyepiece. On the lamp side of this beam splitter is a polarizer (29), and on the photomultiplier side, an analyser (31), which is set to have its plane of polarization at 90° to the polarizer.

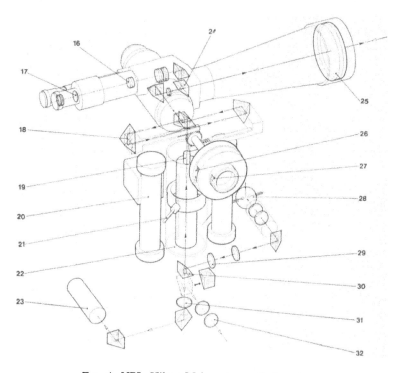

FIG. 4. NPL–Hilger Mekometer, optical system.

16 Telescope-focusing lens	25 Telescope objective lens
17 Telescope graticule	26 Variable-path reading drum
18 Variable path	27 Rounding dial
19 ADP crystal	28 Xenon flash lamp
20 Resonant cavity	29 Polarizer
21 Modulating-cavity tuner	30 Beam splitter
22 Modulating cavity	31 Analyser
23 Photomultiplier	32 Target-viewing eyepiece
24 Field-splitting prism	

Since the principle of the NPL Mekometer distance-measuring system has been fully described by K. D. Froome and R. H. Bradsell, it will be only briefly described here. The xenon lamp is flashed at 100 c/s, each flash being of 1–2 microseconds' duration. The ADP crystal has several thousand volts applied at the modulation frequency for periods of 30 microseconds, commencing just before each flash. The duration of the crystal modulation is more than enough to receive the return flash after reflection from the target. The light from the xenon lamp is plane-polarized before it reaches the ADP crystal where it is elliptically polarized at the modulation frequency. If, on return, the light pulse re-enters the crystal when the modulation is exactly 180° out of phase, the ellipticity is cancelled. Ideally the light is then only plane-polarized and on reaching the analyser it is extinguished, and the output of the photomultiplier is zero. In practice, there is some leakage and a minimum is obtained. To achieve this condition, the total path distance traversed by the light must be an integral number of modulation wavelengths plus half a wavelength, i.e. an odd number of half wavelengths.

In order to provide a readout of distance without ambiguity to 2 km, more than one wavelength is necessary. Our first model is operating with four fixed wavelengths and for the metric system the following have been chosen:

λ_1: 50 cm (f_1)
λ_2: 50 × (20/19) cm $(f_1 - 5\%)$
λ_3: 50 × (400/399) cm $(f_1 - 0.25\%)$
λ_4: 50 × (2000/1999) cm $(f_1 - 0.05\%)$

These wavelengths lead to a distance equation in terms of the four variable path readings, δ_1, δ_2, δ_3, δ_4, as follows:

$$d = 1999(\delta_4 - \delta_1) + 399(\delta_3 - \delta_1) + 19(\delta_2 - \delta_1) - \delta_1$$

The first two terms are approximate solutions in themselves within their ambiguity distances and are, therefore, rounded off to the appropriate multiple of 100 m and 5 m, respectively ('appropriate' since the size and sign of the following term decides whether to round up or down). In calculating this expression the awkward term is $19(\delta_2 - \delta_1)$ which must be rounded off to the nearest 0.25 m. This is done automatically by a 'rounding' dial, concentric with the variable-path reading drum to which it is phased manually after making the δ_1 setting.

The method of operating the instrument is as follows:

1. The cavity is tuned to λ_1, and the variable path is adjusted to 'zero' the photomultiplier reading. λ_1 is read and the 'rounding' dial is set.

2. Variable path settings are made and read in mm with the cavity tuned to λ_3 and λ_4.

3. The cavity is tuned to λ_2, the variable-path setting is made and the rounding dial reading noted.

The distance is then derived quite easily from the above equation.

The other methods of operation used by Froome and Bradsell employing the 'n' count, or 'count of one' have considerable attraction and are of course to be further investigated.

Froome and Bradsell have demonstrated that the Mekometer II system has given an accuracy in distance measurement of a few parts in a million and that their system gives automatic compensation for temperature and pressure changes when the instrument is used on approximately level sights at low altitudes. Small corrections may have to be applied for long steep sights and at high altitudes if maximum accuracy is required.[1]

It is hoped that the first NPL–Hilger model, which is now ready for field testing, will provide further evidence of this, and that the Mekometer will soon find a place in surveyors' and engineers' kits of tools. It is not just another electronic distance measuring instrument but one which promises to fill the gap now existing in the short distance range, i.e. up to a kilometre or two. We foresee that it will be used where surveyors have to provide accurate control for town surveys, and for the provision of control for aerial surveying at large scales. It is also certain to have applications in the building of large engineering structures such as dams and bridges and in mining and tunnelling.

References

[1] Froome, K. D. and Bradsell, R. H. 'Distance measurement by means of a modulated light beam yet independent of the speed of light'; see page 263.
[2] NPL British Patent Application Nos. 919368, 44480/63, 4564/64, 7473/64, 7687/64, 7858/64 and the cognated version of these.

DISCUSSION

on the papers by Dr Froome and Mr Connell

D. A. Rice: How long would the tuning remain constant and how often would calibration be required?

K. D. Froome: The calibration depends on the cavity which is formed on the inside by an invar bar and which should produce a modulation of a constant wavelength providing the air in the cavity is dry. The instrument is provided with a drying agent for this air, and in consequence, should keep its calibration indefinitely.

R. Schöldström: Has any difficulty been experienced with back scatter caused by such things as dust or greasy deposits in the optical system?

K. D. Froome: We did get some trouble with greasy deposits but they were not serious because the objective is not illuminated in the centre by the transmitted light and is placed a distance of half of one modulation wave length from the ADP crystal.

J. W. Wright: How can we be sure that the air in the cavity remains at a constant temperature? Does not the resonance itself cause the air to heat up?

K. D. Froome: Strictly speaking, the cavity should be enclosed in a radiation screen. Even then, there would be some heating, but this is likely to be small and, as $1°$ change means only 1 part in 10^6 error, it is not of major importance.

R. C. A. Edge: The paper gave details of only two base measurements for which the deviation was zero. Might this not be fortuitous?

K. D. Froome: Many more measurements were made and all the results were good, but the instrument was still being experimented with and other results were not quoted as the instrument had been altered. The two results which were quoted were chosen because they were made under very different atmospheric conditions.

H. D. Hölscher: Is there any danger of the temperature in the cavity differing appreciably from the atmospheric temperature?

K. D. Froome: The cavity temperature should be within $1°$ or so of atmospheric, but the accuracy aimed at was better than $1 : 100\,000$; and that represented $10°$ change in cavity temperature and a discrepancy of this magnitude was most unlikely.

J. W. Wright: The current value for the velocity of light used for EDM measures was due originally to Dr Froome and therefore one should pay heed to him. But I venture to say that as far as surveyors go the importance of having an instrument 'that any fool can use' should not be over-estimated.

R. C. A. Edge: I agree that, for repetitive routine operation such as may be undertaken by engineers, a fool-proof instrument may be desirable, but I wonder if at the risk of some complication a truly geodetic instrument with a range of say 10 km at least might not be evolved. Of course, measurement of air conditions and corrections for refraction would be required.

K. D. Froome: You probably could increase the range, but you might well lose simplicity. I suggest that the instrument should be kept simple and used for short distances, leaving the measurement of longer distances to other instruments. As regards corrections for refraction, current standard practice is to observe met. conditions at the two ends of a line only and assume that the mean is representative. The Mekometer system does not seem to involve very much greater probable inaccuracies, even if used over long distances.

F. J. Hewitt: Please will you clarify the use of the system for resolving longer distances. Might not passing vehicles or other interference affect the accuracy of this?

10*

K. D. Froome: I agree that the counting system might give difficulty from interference, but a new system gets over this trouble.

N. A. G. Leppard: The literature on the instrument quotes a hundred measures from one battery charging. Is any inaccuracy introduced by a fading battery and would the failure be indicated to the operator?

K. D. Froome: Perhaps a hundred measures was optimistic. However, there is no effect on the accuracy caused by a failing battery and there is a voltmeter on the instrument which shows when the battery is low.

L. Scott: What is the duration of one battery charge?

K. D. Froome: It would last for about 10 hours' operation.

Range and Accuracy of the EOS Electro-Optical Telemeter

H. RICHTER and H. WENDT

VEB Carl Zeiss Jena

Presented by

H. STROSCHE

In recent years conventional opto-mechanical or mechanical-ly operating telemetric instruments have been supplemented by the range of electronic telemeters which have been developed as a new aid to geodesy. Since the electro-optical principle promises greater accuracy and reliability than the electro-magnetic principle, work proceeded apace at the VEB Carl Zeiss Jena works with the object of developing an electro-optical telemeter with a greater day range, adequate for the bulk of geodetic tasks, together with lower power consumption and operating reliability, even under the extreme measuring conditions applicable to work in open terrain. This development work culminated in series production of the EOS. Reference [1] has dealt exhaustively with the principle and precision of the instrument. However, for better comprehension the principle and precision is outlined briefly again.

1. *Principle*

The intensity of a beam of light is subjected to sinusoidal modulation. This beam passes through the measuring line twice. The modulation phase of the beam coming from the measuring line is measured against the phase position of the initial output beam. This determines the distance to an integral multiple of a scale derived from the modulation frequency and light velocity (half 'modulation wavelength'). In the case of the EOS this amounts to approximately 2·5 m and thus corresponds with a phase difference of 360°. Further scales of approximately 25 m, 200 m, 3000 m are used to eliminate the ambiguity. The relevant frequencies are formed from the difference of each two frequencies lying in the range of 54 Mc/s–60 Mc/s.

The filament of the lamp (6 V 10 W) is pictured on a split diaphragm

0·1 mm in width. The beam which this emits is aligned on a parallel plane by an auxiliary optical system, passes through the modulator and is split to an intensity ratio of 1 : 100 by means of a dividing cube. Both beams are separately incident on two lenses. The diaphragms in the focal points of these lenses allow only the null order of the diffraction spectrum produced in the modulator to pass through. The beam of light of lower intensity then falls direct on to a photo-multiplier, while the beam of greater intensity passes through the transmitting optical system, is reflected on triple mirrors at the end point of the measuring line and travels via an optical receiver to a second photo-multiplier. In the photo-multiplier, the modulated light is converted into high frequency electrical signals which, in turn, while retaining the phase information, are transposed with an auxiliary frequency to 10 kc/s.

Both signals are amplified in selective amplifiers designed for this frequency. In the comparative branch the possibility exists for changing the phase relative to measurement. Both signals impinge upon a discriminator in which the zero indicator can show a phase difference between them of 90° to 270°. Owing to the virtually identical structure of the measuring and comparative branches, phase rotations are compensated to a considerable degree by changing the electrical values of the structural components. However, by incorporating an optical short-circuit line, it is possible to eliminate any phase rotation. If the possible adjustments for 90° and 270° phase differences are used for evaluation the zero deflection of the discriminator is eradicated.

Light modulation takes place through a periodically changing optical phase grating. The maximum phase difference within the range of a grating constant changes periodically from a value of zero (no grating) to a value determined by the maximum degree of modulation. On illuminating this grating with parallel light the intensity of the diffracted and undiffracted light changes with the same periodicity. However both have a relative phase displacement of 180°. By exclusive use of the diffraction image of zero order a modulated beam of light ensues. The phase grid arises in an ultrasonic cell (Debye–Sears effect). In this cell a crystal is excited to ultrasonic oscillations by high frequency control voltages. The resultant diffraction grating builds up and breaks down twice within one period, so that the light is modulated with double the frequency of the ultrasonic wave. By comparison with the conventional Kerr cell modulation system in known instruments, ultrasonic modulation has several advantages to offer. For example, the light loss in the ultrasonic modulator is only marginal owing to the elimination of polarizer and analyser.

Two adjacent Cassegrain systems 7/800 with parabolic main mirror 3·5/420 are used as optical transmitter and receiver. The reflector is located by a telescope with a magnification of 16. The reflector embodies seven small triple prisms so that the reflector alignment need only be approximate.

2. *Accuracy*

The errors of a set of measurements shown in Table 1 represent the mean error of one single distance measurement of 10 to 15 minutes. The mean errors m_i characterizing the inner accuracy were determined for distances of up to 0·7 km from the mean errors $m\phi$ of a single phase measurement. Within the line ranging from 1·1 to 5·6 km the distances were determined for each set of measurements by calculating only the distances resulting from the frequencies f_1 to f_4.

TABLE 1. *Mean errors of a set of measurements*

Distance (km)	Number of sets	Inner accuracy		Outer accuracy	
		$m\phi$ (°)	m_i (cm)	m_a (cm)	m_a/S
0·02	10	±0·3*	(±0·2)	±1·1	1 : 1 800
0·07	10	±0·3	(±0·2)	±1·0	1 : 7 000
0·17	10	±0·2	(±0·1)	±0·5	1 : 34 000
0·68	10	±0·3	(±0·2)	±0·3	1 : 260 000
1·1	20	±0·3	±0·3	±0·9	1 : 120 000
2·5	27	±0·3	±0·3	±0·5	1 : 500 000
3·5	23	±0·4	±0·4	±0·6	1 : 580 000
4·5	27	±0·4	±0·3	±0·8	1 : 560 000
5·6	26	±0·5	±0·4	±0·4	1 : 400 000
7·5	2	—	±0·3	—	—

* $m\phi = 0\cdot1° \triangleq 0\cdot07$ cm.

The variations in these distances against the mean value

$$\tfrac{1}{4} \sum_{i=1}^{4} S_i$$

lead to m_i.

With the electro-optical telemeters the outer accuracy is influenced primarily by the frequencies determined by the scales and their stability, by centring errors and by the atmospheric conditions. The test measurements carried out over a period of four weeks were subject to the most varied external conditions and with fresh setting-up daily in such a manner that all measurable distances could always be observed in continuous change on the same day. It was possible by comparing the overall mean of all measurements with the results of the individual sets measured at various times to determine the mean error m_a of a set of measurements. These showed the outer accuracy to a very close proximity. This is obtained in relation to the distances as per Table 1 with

$$\text{for}\quad S < 200\ \text{m}:\ m_a\ =\ \pm 1\ \text{cm}$$

$$S > 200\ \text{m}:\ m_a\ =\ \pm(0.5\ \text{cm} + 2\cdot 10^{-6}S)\qquad(1)$$

The use of a special reflector in the form of a triple stripe showed, in the range $1\ \text{m} \leq S \leq 170\ \text{m}$, an accuracy of $m_a = \pm 0.5\ \text{cm}$. For distances $> 170\ \text{m}$, the formula (1) also applies.

3. Range

3.1 Theoretical Range

The day range is greatly influenced by imponderables and is thus difficult to summarize. Consequently the following is merely an attempt at estimating the night range which is limited by the noise of the photo-multiplier.

From the equation quoted by Karolus,[2] among others, the boundary range is obtained at

$$S_{\max} = \frac{1}{2\sqrt{}}\left[\frac{C.B.F_S.F_E.\exp(-2kS_{\max})}{I_{0\,\min}}\right]\qquad(2)$$

This equation is derived subject to faultless optical systems and is interpreted as follows:

C = optical transmission factor
B = light density of the lamp
$F_S F_E$ = usable surface of optical transmitter and/or receiver
k = air absorption factor
$I_{0\,\min}$ = minimum light current which still allows an adequate phase measurement.

According to Hartmann and Bernhardt,[3] the following equation applies to the minimum photoelectric current $I_{0\,\min}$ from the photo-multiplier.

$$I_{0\,\min} \approx \epsilon\alpha^2\Delta f\frac{\eta_{\min}^2}{\gamma^2}\left[1 + \sqrt{\left(1 + \frac{\gamma^2}{\eta_{\min}^2}\frac{2I_r}{x^2\epsilon\Delta f}\right)}\right]\qquad(3)$$

where

  ϵ = elementary charge

  α^2 = 1·33 (coefficient dependent on the static character of the secondary emission)

  γ = degree of modulation of the light

  I_r = thermal electron current of the photo-cathode

  Δf = band width of the detector equipment

  η = signal-noise ratio required for the measurement.

(The prerequisites specified in [3] for the validity of this equation are applicable.)

Photoelectric current and light current are connected through the equation

$$\Phi_{0\,mm} = K_E \cdot I_{0\,min} \tag{4}$$

  K_E = cathode sensitivity of the photo-cathode.

With the data for the EOS the following is obtained

$$S_{max} = 1000 \text{ km } e^{-kS_{max}}$$

and for

$$k = 0\cdot1 \text{ km}^{-1} \quad \text{(visual range} \approx 40 \text{ km)}: S_{max} \approx 33 \text{ km}$$

$$k = 0\cdot15 \text{ km}^{-1} \quad \text{(visual range} \approx 26 \text{ km)}: S_{max} \approx 24 \text{ km}$$

3.2 Practical Range

Measurements carried out to date show that, using a reflector with attachment wedges and normal voltage (6 V) in the lamp—lighting service around 100 hours—under good atmospheric conditions, distances of 7 km can be measured reliably by day. If over-voltage is used distances of 10 km can be measured with one reflector. For short distances no focusing is necessary. The range for day measurements when using *one* reflector with attachment wedges can be given as

$$6V\text{: } 20 \text{ m} \leq S \leq 7 \text{ km}$$

$$8V\text{: } 20 \text{ m} \leq S \leq 10 \text{ km}$$

The influence of haze, rain and sunshine are far less disturbing than might have been feared on the basis of existing information on electro-optical distance measurement. For example, with the EOS unit from VEB Carl Zeiss Jena, distances of 5·6 km and 10 km were always measured without difficulty even in sunlight, with a light sky as the reflector background and in light haze. No influence could be traced to the direction in which the sunlight fell.

Periodically the light on the test distances was directly facing or from behind. Neither was there any necessity for screening the separate lighting by means of screens behind the reflector. All of the measurements upon which the report was based were carried out without

TABLE 2. *Results of the range tests*

	Distance (km)	Meteorological visibility (km)	Inner accuracy		Number of reflectors	Lamp voltage (V)	Weather conditions
			$M\phi$ (°) ±	M_l (cm) ±			
Day	7	ca. 15	0·9...1·3	0·6...0·9	1	6	Sunny
	10	,, 20	0·5...0·8	0·4...0·6	1	8	Sunny
	13	,, 30			3	8	Background light
Night	10	,, 15	0·3	0·2	1	6	—
	13	,, 15	—	—	1	8	—
	18	,, 25	0·5...0·8	0·4...0·5	3	8	—
	21	,, 30	0·4...0·5	0·3...0·4	3	8	—

background screens. Distances of 7 km were measurable with 6 V also under light rain. The test distance of 2·5 km could be determined reliably even in heavy rain and haze—meteorological visibility around 2 km, reflector or reflected transmission light no longer visible in the locating telescope. A comparison with the meteorological visibilities which applied during the measurements permits the assumption that the day range (6 V) for $S \leq$ 2·5 km is equal to the meteorological visibility. To the best of our knowledge this is the first time that this has been established for instruments of this type.

Naturally the range can be increased with a reflector station comprising several reflectors. In this manner it was possible with the aid of a combination of three reflectors, to measure a distance of 13 km by day and 21 km by night.

In Table 2 a comparison is made of the measurement conditions which applied for the results achieved from special range tests; $m\phi$ is the mean error of the average of a phase measurement repeated five times in two phase positions.

Tests for the measurement of distances exceeding 13 km by day and 21 km by night have not so far been undertaken. During the period in which the specified distances were measured no greater distances were available. The external conditions which applied during the range tests and the reserves inherent in the instrument would, however, permit assumption of the following ranges subject to very good atmospheric conditions:

	6 V/1 reflector	8 V/1 reflector	8 V/3 reflectors
Day	7 km	10 km	15 km
Night	10 km	15 km	25 km

References

[1] Richter, H. & Wendt, H., 1965. EOS opto-electrical telemeter of VEB Carl Zeiss Jena. *Vermessungstechnik* (Berlin), 13th Year No. 4, 124–5.

[2] Karolus, A., 1958. The physical fundamentals of opto-electrical distance measurement. (Munich, Verlag der Bayr). Akademie der Wissenschaften.

[3] Hartmann, W. & Bernhardt, F., 1957. Photographic duplicators and their application in nuclear physics. (Berlin, Akademieverlag).

DISCUSSION

L. E. Wood: Would the addition of a different light source add to the range?

H. Strosche: Tests are being made with a mercury vapour lamp. I would guess that a range of 40 km might be possible at night.

J. W. Wright: Do you plan to re-design the instrument with the transmitter and receiver coaxial?

H. Strosche: Not for an instrument of this size.

Air and Satellite Borne Systems

Shiran*

G. R. WOODRING
U.S.A.F.

A general description is given of the operation of the
SHIRAN system and its advantages over the HIRAN system.
Fifteen lines from 100 to 500 miles in length were
measured with a probable error for a single measure of
about 7 ft., i.e. at 300 miles, this is approximately 1/226 000.

The SHIRAN system utilizes an aircraft to transport the interrogator, permitting the measurement of lines between ground transponders which are hundreds of miles apart. It serves two purposes: one, to establish geodetic ground control by trilateration, and two, to establish the position of the aircraft at the instant of exposure of aerial mapping photographs. The interrogator transmits a data-modulated, encoded, continuous wave in the S-band frequency range to transponders at four ground stations. The returned signals are decoded by the interrogator; the phase shift in the modulation frequencies is measured, translated into distance, and the information stored directly on to magnetic tape.

The advantages in the operation of SHIRAN over HIRAN are as follows:

HIRAN	SHIRAN
Two-station operation	Four-station operation
Data recorded on 35-mm film (manually processed)	Data recorded on magnetic tape (computer processed)
Two operators required	One operator required
No in-flight maintenance facility	In-flight maintenance features built-in
Slow data acquisition—records from two stations every two seconds	Rapid data acquisition—ranges from four stations recorded five times per second

As used at present, HIRAN line crossings are repeated twelve times, six crossings at each of two altitudes. It is expected that SHIRAN will need only two line crossings instead of twelve.

When used to control aerial photographs the four ground stations provide solutions for six nadir points, but only four are strong.

* The paper on 'Electronic Airborne Systems' by M. J. Pappas was not available. Instead an informal description of the SHIRAN system was given by Mr Woodring.

The system has been tested both to extend ground control and to control photography in a test area in Arizona, using U.S. Coast and Geodetic first-order stations as control. The shortest side of the trilateration net was about 100 miles long while the longest was about 500 miles long. Fifteen lines were measured, the probable error of a single measurement being about 7 ft. (2·1 m).

An area about 30 miles by 34 miles in the centre of the net was used to test the photogrammetric control. The circular probable error and the 90 per cent distribution limits for nadirs determined from the four corners of a quadrilateral taken two at a time were:

Control points	1–2	2–3	3–4	4–1
C.P.E. (metres)	±3·36	±3·38	±3·97	±3·77
90% (metres)	±6·13	±6·16	±7·24	±6·88

Since the requirement for 1 in 50 000 scale maps is that 90 per cent of the points should lie within 83 ft. (25 m) of their true position, it looks as though SHIRAN will provide a very good system for large scale mapping.

SHIRAN is only part of an airborne mapping and survey system known militarily as AN/USQ–28. The other elements included are:

Terrain profile recorder (or perhaps, later, a Laser altimeter).

Inertial reference unit capable of vertical read-out accurate to 30 in. for 90 per cent of the time.

Two KC–6 aerial mapping cameras.

A point light source (for azimuth determination), either flashing for use with ballistic cameras using a star background, or steady for use with photo recording theodolites, terrestrially oriented.

All these items are mounted in an RC–135 aircraft (similar to the Boeing 707).

The main disadvantage of the new system is due to increasing the frequency from 300 megacycles (HIRAN) to 3000 megacycles (SHIRAN), because the radiation cannot now penetrate foliage nor strong temperature inversions (at low incidences) in some instances. The terrain must therefore be cleared of trees around the ground stations in all transmission directions.

DISCUSSION

L. Asplund: What is the distance between the control points used for the determination of the errors shown for the nadirs of the photo stations?

G. R. Woodring: About 110 miles.

J. Kelsey: I see that for the determination of the nadir points you have only used two stations at a time. Would it not be better to use four thereby improving the accuracy by about a factor of two?

G. R. Woodring: Yes, it would be better to do this and such is our intention. The use of only two stations at a time was to indicate the consistency of the results.

P. J. Carmody: How critical is the angle of intersection of lines to the control points?

G. R. Woodring: Using four channels this is never critical since the system is used in continuous trilateration to establish control triangles whose sides are about 200 miles long. You then always get good intersections for controlling photography from at least three stations.

J. Kelsey: Did you have any difficulty in obtaining clearance for the use of 3000 Mc/s?

G. R. Woodring: No, the difficulty was when we used 300 Mc/s and this was one reason for increasing it.

N. J. D. Prescott: Does the stabilized mounting for the camera give any trouble when making sharp turns?

G. R. Woodring: No, although the inertial mountings do experience difficulty, they are caged during such manoeuvres.

A. R. Robbins: To what extent is the camera stabilized and what is the maximum range for line crossing?

G. R. Woodring: The camera is stabilized to an angle of the order of a degree or two, but the vertical is recorded independently to about ± 30 in. The maximum range is limited only by aircraft height. At about 40 000 ft it should be possible to measure a line about 900 miles long with large negative take-off angles at the ground stations.

F. J. Hewitt: I am surprised that temperature inversion ducts cut off the signal and would be interested to hear details.

G. R. Woodring: This occurred in a test over water off the Californian coast, where temperature inversions are common. In the case in question there was a strong inversion between 5800 ft and 6200 ft. Signals were satisfactory at about 4500 ft but for greater altitudes signals were either not getting through at all or were very noisy (indicating reflections within the duct) until at about 12 000 ft signals were again consistent.

F. J. Hewitt: I should think that in this case bending of the ray might be significant.

G. R. Woodring: The meteorological data are measured at 500-ft intervals up to 5000 ft and then at 1000-ft intervals. This gives the propagation velocity, but it is realized that this cannot be determined to an accuracy better than 1 in 200 000 and this sets the limit to the system.

Radio Ranging on
Artificial Earth Satellites

FRANK L. CULLEY

Army Map Service, U.S. Army Corps of Engineers

The length of the modern geodetic measuring tape is
unlimited. It is the high-frequency electromagnetic wave
which is graduated with wavelengths of low, modulating
frequencies. The number of wavelengths is determined
by the phase differences of the outgoing and incoming
modulating waves. The artificial Earth satellite provides a
beacon carrying a transponder to relay the modulating
frequencies back to ground stations from which they were
emitted, permitting instantaneous measurements in rapid
sequence from several widely-separated ground points.
The position of a ground point of unknown co-ordinates is
determined from three points of known co-ordinates in two
steps. Three imaginary mathematical spheres, with ground
stations at their centres and with distances determined by
radio ranging from ground stations to satellite as their radii,
intersect in two points of which one is the satellite position.
Next, three spheres, with three determined satellite
positions as centres and with distances measured from a
fourth station of unknown coordinates as the radii,
intersect in two points of which one is the position of the
fourth station. This surveying method may be called
appropriately *electronic trispheration*. The operation of the
equipment is described, and an example of mathematical
treatment is given on a theoretical basis.

Geodesists concern themselves with the geometric relations—distances
and directions—between points on the Earth. They usually find these
distances and directions by making ground surveys such as by running
a traverse or an arc of triangulation between the points.

Often surveyors encounter physical barriers, such as wide ocean
areas, and politically inaccessible areas which prevent the use of these
conventional ground methods. Geodesists have adopted techniques
from other disciplines to overcome or bypass these obstacles. In recent
years, hitherto unfamiliar terms and techniques such as astrogravi-
metric levelling, occultations, and satellite geodesy have become
common. Precise radio ranging has been developed and put into
operation as a new geodetic tool for satellite geodesy.

This surveying system consists of four radio-ranging ground stations and a satellite bearing a transponder, or radio relay. Each ground station measures the distance to the satellite and back in sequential order twenty times per second. The trade name given the system by its developer is SECOR, an acronym for Sequential Collation of Ranges. Although measurements are not made strictly simultaneously, corrections are applied to correct the ranges for a given instant within a 50-millisecond period. In this discussion, these are considered simultaneous measurements.

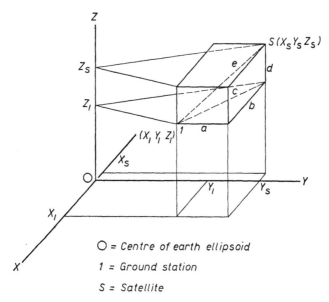

O = Centre of earth ellipsoid
1 = Ground station
S = Satellite

FIG. 1. Three-dimensional co-ordinate system.

The electronic equipment in the system is highly complex. The mathematical computation is explainable by solid geometry. The quantity of data is formidable and highly redundant.

We need only the Pythagorean theorem which states that the square of the hypotenuse of a right-angled triangle is equal to the sum of the squares of the other two sides. We use a three-dimensional, spheroid-centred, Cartesian co-ordinate system for computations, finally transforming them into geodetic co-ordinates of latitude, longitude, and heights above the spheroid. We extend the Pythagorean theorem to a rectangular parallelepiped, the edges of which are parallel to the X, Y, and Z axes of the co-ordinate system (see Fig. 1).

One corner represents a ground station (1) of which the co-ordinates X_1, Y_1, and Z_1 are unknown. The opposite corner of the parallelepiped is at a given instant the position of the satellite of which the co-ordinates X_s, Y_s, and Z_s are also unknown. The diagonal line, e, is the distance $d(S\text{-}1)$ between the satellite and the unknown ground station. Here

$$e^2 = a^2 + b^2 + d^2$$

or

$$\overline{d\,(S\text{-}1)}^2 = (X_s - X_1)^2 + (Y_s - Y_1)^2 + (Z_s - Z_1)^2 \qquad (1)$$

The ground station measures the distance to the satellite. We then have only one known value in equation (1), $d\,(S\text{-}1)$, and six unknowns: X_s, Y_s, Z_s, X_1, Y_1, and Z_1.

At the same time that Station 1 is ranging on the satellite, Station 2 is also measuring the distance from itself to the satellite. Station 2 occupies a point for which the geodetic co-ordinates on a given datum are known. We convert these co-ordinates to the spheroid-centred, Cartesian co-ordinate system giving the location of Station 2 as X_2, Y_2, and Z_2.

From this we get a second equation

$$\overline{d\,(S\text{-}2)}^2 = (X_s - X_2)^2 + (Y_s - Y_2)^2 + (Z_s - Z_2)^2 \qquad (2)$$

This gives us two equations with the same six unknown values:

$$\begin{matrix} X_s & Y_s & Z_s \\ X_1 & Y_1 & Z_1 \end{matrix}$$

and five known values:

$$d\,(S\text{-}1),\ \text{measured}$$
$$d\,(S\text{-}2),\ \text{measured}$$
$$X_2,\ Y_2,\ \text{and}\ Z_2$$

In order to solve these equations, we must have as many equations as there are unknowns. Let us add two more ground stations, No. 3 and No. 4, on points for which geodetic co-ordinates on the same datum as that of Station 2 are known. If these stations measure distances between themselves and the satellite simultaneously with Station 1 and Station 2, we get

$$\overline{d\,(S\text{-}3)}^2 = (X_s - X_3)^2 + (Y_s - Y_3)^2 + (Z_s - Z_3)^2 \qquad (3)$$

and

$$\overline{d\,(S\text{-}4)}^2 = (X_s - X_4)^2 + (Y_s - Y_4)^2 + (Z_s - Z_4)^2 \qquad (4)$$

in which we have no new unknown values added to our six unknown, but we have eight new known values:

$$d\,(S\text{-}3),\text{ measured}$$
$$d\,(S\text{-}4),\text{ measured}$$
$$X_3,\ Y_3,\ Z_3,\ X_4,\ Y_4,\text{ and }Z_4$$

Now we have four equations with six unknown values.

In the three equations, Nos. (2), (3), and (4), we have only three unknown values: X_s, Y_s, and Z_s. These are the coordinates of the satellite which we can now determine by solving equations (2), (3), and (4).

We have made some progress; at least, we have the coordinates of the satellite when ranged upon simultaneously by the ground stations.

We see that equations (1) through (4) are equations of spheres with the ground stations as centres and ranges to the satellite as radii. We know that two intersecting spheres form a circle and that a circle intersects a sphere in two points. The three spheres centred on Stations 2, 3, and 4 intersect in two points (see Fig. 2). The satellite must be in the surface of each sphere having a ground station as its centre and the corresponding range to the satellite as its radius. Therefore the satellite must be at one of the intersections of the three spheres. To imagine it to be at the other intersection is obviously ridiculous as this point would be below the surface of the Earth.

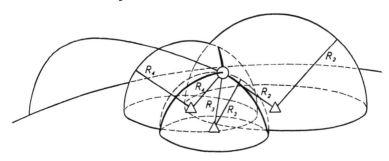

○   Satellite is at intersection of 3 spheres

△   Ground station on known position

R   Radius of sphere or range from ground station to satellite

Heavy lines are intersections of spheres

Fig. 2. Trispheration.

Likewise, the solution of the three simultaneous quadratic equations (2), (3), and (4) gives the co-ordinates of these two points, those values for one being the correct position of the satellite and those for the other being a ridiculous position.

We measure neither angles nor trilaterals in this type of survey. We do not compute coordinates of unknown ground positions directly from coordinates of known ground positions and from direct measurements between known and unknown ground points. Two steps are required: first, satellite positions are found from known positions; secondly, the unknown position is found from the satellite positions. Each new point of satellite position or ground station is at the intersection of three spheres. We can, therefore, call this *trispheration*. The name has a respectable philological origin.

Let us look at equation (1) again. We do not know the position of Station 1, but we have found the position of the satellite. Let us use the satellite as the centre and let Station 1 lie on the surface of the sphere of radius d (S-1). Let us get another satellite position a few hundred kilometres farther along on the same orbit. Then if we get a third satellite position in the same way, except on another orbit to avoid having nearly collinear centres, we have a solution.

Let us get back to the Pythagorean approach which really gives us the equations of the spheres. Designating the coordinates of the second satellite position along the same orbit as X_{s2}, Y_{s2}, and Z_{s2}, and distances from ground stations to the new satellite position as d (S_2-1), d (S_2-2), d (S_2-3), d (S_2-4), we get

$$\overline{d\,(S_2\text{-}1)^2} = (X_{s2} - X_1)^2 + (Y_{s2} - Y_1)^2 + (Z_{s2} - Z_1)^2 \qquad (5)$$

$$\overline{d\,(S_2\text{-}2)^2} = (X_{s2} - X_2)^2 + (Y_{s2} - Y_2)^2 + (Z_{s2} - Z_2)^2 \qquad (6)$$

$$\overline{d\,(S_2\text{-}3)^2} = (X_{s2} - X_3)^2 + (Y_{s2} - Y_3)^2 + (Z_{s2} - Z_3)^2 \qquad (7)$$

$$\overline{d\,(S_2\text{-}4)^2} = (X_{s2} - X_4)^2 + (Y_{s2} - Y_4)^2 + (Z_{s2} - Z_4)^2 \qquad (8)$$

A third satellite position is observed on an orbital pass selected so that the third position could form a good-shaped triangle (approximating equilateral) with the other two positions. Designating the co-ordinates of this position as X_{s3}, Y_{s3}, and Z_{s3}, and distances from the ground stations to the third satellite position by d (S_3-1), d (S_3-2), d (S_3-3), and d (S_3-4), we get

$$\overline{d\,(S_3\text{-}1)^2} = (X_{s3} - X_1)^2 + (Y_{s3} - Y_1)^2 + (Z_{s3} - Z_1)^2 \qquad (9)$$

$$\overline{d\,(S_3\text{-}2)^2} = (X_{s3} - X_2)^2 + (Y_{s3} - Y_2)^2 + (Z_{s3} - Z_2)^2 \qquad (10)$$

$$\overline{d\,(S_3\text{-}3)^2} = (X_{s3} - X_3)^2 + (Y_{s3} - Y_3)^2 + (Z_{s3} - Z_3)^2 \qquad (11)$$

$$\overline{d\,(S_3\text{-}4)^2} = (X_{s3} - X_4)^2 + (Y_{s3} - Y_4)^2 + (Z_{s3} - Z_4)^2 \qquad (12)$$

We have added to the first four equations eight more containing six more unknown values:

$$\begin{matrix} X_{s2} & Y_{s2} & Z_{s2} \\ X_{s3} & Y_{s3} & Z_{s3} \end{matrix}$$

and eight more known values, all measured:

$$\begin{matrix} d\,(S_2\text{-}1) & d\,(S_2\text{-}2) & d\,(S_2\text{-}3) & d\,(S_2\text{-}4) \\ d\,(S_3\text{-}1) & d\,(S_3\text{-}2) & d\,(S_3\text{-}3) & d\,(S_3\text{-}4) \end{matrix}$$

Now we have twelve simultaneous, quadratic equations with twelve unknown values and twenty-one known values. The left members of all twelve equations are known because we measured their square roots. The nine co-ordinate values of three ground stations are known from previous geodetic surveys.

We can continue a step-by-step solution by solving equations (6), (7), and (8) for X_{s2}, Y_{s2}, and Z_{s2}, and by solving equations (10), (11), and (12) for X_{s3}, Y_{s3}, and Z_{s3}. By substituting the values of X_s, Y_s, and Z_s in equation (1), values of X_{s2}, Y_{s2}, and Z_{s2} in equation (5), and values of X_{s3}, Y_{s3}, and Z_{s3} in equation (9), we can now solve these three equations for X_1, Y_1, and Z_1. We can also solve all twelve equations simultaneously with an electronic computer.

We have been attacking this problem as if the observed ranges were the shortest, or straight-line, distances between ground stations and satellite. Actually we have to make corrections to the ranges for aberration, for Doppler effect, for delays within the ground equipment and the transponder, and for atmospheric refraction.

The correction for aberration is for the change in apparent position of the satellite due to the combined effect of the motion of the ranging signal and the motion of the satellite relative to the ground station. It is small enough to be neglected at present.

The Doppler effect on the range is small. While the satellite is approaching the ground station, the frequency of the ranging signal is higher and the wavelength is shorter than when it is moving away. The effect lessens as it comes nearer because it is approaching the zenith where its velocity vector and the vector of the ranging signal would form a right angle. The effect increases as it approaches the horizon where the vectors would be nearest to collinearity. A Doppler loop is built into the ground station to correct for this effect automatically.

The signal spends some time going through the circuits of the ground station and of the transponder. The distance it would travel in free space during this time must be subtracted from the observed range. This is called the calibration correction. It is measured in the transponder before launching. It is measured in each ground station before and after each pass of the satellite by ranging on a transponder at a known distance from the ground station and subtracting the known distance from the observed range.

As the radio waves from the transponder enter the Earth's atmosphere, they first encounter the ionized layers (ionosphere) which, in effect, speed them up, bending those which enter obliquely even farther from the normal to the ionosphere. When they leave the ionosphere, they enter the troposphere which bends their path back toward the perpendicular to the troposphere. This results in a crooked path which is longer than the straight-line distance. The signal transmitted from the ground to the satellite follows the same path (see Fig. 3).

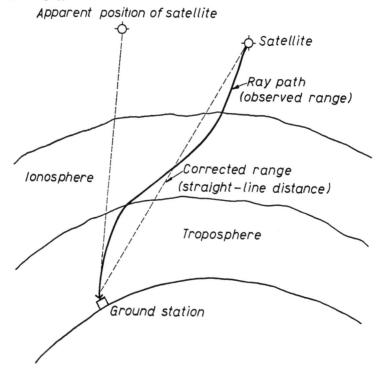

FIG. 3. Effect of the troposphere and ionosphere on the bending of the ray path.

Corrections for the curvature in the troposphere are fairly simple. The troposphere does not change much; observations have been made through it for many years, and radio waves behave much like light rays in it. On the other hand, the ionosphere changes its depth, its extent, and its nature from season to season, from day to day, and from hour to hour, and with latitude. It is also affected by sunspots. We cannot apply much of what is known to make reliable corrections. Hence the SECOR system estimates the ionosphere's effect at each instant by measuring the same range at two different carrier frequencies. This system provides for the transponder in the satellite to retransmit the highest-frequency modulation signal simultaneously on a high-frequency carrier, 449 megacycles per second, and on a low-frequency carrier, 224·5 megacycles per second. The refraction, or curvature, in the ionosphere has a simple relationship to the frequency, decreasing as the frequency increases. The difference in observed ranges, or lengths of paths, of the two frequencies enables us to make a correction for ionospheric refraction. For example, if the observed distance shortens 8 m when the frequency is doubled, about 3 m subtracted from the distance measured by the higher frequency and of about 11 m from that measured by the lower frequency should give the straight-line distance.

We have considered the mathematics and the physical environments of the system. Now let us look at how the system operates.

The ground station transmits a carrier wave with a frequency of 420·9 megacycles per second. This carrier frequency is modulated with four ranging frequencies. The very fine ranging frequency is greatest, the wavelength being 512 m. The fine ranging frequency is one-sixteenth that of the very fine, and the wavelength is sixteen times longer, as shown in Table 1.

TABLE 1. *SECOR frequencies, wavelengths and ranges.*

Ranging frequency	Ranging wavelength (λ)	Non-ambiguous range ($\lambda/2$)	Resolution
kc/s	metres	metres	metres
585·530 (very fine)	512	256	0·25
36·596 (fine)	8 192	4 096	16
2·287 (coarse)	131 072	65 536	256
0·286 (very coarse)	1 048 576	524 288	2048
20 c/s	15 000 km	7 500 km	41·83 km

The ground station measures and records on magnetic tape in binary digital form the difference in phase of the outgoing and incoming ranging signal on each ranging frequency (see Fig. 4). We get only a measurement of the fractional half-wavelength which, if added to the length of the unknown integral number of wavelengths, should give the distance from ground station to satellite. But we are limited to measuring no more than 256 m at the very fine frequency although the resolution is 0·25 m.

Suppose our phase difference at this very fine frequency is 270°. Twice the distance would be

$$2D = (V + \tfrac{3}{4})\lambda$$

or

$$D = (V + \tfrac{3}{4})\lambda/2$$

$$= (256V + 192) \text{ m} \tag{13}$$

where

$$V = \text{integral number of wavelengths}$$
$$\lambda = \text{wavelength.}$$

Using ranging wavelengths of 8192 m, 131 072 m, and 1 048 576 m, and representing the integral numbers of wavelengths by W, X and Y, suppose we get

$$D = (W + 0·047) 4096 \text{ m} \tag{14}$$

$$= (4096W + 193) \text{ m}$$

$$D = (X + 0·940) 65 536 \text{ m} \tag{15}$$

$$= (65 536X + 61 604) \text{ m}$$

and

$$D = (Y + 0·993) 524 288 \text{ m} \tag{16}$$

$$= (524 288 Y + 520 618) \text{ m}$$

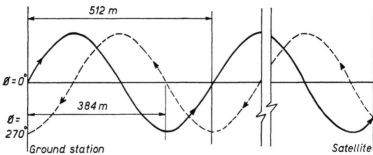

Fig. 4. Phase measurement. D = Distance from ground station to satellite. $\Delta\phi = 270°$, phase difference between signal transmitted and returned. $2D = (V + \tfrac{3}{4})512$ m.

We do not know the values of V, W, X, and Y. We must get rid of these ambiguities. We can get a rough idea of D from the extended range by measuring the time for the signal to make the trip to the satellite and back. An electronic clock ticking off 20 cycles per second is the measuring device. With a frequency of 20 pulses per second, we may say that we have a wavelength of 15×10^6 m, or a half wavelength of 7.5×10^6 m. Suppose that we get a reading of 0·139 twentieths of a second. The distance must then be of the order of

$$D = 0.139 \times 7\ 500\ 000 \text{ m} \qquad (17)$$
$$= 1\ 042\ 500 \pm 41\ 830 \text{ m}$$

Dividing this last value by 524 288, we get

$$Y = 1$$

Substituting 1 for Y in equation (16) we get

$$D = 1\ 044\ 906 \pm 2048 \text{ m}$$

Dividing this value by 65 536, we get

$$X = 15$$

and from equation (15)

$$D = 1\ 044\ 644 \pm 256 \text{ m}$$

Dividing by 4096, we get

$$W = 255$$

and from equation (14)

$$D = 1\ 044\ 673 \pm 16 \text{ m}$$

Dividing by 256, we get

$$V = 4080$$

and, finally, from equation (13) we get

$$D = 1\ 044\ 672 \pm 0.25 \text{ m}$$

We have considered the operation wherein the satellite is ranged upon simultaneously from four stations, three of them on known points, and one on an unknown point. There will be situations in which an unknown station will be so far removed from the nearest known points that it cannot measure ranges simultaneously with the other three stations on known points (see Fig. 5). In this situation, the three stations on known points will be used to determine small segments of several orbits of the satellite. While the satellite is in these same orbits and in electronic view of the ground station on the unknown point, that station will measure ranges to it. We can now extrapolate the orbits from the computed segments to include the segments where the unknown station ranged on the satellite. Accurate timing is important in this situation, for we must know the satellite position at the time it was ranged upon from the unknown point.

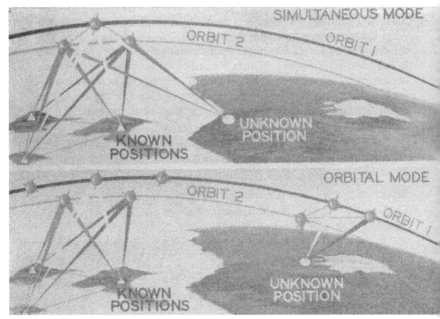

FIG. 5. Simultaneous and orbital tracking modes.

Knowing the co-ordinates of the satellite positions at definite instants, we can use these positions as centres of the spheres with corresponding ranges as radii to compute the position of the unknown point.

Redundancy of data enables us to select the best geometric figures to make our computations and to decrease statistical probable errors.

Data from radio-ranging operations should provide a valuable by-product, an increase of our knowledge of the atmosphere, particularly the ionosphere. This system should enable us to increase greatly the distances over which we can make geodetic ties. It should give us a better knowledge of the shape and size of the Earth and of the external gravity field of the Earth. Attempting to separate the effects of air drag, solar-radiation pressure, and geomagnetism from effects of gravity anomalies will be an interesting occupation.

Trispheration is a useful precise method for three-dimensional geodesy. Except for extreme conditions such as typhoons it can be operated in all kinds of weather and at any time of day or night. It can provide a good scale for the photogrammetric satellite geodesy. We expect it to play a major role in connecting the major geodetic datum points to form a single, world-wide geodetic network.

Fig. 6. Secor ground station.

DISCUSSION

For the discussion on this paper see that at the end of Major Prescott's paper.

Experiences with Secor Planning and Data Reduction

N. J. D. PRESCOTT

British Military Survey

The United States Army Corps of Engineers acquired the SECOR (Sequential Collation of Range) electronic distance measuring equipment as an all-weather system for providing a world-wide geodetic network by means of artificial satellites. Used mainly in a geometrical role by Army Map Service the equipment has produced results which show remarkably high accuracy. The planning and data reduction techniques are described in this paper. Taking into account satellite heights and other relevant considerations station configurations are selected. After observations are made the measured range is corrected for calibration, troposphere and ionosphere. A large number of satellite positions are fixed in all available orbits and one, or more, least-square solutions are used to solve for the best position of the unknown station. Examples of the data printed out at each stage and a typical solution output are included. The latest results obtained in the Pacific Operation are listed. Possible improvements in data handling and other geodetic applications of SECOR are discussed.

Introduction

Artificial satellites and precise electronic distance measuring equipment are both products of the last decade. It is to the credit of the United States Army Corps of Engineers that they foresaw the advantages of combining them in a system capable of making geodetic ties across oceans between the world's triangulation datum points. The SECOR (SEquential Collation Of Range) all-weather electronic distance measuring equipment was developed by a contractor, Cubic Corporation, and the system's first successful satellite was launched in January 1964. Since then Army Map Service (AMS) has obtained good results during a series of tests in the U.S.A., and their more recent operation in the Pacific is making excellent progress. Owing to the keen interest shown in the project by the Director of Military Survey, a British team was invited to man one of the tracking stations. The writer was fortunate in commanding it for twelve months, and since then has also spent a year at AMS, Washington, D.C., assisting with operational planning and data reduction.

This paper describes some of the factors to be considered in SECOR planning and then deals at length with the data reduction. The latest results from the Pacific Operation are given and possible improvements in computational methods and observation techniques are discussed. Details of the data reduction have been relegated to appendixes.

Planning Considerations

During the test period in the U.S.A., experiments were made with an orbital technique but results were not very satisfactory. It was, therefore, decided to avoid possible errors that might be caused through uncertainties in the gravity field by using SECOR in a geometric role only, during the present operations in the Pacific.

In this mode of observation four ground stations (one designated as master and three as slaves) make range observations, that are effectively simultaneous, to a transponder in a satellite. Three of the stations must occupy known points that have been previously established and the fourth station may be placed on an unknown position. Each set of ranges from the known stations fixes a satellite point in space. A known range is also available from the unknown station to each fixed satellite point. In theory, ranges from a minimum of three well-chosen satellite positions will locate the unknown station. In practice, the satellite is tracked over many passes and numerous satellite points are fixed in a wide area above the ground stations. The position of the unknown station is then solved by least squares.

Good geometry has a great bearing on the accuracy of a SECOR simultaneous solution. The three known stations should form a well-conditioned triangle with side lengths of about one to two times the satellite's height. The satellite point is fixed most strongly when it is over the centre of the triangle and accuracy falls off as it moves away. The strongest possible satellite position would occur if the three lines from the known stations intersected at right angles (a situation that is almost impossible to achieve in practice). It is, however, advantageous if the geometry avoids acute angles at the range intersections. The accuracy of SECOR range measurements is discussed below.

The same applies to the fix at the unknown station. The following are some of the considerations that affect the selection of stations in a framework, apart from those of logistics:

(*a*) *Available satellite height*—Appendix A shows the approximate distances that may be spanned with satellites of different heights.

(*b*) *Minimum elevation angle*—The ionospheric correction and, to a lesser extent, the tropospheric correction to the range increase fairly sharply at low elevation angles. For planning purposes, it is considered that only ranges measured at elevation angles of 15 or greater can be reduced with acceptable accuracy.

(c) *Good intersections*—An arc, indicating the 15° minimum elevation angle for the satellite used, is drawn around each station. Good simultaneous observations may be made in the area enclosed by the four arcs (an example is given in Appendix A, p. 321). It is necessary for this area to be sufficiently large so that lines from satellite points near the edge will intersect at reasonable angles (i.e. not too acute) if a weak fix is to be avoided.

(d) *Extension of control*—Each new station should be suitably placed so that, when established, it plays a useful part in fixing the next unknown point.

When the programme has been tentatively planned, the strength of each figure is checked by running an error simulation programme on the computer. This enables various alternatives to be compared and shows when the geometry is being stretched too far.

Secor Ranging Errors

The main sources of error in SECOR range measurements are caused by calibration, tropospheric and ionospheric refraction. The corrections applied for these are described in Appendix C (p. 324). Gross timing errors and ambiguities will result in mistakes but small timing errors should not harm simultaneous observations (Appendix B, paragraph 3, p. 323). Electronic noise in the system causes a small random error in each measurement. A constant scale error not exceeding one part per million might be expected from uncertainties in the velocity of light and the frequency standard. The writer estimates that these sources might contain the following probable errors using a satellite at a height of 500 nautical miles above the Earth's surface:

Probable Error of a Single Range Measurement

		Elevation (slant range)	
		60°–90°	15°
		(1000 km)	(2230 km)
(a)	Calibration—satellite transponder	1 m	1 m
(b)	Calibration—station	2 m	2 m
(c)	Tropospheric correction	0·25 m	1 m
(d)	Dual frequency ionospheric correction —night	0·25 m	1 m

		Elevation (slant range)	
		60–90°	15°
		(1000 km)	(2230 km)
(e)	Dual frequency ionospheric correction —day	1·5 m	6 m
(f)	Random electronic noise	1 m	1 m
(g)	Frequency and light propagation errors	1 m	2 m
	Combined (day) error	3·0 m	6·9 m
	Combined (night) error	2·7 m	3·5 m

This indicates that the probable error in daytime observations is proportional to the slant range with a value of about 3 parts per million (ppm). At night the constant errors predominate and a probable error of about 2·7 ppm near the zenith improves to about 1·6 ppm at 15° before increasing fairly sharply at lower elevations.

The errors, with the exceptions of errors (a) and (g), may be considered as random over a number of satellite passes. A significant phase shift in a satellite should be observed by test stations in the U.S.A. and will also show up as a discrepancy amongst data tracked from other satellites.

It must be emphasized that these estimates of probable error are very approximate, since it has been extremely difficult to check the precise accuracy of SECOR range measurements. During the tests in the U.S.A. the ground triangulation system was possibly a greater source of error than SECOR. The geometry was often poor and the less accurate analytic model corrections were used for the ionosphere as technical difficulties prevented dual frequency data from being available from all stations.

The writer considers that higher satellites should result in still greater proportional accuracy in the SECOR system, since most sources of error can be expected to remain fairly constant or increase relatively gradually. It is, therefore, to be hoped that higher SECOR satellites will soon be available.

Data Handling

During a tracking operation all the relevant ranging and timing information is recorded on seven-channel magnetic tape, except for calibration and meteorological readings (barometer, wet and dry bulb), which are recorded on a form. A number of other forms concerning the electronic performance of the station are also completed and the tape, together with all the paper work, is immediately sent back to AMS by air mail. The following is a general description of the computing process used. A detailed description is given later in the paper.

The tapes are immediately run on the Honeywell 800 computer and a *raw data print out* (Appendix E, paragraph 1) is produced. At this stage the range parts measured on the five channels are put together to form one complete range word (Appendix B, paragraph 2).

Provided all four stations have potentially good simultaneous data, the *pack* and *edit* programmes (Appendix E, paragraph 2) are run on the computer. Thus, all usable data relating to a single track are concentrated in convenient and economical form for computations and storage.

When two or three passes with good data are available, solutions may be started. Satellite points are computed using the measured ranges from the known stations. In the least-squares solution used at present, these points are held fixed in solving for the position of the unknown station. A detailed description of the method is given in Appendix D, paragraph 5 and an alternative method of adjustment is discussed later in this paper. At first, small independent solutions are the most useful for checking passes for ambiguities and obtaining an approximate position. Once three independent solutions agree, it is a relatively easy matter to identify and remove an ambiguity from a pass, provided only one set of the range measurements contains an ambiguity (Appendix E, paragraph 4).

Input to the solutions can be by punched cards. This enables up to a hundred passes incorporating a maximum of 1000 satellite points to be used in a single solution. As an alternative, input can be by magnetic tape when there is no absolute limit to the amount of data that may be used in a solution. In general, card input is used to begin with but tape input has been found more satisfactory for the final solution.

Results of the Secor Operation in Pacific

Starting from three known stations on Tokyo datum (Bessel Spheroid), unknown stations on the following islands have been fixed:

	Probable error in metres		
	Lat	*Long*	*Ht*
Minami Daito Shima	± 1·6	± 1·6	± 1·0
Iwo Jima	± 2·1	± 2·2	± 2·8
Guam	± 1·3	± 1·5	± 2·7
Marcus	± 3·7	± 2·2	± 0·5

The probable error has been obtained from the internal agreement between the solutions.

Work is now progressing well and it is anticipated that the next two islands, Truk and Woleai, will be fixed shortly.

It is expected that the tie will be carried through to Hawaii by the summer of 1966.

Possible Improved Techniques

Alternative Methods of Adjustment

An alternative method of adjusting SECOR results has been considered at AMS. This allows the satellite position as well as the unknown station to adjust in the solution. The argument put forward against it is that such a solution would be highly correlated owing to the geometry. It is possible that orbital constraints properly applied to the satellite positions might overcome the problem, but no method has yet been worked out for this, though a computer programme, which will allow satellite positions as well as the unknown station to adjust, has been prepared by AMS. This method of adjustment is to be recommended if the danger of correlation can be overcome as it lends itself to weighting of the observed quantities (ranges). Weighting might be an advantage if day and night observations were combined in the same solution and it would also allow data from satellites at different heights to be used together.

At present only data with good simultaneous observations from four stations are being used. The Research and Analysis Division of AMS is working on orbital computational techniques that will enable any good data in a pass to be used. This study may also lead to the use of semi-orbital and orbital solutions in fixing unknown stations.

Observational Methods

The principal world-wide control system is planned as a 36-station network to be observed by the BC-4 cameras. Work on this will start about the middle of 1966 after the PAGEOS (100 ft diameter) balloon satellite has been launched to a height of 4250 km above the Earth's surface. SECOR has been allotted a future role of densification of control. The 36-station net is to be considered as known datum and three SECOR stations will occupy each triangle in turn, while other stations move to unknown points that need to be tied into the main framework.

Though SECOR may be well suited to this task, there is a strong argument against restricting it solely to densification and allowing the BC-4 cameras to be completely responsible for the 36-station net. There are many differences of opinion over the relative accuracy of SECOR in distance measurement and various camera systems in the determination of direction. However, it should be noted that whereas a

11*

camera will retain much the same proportional accuracy in observing to higher satellites, the random and systematic biases in SECOR will remain fairly constant and the system will, thus, become relatively more accurate at a greater range. Therefore, it is probable that, using SECOR, range observations could be made to a satellite at a similar height to that proposed for Pageos with considerably greater proportional accuracy than would be possible with the BC-4 cameras.

However, this is not a sound argument for suggesting that SECOR stations should replace the BC-4 cameras as the system for observing the 36-station network. The geometry used by SECOR to fix a fourth station is sound with the satellites now in orbit, but owing to the increased effects of the Earth's curvature, with stations as far apart as those of the 36-station net, the geometry for a simultaneous solution would be badly stretched unless an extremely high satellite were used.

In the writer's opinion there is a very strong case for observing the 36-station net with combined BC-4 camera/SECOR stations. There may be practical difficulties but these are small in comparison with the problems already overcome in the development of both systems. A SECOR transponder and its antennae would have to be positioned so that measured ranges would always relate to either the centre or the nearest part of the surface of the optical target. There would be no need for the camera and SECOR observations to be synchronized exactly provided the timing systems were combined. For example, if the times when the SECOR master pulse was received at each station were recorded on the BC-4 timing equipment, two purposes would be served: the times of range and direction observations at each station would effectively be tied together, and a most accurate time synchronization between all the camera stations would be provided.

With a combined system a more flexible observing plan would be possible and more satellite passes would become usable. The main disadvantage of an optical system is its dependence on good weather conditions. Whereas a two-station simultaneous optical observation provides usable data, a single camera observation combined with SECOR would do so as well. It should also be noted that a two-station camera/SECOR observation immediately provides a position for the unknown station. Some modifications will be necessary in the data reduction methods of both systems, but it should be possible to obtain a very strong and uncorrelated solution for the unknown point. The advantages of having scale directly observed into each figure should not be overlooked.

There may be some other valid arguments against the use of a combined optical/distance measuring system. However, in the absence of such reasons, it would be unfortunate if the BC-4 cameras alone were to observe this 36-station network on which so much subsequent control will depend.

Summary

The accuracy of simultaneous SECOR solutions depends very much on good geometry. An observing plan, therefore, needs to be carefully considered and each figure tested with an error simulation programme. Range accuracy will improve proportionally if higher satellites are used.

A raw data print-out is immediately made at AMS from all magnetic tapes received from the field. Good simultaneous data are edited and packed on new tape in a more convenient and economical form for storage and computations. Selected data are then used for station solutions.

Four islands have now been fixed in the Pacific operation and two others should be located very soon.

AMS is working on more sophisticated methods of data reduction. The study of orbital techniques may lead to better use being made of SECOR data.

There is a strong case for developing a combined BC-4/SECOR station as the system to be used for observing the 36-station network.

Acknowledgement: The writer is grateful to those concerned with the SECOR Project in the Departments of Geodesy and Computer Services, Army Map Service, who have helped by checking and criticizing this paper.

APPENDIX A

Secor Planning

1. Whereas the quality of figures in a triangulation system may be readily estimated at a glance, the complex three-dimensional SECOR geometry needs more careful analysis. The known stations ideally will form a figure not too far removed from an equilateral triangle, with side lengths of one to two times the satellite height. The exact height/side length relationship does not vary directly owing to the increased effects of the Earth's curvature at greater distances.

2. In any four-station configuration, simultaneous tracking must be possible with acceptable elevation angles over a sufficiently large area. The limiting factor is generally the distance between the unknown station and the known station most remote from it. The writer has found a simple nomogram, such as Fig. 1, to be a useful planning aid. This illustrates the geometric strength of fixes over various distances

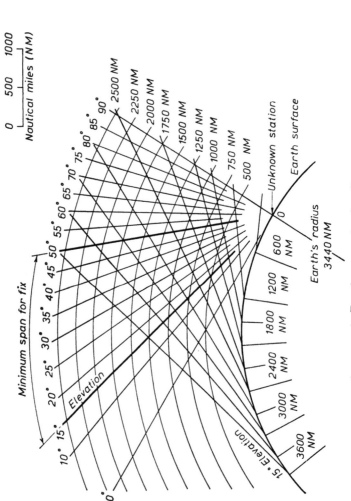

APPENDIX A. FIG. 1. SECOR planning diagram.

for different satellite heights, provided only elevation angles of 15° or greater are used. The view presented is the fix of the unknown station in its weakest direction. An intersection angle of 35° is considered the absolute minimum that can be tolerated, but for a strong fix this angle should be about 60°. Therefore, the unknown station is able to observe satellite positions at elevation angles from 15° to 50° and preferably up to 75°. Fig. 1 caters for satellites of heights varying from 500 to 2500 miles above the Earth's surface and also shows the distance to the sub-satellite point for different elevation angles.

3. The next stage in planning is to prepare a drawing similar to Fig. 2 for the stations likely to be used in each fix. This shows the extent of the area available for simultaneous tracking, using the current (500-nautical mile) satellites, from a different viewpoint. There is generally no problem in observing satellite positions from the unknown station through an arc of at least 60° in azimuth, unless the known stations were badly placed. In the example (Fig. 2), the satellite can be usefully

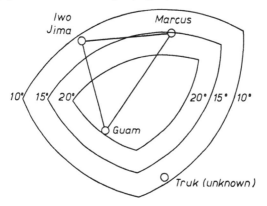

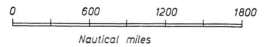

APPENDIX A. FIG. 2. Areas of simultaneous tracking. This is an example of the fix of Truk in the Pacific, using Iwo Jima, Marcus, and Guam as known stations. Areas indicating minimum elevation angles of 10°, 15°, and 20° are shown. Satellite passes within the 15° area would be considered as usable data. This refers to present satellites, which are 500 nautical miles above the Earth's surface.

observed from the unknown (Truk) at elevations of 15°–75° and through an azimuth arc of at least 90°. This is considered fairly good geometry, which should result in a strong solution for Truk.

4. AMS have an error simulation programme for their computer to test the strength of any figure, which provides a basis for comparison between different configurations. It is not proposed to go into details of the techniques that have been used in error simulation studies.

APPENDIX B

Secor System Range Measurement

1. *Modulation Frequency*

The modulation frequencies are all generated by a James Knight oscillator with a stability of 1 part in 10^9 which produces a basic frequency of 1171·065 kc/s. This reference frequency is divided a number of times as follows:

Ref 1171·065 kc/s

Very fine D_1 585·533 kc/s (reference divided by 2)
Fine   36·596 kc/s (reference divided by 2^5)
Coarse   2·288 kc/s (reference divided by 2^9)
Very coarse   0·286 kc/s (reference divided by 2^{12})

As these modulation frequencies cover too wide a range to be transmitted in this form, it is necessary to mix them into the following four signals, suitable for transmission:

D_1 585·533 kc/s
D_2 548·937 kc/s   Very fine–fine
D_3 583·245 kc/s   Very fine–coarse
D_4 549·223 kc/s D_2+very coarse

Measuring signals received from the satellite are compared with a sample of the outgoing signal, the phase shift is extracted and the original measuring frequencies are recovered. As one oscillator produces all the modulating frequencies, serious biases should not occur to individual channels, though small random biases have unfortunately often been present in the data.

2. *Ranging Channels*

The phase difference in each measuring channel is recorded in metres. A fifth channel, the extended range, also contributes towards the full range word. Values for the extended range are obtained from the transit time to and from the satellite of each measuring pulse and not from phase shift. The 'Extended Range' is not particularly accurate but this is unimportant, since 'Very Coarse' gives a non-ambiguous range of up to 524 288 metres, and multiples of this can easily be determined from the prediction data.

Figure 1 illustrates how the five channels overlap to make up the full-range word.

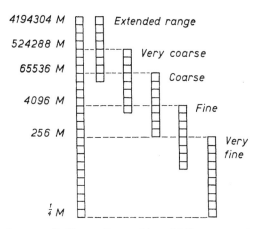

APPENDIX B. FIG. 1. Composition of full-range word.

3. *Timing of Observations*

In the SECOR system, four ground stations measure range in sequence at $12\frac{1}{2}$-millisecond intervals at a rate of twenty measurements each per second. The range on each channel reaches its analogue servo at these intervals. The servo is, however, electronically damped so that it keeps a smooth and continuous record of phase angle. Each analogue servo is monitored by a digital servo which keeps a binary representation of the range in terms of metres. To achieve simultaneous range measurements from the four ground stations, one station, previously designated as master station, emits a special pulse every 50 milliseconds with its ranging signal. This pulse, retransmitted by the satellite, is received at all stations including the master. On reaching the stations the pulse causes the contents of each digital servo to be immediately recorded on the magnetic tape. These range readings all correspond to the same time at the satellite.

Time is also recorded with each range. The stations have crystal-controlled frequency standards which operate time-code generators. Radio checks are carried out every day and the propagation time of the radio signal is allowed for. In these circumstances it is unlikely that the maximum error ever exceeds 2 milliseconds. Using the simultaneous method of observation small clock errors at a station are of no consequence, since time is only needed to match the four sets of ranges to the correct 50-millisecond interval. Signal propagation delays from the satellite cause slight apparent time variations between stations, but these are unlikely to exceed 6 milliseconds. The computer programme which matches the ranges accepts variations of up to 10 milliseconds (Appendix E, paragraph 2).

It should be noted that range readings will not occur at exactly 50-millisecond intervals as the satellite range from the master shortens the intervals between successive readings to slightly less than 50 milliseconds. In the same way, when the satellite range is increasing, the time intervals become slightly greater. In a simultaneous solution this does no harm, but a correction has to be applied to the data before they can be conveniently used in orbital computations.

APPENDIX C

Corrections to Observed Range

1. *Calibration*

It has been found that the bias in a system as complicated as SECOR does not remain constant for any length of time. The main sources of bias are the ground station and antenna circuitry. The satellite transponder also contributes a significant delay to the system. The various sources of bias are calibrated as follows:

(*a*) *Satellite Transponder*. A transponder calibration unit was used to test the satellite transponder and its antenna system. Different orientations of the satellite and its antenna showed variations of 3·75 m. It was felt that the satellite's motion around its axis would tend to average this variation so a mean value was accepted. The satellite launched in January 1964 required the following calibration corrections:

High frequency	449 Mc/s	− 12·25 m
Low frequency	224·5 Mc/s	− 8·50 m

It was considered that these values were unlikely to change significantly during or after the launch.

(*b*) *Ground Station*. The non-symmetrical twin-dish antenna in current use does not lend itself to a straightforward and reliable method of internal self-calibration. After numerous experiments during the tests in the U.S.A., the best results were obtained by ranging on a discone antenna 100 metres away which was connected to a test transponder. This involved having an extra piece of equipment at each ground station which had to be checked by a transponder test unit periodically. Calibration of the ground station against the test transponder is carried out before and after every satellite track. Generally the two values agree fairly closely.

2. Tropospheric Refraction Correction

The present computer programme uses an empirical formula which can be expected to be accurate to within about 10 per cent of the correction.

$$\Delta R = \frac{K_1 \left(1 - e^{-ZR}\right)}{\sin E_0 + K_2 \cos E_0} \qquad \text{(The correction to the observed range is always negative.)}$$

where

K_1 = zenith refraction value (2·6 metres)
K_2 = control constant (0·0236)
Z = control constant (1/6858)
R = slant range in metres
E_0 = elevation angle

Using this formula a correction of 9·2 m is applied to a line with an elevation angle of 15°. As the elevation angle increases, the correction decreases until it is 2·6 m at the zenith. At 15° elevation the range to the satellites which are in use now (both at a height of 500 nautical miles) is about 2 230 000 m. The maximum likely error to the range corrected by this formula is about 0·4 ppm at this elevation and 0·2 ppm near the zenith, which is of no great significance.

3. Correcting for Effects of the Ionosphere

A radio signal of sufficiently high frequency to penetrate the ionosphere instead of being reflected is subject to some degree of retardation. The amount of retardation depends on the electron density profile of the ionosphere, the elevation angle, the length of the line and the frequency of the carrier signal.

The width of the F2 layer of the ionosphere and its electron density vary with the time of day, the latitude, and with the amount of sunspot activity. The maximum density occurs around mid-day when an error of as much as 80–120 m may be caused in a measured SECOR range. The density decreases to a minimum at local midnight with very little increase for at least three hours before or after. At this time range errors should not exceed 5–20 m, so this is clearly the optimum time for observations. The electron density is at its greatest around the equator and considerably less dense in medium and high latitudes. Sunspot activity was relatively low in 1964–5, but will have increased considerably by 1969–70. Around that date uncorrected range errors may be many times as great as at present, and it will be very necessary to restrict observations to night time.

The elevation angle, and hence the length of line passing through the ionosphere, affects the amount of retardation, though not to the same extent as the diurnal changes, providing elevation angles below about 15° are avoided.

The range error caused by ionospheric retardation also decreases as the frequency of the carrier signal increases. The high frequency signal in the SECOR equipment is 420·9 megacycles for the outgoing carrier while the return signal from the satellite uses 449 megacycles. A second carrier frequency of 224·5 megacycles is also transmitted by the satellite. To a first approximation the retardation varies inversely with the square of the frequency. This fact is put to good purpose in SECOR by using the difference between the ranges measured on the high- and low-frequency carriers to compute a correction for the ionospheric retardation. This difference is first corrected for calibration errors in the high- and low-frequency very-fine-measuring channels. The resulting value is then multiplied by a constant determined from the three carrier frequencies involved in the system.

$$\text{Range correction} = K\,(\text{1C-D1 corrected for calibration})$$

$$\text{where } K = \frac{f_2^{-2} + f_1^{-2}}{f_2^{-2} - f_3^{-2}}$$

For SECOR

$$f_1 = 420\cdot9 \text{ Mc/s}$$
$$f_2 = 449 \quad \text{Mc/s} \qquad K = -0\cdot7125$$
$$f_3 = 224\cdot5 \text{ Mc/s}$$

Note: The correction to the range must always be negative.

The dual frequency method of correcting for the ionosphere may leave small residual errors but it is not considered that these should amount to more than 5 per cent of the correction. Therefore, satellite passes within three hours of midnight should only contain a maximum uncorrected error of about one metre. The use of a third frequency would lead to an improved correction but the extra electronics needed would make this a difficult modification for the SECOR system, and would not appear to be necessary.

Owing to past problems with interference on the low-frequency carrier signal, it was often not possible to obtain the second range measurement at one or more of the stations. An analytic method of determining the ionospheric correction was therefore produced and used on some of the earlier preliminary computations. This formula was not very satisfactory as it took no account of the variations in electron density due to the solar zenith and variations in latitude. Improved observation techniques now enable both frequencies to be recorded at all stations during most tracks and an analytic correction is seldom required. A more sophisticated analytic ionospheric model for the few occasions that it will be needed is expected to be available soon.

4. *Doppler Effects*

A special Doppler loop circuit in the servos compensates electronically for Doppler effects on the wavelength. As a result no computational correction to the observed range is necessary.

APPENDIX D

Outline of Mathematical Steps

1. The mathematical formulae and computational steps used in a SECOR solution are described in great detail in [1] and [2]. The following is a general description of the mathematical steps taken in a SECOR simultaneous solution.

2. *Known Data*

(a) Latitude, longitude and height of the three known stations (Stations 1, 2, and 3).

(b) Observed ranges to the satellite from the known and unknown stations.

(c) The approximate (trial point) co-ordinates of the unknown station (Station 4).

(d) Prediction data which provide approximate azimuth, elevation and range to the satellite at different times from each ground station.

3. *Cartesian Co-ordinates*

Station co-ordinates are converted to an Earth-centred origin Cartesian system for the computations using the standard formula:

$$\begin{bmatrix} x \\ y \\ z \end{bmatrix} = (N + h_s) \begin{bmatrix} \cos \varphi \ \cos \lambda \\ \cos \varphi \ \sin \lambda \\ \sin \varphi \end{bmatrix} - e^2 N \begin{bmatrix} 0 \\ 0 \\ \sin \varphi \end{bmatrix}$$

where x, y, z are the Cartesian co-ordinates of the station, with the origin at the Earth's centre.

$$N = a(1-e^2 \sin^2 \varphi)^{-1/2}$$

a = semi-major axis of Earth
e = eccentricity of Earth
φ = geodetic (spheroidal) latitude
λ = geodetic (spheroidal) longitude
h_s = height of station above spheroid

(Height above the geoid must be corrected for the separation between geoid and spheroid.)

4. *Solution for Satellite Positions*

The basic formulae for the solution of a satellite position from ranges measured from three ground stations are:

$$r_1^2 = (x_s - x_1)^2 + (y_s - y_1)^2 + (z_s - z_1)^2$$

$$r_2^2 = (x_s - x_2)^2 + (y_s - y_2)^2 + (z_s - z_2)^2$$

$$r_3^2 = (x_s - x_3)^2 + (y_s - y_3)^2 + (z_s - z_3)^2$$

where

$$\left.\begin{matrix} x_1\, y_1\, z_1 \\ x_2\, y_2\, z_2 \\ x_3\, y_3\, z_3 \\ x_s\, y_s\, z_s \end{matrix}\right\} = \left\{\begin{matrix} \text{Cartesian co-ordinates} \\ \text{of the three ground} \\ \text{stations and the} \\ \text{satellite} \end{matrix}\right.$$

$r_1\, r_2\, r_3 \equiv$ Range from Station 1, 2, 3 to the satellite

The solution of these equations in their present form is laborious and it is more convenient to transform to temporary co-ordinates $x'\, y'\, z'$. The origin should be one of the known stations; in this example let it be Station 1. The axes are rotated so that the three ground stations all lie in the xy plane, and the y axis is aligned so that it passes through one of the other stations, in this example Station 2. The temporary Cartesian co-ordinates for the stations will then be:

	x'	y'	z'
Station 1	0	0	0
Station 2	0	y_2'	0
Station 3	x_3'	y_3'	0

The equations for the satellite positions will then become

$$r_1{}^2 = x_s'^2 + y_s'^2 + z_s'^2$$
$$r_2{}^2 = x_s'^2 + (y_s' - y_2')^2 + z_s'^2$$
$$r_3{}^2 = (x_s' - x_3')^2 + (y_s' - y_3')^2 + z_s'^2$$

These three equations can be relatively easily solved to obtain satellite positions with a positive value for z_s'. In the computer programme many satellite positions are solved for and the values transformed back into the Earth-centred Cartesian co-ordinates.

5. *Solution for the Unknown Station*

In theory, the position of the unknown station might be solved for from only three suitably placed satellite points. The procedure would be the same as for fixing the satellite position except that a negative value for the co-ordinate z_4' would have to be obtained. In practice, the trial point co-ordinates available for the unknown station are used as a first approximation and its precise position is solved for by least squares. To date the policy has been to hold satellite positions fixed, allowing only the unknown station to adjust. The solution is obtained as follows:

Unknown station trial point co-ordinates

$$x_4{}^t \quad y_4{}^t \quad z_4{}^t$$

Satellite positions 1 to n

$$x_{s1} \quad y_{s1} \quad z_{s1}$$
$$\cdot\cdot \quad\quad \cdot\cdot \quad\quad \cdot\cdot$$
$$x_{sn} \quad y_{sn} \quad z_{sn}$$

Measured ranges from satellite points to unknown $r_{14}^m, r_{24}^m \cdots r_{n4}^m$
Computed ranges from satellite points to trial points $r_{14}^c, r_{24}^c \cdots r_{n4}^c$
Direction cosines (l, m, n) are calculated for the lines from the satellite points to the trial point:

$$l_{14} = \frac{x_4{}^t - x_{s1}}{r_{14}}$$

$$m_{14} = \frac{y_4{}^t - y_{s1}}{r_{14}}$$

$$n_{14} = \frac{z_4{}^t - z_{s1}}{r_{14}}.$$

Observation equations are then written:

$$
\begin{matrix}
(A) & (X) & (K)
\end{matrix}
$$

$$
\begin{bmatrix}
l_{14} & m_{14} & n_{14} \\
l_{24} & m_{24} & n_{24} \\
l_{34} & m_{34} & n_{34} \\
\cdot\cdot & \cdot\cdot & \cdot\cdot \\
l_{n4} & m_{n4} & n_{n4}
\end{bmatrix}
\times
\begin{bmatrix}
dx_4 \\
dy_4 \\
dz_4
\end{bmatrix}
=
\begin{bmatrix}
r_{14}^m - r_{14}^c \\
r_{24}^m - r_{24}^c \\
r_{34}^m - r_{34}^c \\
\cdot\cdot \quad \cdot\cdot \\
r_{n4}^m - r_{n4}^c
\end{bmatrix}
$$

The normal equations become a 3×3 matrix, which is easily solved by obtaining the inverse: $AX = K$

$$X = (A^T A)^{-1} A^T K.$$

dx_4, dy_4 and dz_4 are the corrections to be applied to the trial point co-ordinates x_4, y_4 and z_4. If the corrections are fairly large the process may be iterated using $(x_4 + dx_4)$, $(y_4 + dy_4)$ and $(z_4 + dz_4)$ as the new trial co-ordinates.

References

[1] AMS Geodetic Memorandum No. 1536 A, Range Methods for Geodetic Adjustment.
[2] Cubic Corporation, Final Technical Report, Mathematics of Geodetic SECOR Data Reduction.

APPENDIX E

Data Reduction Processes

1. Raw Data Print-out

AMS receives a magnetic tape from each station for every satellite pass tracked. These tapes are fed into the Honeywell 800 computer via a data translator. A raw data print-out (Fig. 1) is produced first. This is designed to be of especial value to the electronic engineers and much of the information given is not needed for the data reduction. The following is an explanation of the information recorded in each column:

'Q' Column—This gives an indication of data quality. A zero shows that all the electronic servos at the tracking station were locked and that the data should therefore be good. A figure 1 indicates that one or more of the five servos were out of lock and therefore the data are bad (though possibly recoverable).

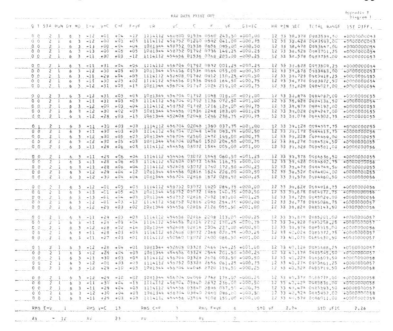

APPENDIX E. FIG. I. Example of raw data print-out.

'*T*' *Column*—A figure 1 is printed out every second (every twentieth measurement), otherwise 0 is shown.

STA—The station number, in this case Station 2.

RUN—Each run or track recorded during a day is numbered.

DY and MO—shows the day and month. In the example, 6 March.

E-V, V-C, C-F and F-VF—These columns show the numerical value of the discrepancy between the overlapping parts of the five measuring frequencies that make up the total range word. The figures and signs are a little confusing. With the exception of *V–C* (the overlap between very coarse and coarse), four binary bits overlap between the frequencies (Appendix B, paragraph 2). In these columns a value of 7 or 8 indicates an almost completely ambiguous reading. A value of 12 indicates a numerical discrepancy of 4. In the *V–C* column, where there are five bits of overlap, a value of 31 is the same numerically as 1. A major ambiguity in a channel cannot be detected in these columns.

ER, VC, C, F, VF—Ranges in metres are given for the Extended Range, Very Coarse, Coarse, Fine and Very Fine channels respectively. The Extended Range channel is very unreliable. It is based upon the transit time of a measuring pulse, and not on phase shift. However, this is not important as Very Coarse gives a non-ambiguous range of up to 524 288 m.

*D*1–1*C*—Gives the difference in range between the high and low frequency carriers' (very fine) ranges. In the example, the difference is very small and the fact that positive and negative signs appear is due to noise in the servos. Theoretically, a shorter range must always be measured on the high frequency and the ionospheric correction (computed later) is always negative.

HR, MIN, SEC—Clock time of instant at which data are recorded.

TOTAL RANGE—Total range word in metres is given.

1st *DIFF*—The first difference to the nearest metre between successive ranges is shown.

2. *Pack and Edit Processes*

The data from four stations for a pass are time synchronized and ranges are edited and re-recorded in much more compact form on fresh tape. This is then suitable for future computations.

The packing process matches data from the stations together on a common time base accepting discrepancies of up to ten milliseconds. The stations allow for propagation delay in rating their time code generators. During a track, differences in range to the satellite account for small discrepancies between the recorded times at each station, but these are unlikely to exceed six milliseconds.

Editing is designed to remove intermittent ambiguities from the data. However, it is possible that the resulting ranges may all contain the same ambiguity. Editing is achieved by comparing the measured first difference against a predicted first difference, computed from a span of previously edited data using linear extrapolation. The acceptable difference between the predicted and measured values must allow for random noise.

Three or four standard deviations of the random noise variations may be programmed as the limit for determining whether a bad sample should be rejected. When a discrepancy occurs multiples of 256 metres are added to, or subtracted from, the new range to bring it within the required tolerance. The following is an explanation of the columns shown on the editing sheet (Fig. 2):

Time—The master station time is recorded. The ranges at each

APPENDIX E. FIG. 2. Example of an editing sheet.

station are read on a command pulse initiated by the master and retransmitted to all stations by the satellite. As the pulse leaves the satellite at one time the ranges recorded at the stations all refer to that instant even though the pulse transit time varies with the range to the satellite.

Edited Ranges (station names listed below)—The edited ranges given in metres are listed for the four stations.

Edited Corrections—Edited corrections to each range are shown. The 524 288 m value, which is shown as a correction to the range of the first station, would have been determined from the prediction data. This is a necessary check because the extended range channel is of such poor accuracy. If the extended range were to be ignored altogether, adding multiples of 524 288 m would be similar to adding multiples of ten miles to a Tellurometer measurement.

Edited 1st Diff—First differences to the edited ranges to the nearest metre are given.

1C–D1—These columns record the difference between the ranges measured on the high and low frequency carriers. This value should be negative indicating that the high frequency measured a shorter range than the low frequency. It should be noted that occasionally large positive values are printed out, for example, in the fourth station (fifth line) 239 is shown. The low frequency channel only records values of up to 256 m. Therefore, 256 m should be subtracted from 239 making −17.

3. *Solution*

The example headed 'Overdetermined Solution for Pacific Quad, shows a solution for Station 6 (Fig. 3).

Spheroid Constants are listed at the head of the page, *AO* and *BO*, semi-major and semi-minor axes.

Input Site Co-ordinates. All stations are listed with co-ordinates in latitude, longitude (Bessel spheroid, Tokyo Datum) and height above the spheroid. *X*, *Y* and *Z* Cartesian co-ordinates are also tabulated for each station. The unknown stations are given approximate co-ordinates.

Blocks 1–6. Six blocks follow, each representing a different orbit. They are all laid out in the same way and deal with the corrections that must be applied to the measured range before it can be used in the computations. Each block lists a number of simultaneous ranges from all four stations.

Block 1. The top line gives the orbit number and the name of the unknown site. On the right, information about the ionospheric correction is given.

IONO 1. A figure of 1 after IONO indicates that the measured ionospheric correction (IM) has been applied to the measured range (RM).

F MAX, CK and HT are parameters that have been used in computing an analytic ionospheric correction. *F MAX* is the maximum electron density of the *F*2 layer of the ionosphere. If the value given, 0·150, is multiplied by 10^{12} it represents electrons to the cubic metre. *CK* is the density gradient and *HT* the approximate height of the satellite for the computation. This correction is now seldom used.

Immediately below, information relating to each of the four stations is tabulated. Under the station name the total calibration corrections (Appendix C, paragraph 1) are given. The first value is for the high and the second for the low frequency carrier signals. Ambiguities that have been removed are shown beneath the high frequency calibration correction.

R NO—Each range used in the solution is numbered.

```
                              OVERDETERMINED SOLUTION FOR PACIFIC QUAD                      APPENDIX E
  7   2   2   3   1                                                                         DI-GRAN 3

SPHEROID CONSTANTS AO 6377397.160   BO   6356078.963                          FPEQ   42.00

INPUT SITE COORDINATES
                              LATI         LONG           HGT        X EQ        Y EQ        Z EQ
  0   1    (Station         (Geographical             (Cartesian
  0   2     Names)          Co-ordinates)             Co-ordinates)
  0   3
  0   4
  0   5
  0   6
  1   7

1 BLOCK   1   ORBIT  5703   UNKNOWN SITE (Sta 6)            IONO  1 FMAX  .150  CK  .050  HT  92000.00
                    Sta 1            Sta 4                          Sta 5                     Sta 6
CAL          -31.00   -33.00              -21.25   -18.59              -26.75   -23.00              -22.25   -18.75
AMB            .00                         -4096.00                      .00                          .00

R NO   RC(1)   RM(1)  EL  TR IM  IA   RC(4)   RM(4)  EL  TR  IM  IA   RC(5)   RM(5)  EL  TR IM  IA   RC(6)   RM(6)  EL  TR  IM  IA
  1  1001729 1001762  64   2 -1   3  1347096 1351226  38   4   7   5  1641695 1641731  28   5   3   6  2061089 2061134  18   7  15   9
  2   999792  999826  65   2  0   3  1256332 1260460  43   3   6   4  1641528 1641564  28   5   3   6  1975634 1975677  20   7  13   8
  3  1015923 1015959  63   2  1   3  1173964 1178091  48   3   6   4  1652332 1652369  27   5   4   6  1895725 1895768  21   6  14   8
  4  1049283 1049319  59   3  1   3  1101899 1106024  53   3   3   4  1673884 1673925  27   5   8   6  1822175 1822175  23   6  12   7
  5  1098292 1098328  54   3  1   4  1042321 1046447  59   3   5   3  1705775 1705816  26   5   8   7  1755669 1755710  25   6  12   7
  6  1160951 1160988  49   3  2   4   997499 1001624  65   2   4   3  1747417 1747458  25   6   8   7  1697200 1697241  26   5  13   7
  7  1235165 1235203  44   3  3   4   969500  973624  70   2   4   3  1798113 1798157  24   6  11   7  1647597 1647639  28   5  13   6
BLOCKS 2-6 OMITTED

                              OVERDETERMINED SOLUTION FOR PACIFIC QUAD
                         SATELLITE SOLUTION AND INTERNAL COMPARISON
BLK   ORB   NO.      X EQ         Y EQ        Z EQ         LATI         LONG        HGT      DR(1)   DR(4)   DR(5)   DR(6)
  1  5703    1   -3629588.    497436P.    3929793.   32.6972650  126.1186568  933688.1    1.3    -1.0    -1.2    1.0
  1  5703    2   -3712904.   5001899.    3815318.   31.6356856  126.5864545  933318.0     .6     -.5     -.5     .4
  1  5703    3   -3794633.   5027837.    3699287.   30.5715874  127.0438548  932955.5    -9.6    8.6     8.5    -7.9
  1  5703    4   -3874763.   5051566.    3581733.   29.5049774  127.4897219  932600.0    -4.8    -4.7    -4.3    4.3
  1  5703    5   -3953240.   5073676.    3462707.   28.4360766  127.9245424  932245.3    6.1    -6.7    -5.7    5.9
  1  5703    6   -4030044.   5093950.    3342268.   27.3650682  128.3490483  931896.6    3.8    -4.6    -3.7    4.0
  1  5703    7   -4105159.   5112371.    3220461.   26.2920338  128.7639539  931556.1    3.3    -4.3    -3.3    3.8
BLOCKS 2-6 OMITTED

                                                          SOLUTION OUTPUT
  1        X EQ   Y EQ   Z EQ   DX   DY   DZ      LATI      LONG      HGT        D LATI         D LONG        DH
  2    (Station        (Cartesian         (Corrections           (Geographical               (Corrections
  3     Names)          Co-ordinates)       to                    Co-ordinates)                 to
  4                                         Co-ordinates)                                       Co-ordinates)
  5
  6
  7
EDITED RANGE RESIDUALS
   1    -8.92 0     2    -8.72 0     3   -16.17 0     4    -3.04 0     5     -.28 0     6     -.94 0
   7      .16 0     8   -12.29 (     9    -1.68 0    10    -2.81 0    11    -6.86 0    12    6.87 0
  13     5.24 0    14     3.32      15    10.45 0    16     4.13 0    17     8.18 0    18    -3.28 0
  19      .17 0    20     -.91 0    21    -3.98 0    22    -7.99 0    23    -2.28 0    24     5.14 0
  25   -13.61 0    26     9.09 0    27     8.70 0    28    10.35 0    29     5.71 0    30    6.90 0
  31     8.76 0    32    15.37 0    33     -.63 0    34    -1.77 0    35     3.30 0    36   -2.26 0
  37   -10.10 0    38   -12.55 0    39   -44.00 1    40    35.63 1    41   -37.77 1    42  -151.73 1

NO. PTS  42   NO. BAD PTS   4   SICR       7.64   RESIDUAL AVG       -.24
  6 (Sta Name)       .77      .29    .38    1.00    -.41   -.36
                                           -.41    1.00   -.36
                                           -.36    -.36   1.00
```

APPENDIX E. FIG. 3. Over-determined solution for Pacific Quad.

RC (1) *and RM* (1)—Corrected and measured ranges. The print-out has been rounded off to the nearest metre. The measured range after being corrected for calibration, troposphere and ionosphere corresponds to the corrected range.

EL—Elevation angle in degrees.

TR—Tropospheric correction in metres (Appendix C, paragraph 2).

IM—Measured ionospheric correction in metres (Appendix C, paragraph 3).

IA—Analytic ionospheric correction in metres computed using the parameters listed in the top line of the block.

4. Satellite Solution and Internal Comparison

Ambiguities have complicated the work involved in reducing SECOR data. This block is designed to reveal any remaining in a pass.

BLK, ORB and NO—Block, orbit and satellite position numbers.

X EQ, Y EQ and Z EQ—are the satellite positions in Earth-centred origin Cartesian co-ordinates, to the nearest metre.

LATI, LONG and HGT—gives the same satellite positions in values of latitude, longitude and height with reference to Bessel spheroid, Tokyo Datum. These satellite positions are those fixed from the three known stations. Alternative satellite positions are also computed in three other combinations using the best value available for the unknown station and treating it as known, but their co-ordinates are not printed out.

DR (1), *DR* (4), *DR* (5) *and DR* (6)—The four satellite positions were established using different combinations of the four stations. For each combination a computed range is obtained to the satellite from the station not used in solving for the satellite position. This computed value is subtracted from the observed range. In this way stations (1), (4), (5) and (6) are each treated as the known in turn and the values tabulated in the four columns are residuals in metres. Provided preliminary solutions have fixed the original unknown station reasonably accurately, an ambiguity, of for example 4096 m, in a block of data shows up as unusually large residuals in all four combinations. In one set the residuals would be approximately 4096 m and reasonably consistent. This indicates the ambiguity is in those measured ranges and not in the three sets of ranges used to fix the satellite positions in that combination. The other combinations can be expected to show rather smaller residuals, perhaps 2000–4000 m, which are likely to increase or decrease progressively along the pass. In this way, provided only one set of ranges is ambiguous and the position of the unknown station is approximately known, the data can be quickly corrected and used in the solution.

5. Solution Output

This starts with a repeat of the input co-ordinates for the stations. Here the Cartesian co-ordinates are given first and the geographical coordinates second. Also, each set of co-ordinates is followed by a column which indicates a shift required for the unknown station (*DX, DY, DZ*, and *D LATI, D LONG, DH*).

The Edited Range residuals are listed by number. Each number refers back to the satellite position and range block lists. Ranges with residuals greater than two standard deviations are rejected and this is shown by a figure 1 after the residual. A zero indicates that the range has been accepted.

The line immediately below the edited range gives the number of good and bad points (in this case 42 and 4 respectively), the standard deviation for a single range (SIGR 7·64 m), and the residual average (−0·24 m).

The last line gives the station number and name. The values 0·77, 0·29 and 0·38 give the standard deviation of the unknown station in X, Y and Z respectively as determined from the geometry of the solution. Finally, a 3×3 correlation matrix is given. This shows the geometric correlation of the solution between the X, Y and Z axes. The example is quite good as there is relatively little indication of correlation.

DISCUSSION

Discussion on the two preceding papers by
F. L. Culley and N. J. D. Prescott

R. C. A. Edge: Both papers are extremely interesting and encouraging and I strongly support the proposal for measuring range and direction simultaneously. Since the correction due to the ionosphere is the biggest source of error, and since this error decreases with the square of the frequency, would it not be advisable to increase the frequency?

N. J. D. Prescott: Theoretically this is so, but there are practical difficulties. Increasing the frequency makes the waves more sensitive to moisture in the atmosphere and thus the all weather capability would be reduced. AMS consider that there might be some advantage in increasing the frequency from about 500 Mc/s to say 1000 or 1500 Mc/s but cannot yet commit themselves.

L. Asplund: I support General Edge's plea for combining range and direction measurements. Is there any intention of making simultaneous observations with the Coast and Geodetic BC4 cameras, or at least to operate from the same stations?

F. L. Culley: While the AMS is involved in the BC4 operations, GIMRADA* will conduct experiments using the Baker–Nunn camera of the Smithsonian Institute. These observations will be on the satellite launched in August 1965, which has a pronounced elliptical orbit, the height varying between 611 and 1309 nautical miles.

G. C. Weiffenbach: Simultaneous SECOR and optical measurements were attempted with the ANNA satellite, but failure of the SECOR prevented measurements. It is hoped to try again with PAGEOS.

A. R. Robbins: The estimates of accuracy have been made from internal evidence only; is there any evidence from external sources? I should also like to know what accuracy you would expect to obtain using orbital measurements, since I think this method would give results which were poor compared with your present figures.

F. L. Culley: A comparison of SECOR measurements with U.S. Coast & Geodetic points was made for a quadrilateral whose sides were about 800 miles long. Holding three corners fixed, the discrepancy at the fourth corner was about 6 m. I agree that the orbital method is difficult and in the only test so far made the error was 55 m. However one station

* Geodesy Intelligence and Mapping Research and Development Agency, U.S.A.

was not operating too well which may at least partly explain such a poor result.

J. Kelsey: According to estimates given in the paper, the errors in x, y, and z, are all consistent, although previously published work has suggested that the error in z is substantially worse. What is the explanation for this?

N. J. D. Prescott: The error depends on the geometry. If you go out too far there will be a resultant large error in z. There will also be a large error in the horizontal plane in the direction in which the ray is elongated.

J. Kelsey: In the test just mentioned by Mr Culley, there was a larger error in height than the 6 m quoted for planimetry.

N. J. D. Prescott: This was compared with the geoidal height and it is more likely that this figure was wrong than that the SECOR heights were wrong.

J. A. Weightman: Is there any independent source of information to correlate the 50-millisecond data bursts at the various ground stations, or does one rely solely on the times recorded by each ground-station clock in the data print-out to avoid a mis-match of data?

N. J. D. Prescott: Yes, one does rely on the data print-out for time correlation, but the clocks are checked independently.

K. Bretterbauer: One source of error is uncertainty in the velocity of propagation. If four known stations are used instead of three, this error would be eliminated.

F. L. Culley: This would give redundancy, which is very desirable, since the redundancy obtained from the multiplicity of observations from three stations is of no help if one of the stations themselves is in error. However with the present system, the maximum number of stations that can be operated simultaneously is only four.

Tropospheric and Ionospheric Propagation Effects on Satellite Radio-Doppler Geodesy*

G. C. WEIFFENBACH

The Johns Hopkins University
Applied Physics Laboratory

Radio-frequency transmissions between orbiting satellites and ground stations have proved to be a very effective means of obtaining geodetic information. A basic limitation on the accuracy of such data is introduced by the influence of the atmosphere on the propagation velocity of radio signals. Indeed, these atmospheric effects can result in errors which are quite serious in geodetic applications. This paper discusses the magnitude of these errors, and methods which are being employed to correct or eliminate them in the specific case of radio Doppler measurements, where the residual errors from this source are estimated to be about 5 m at this time.

Introduction

The influence of the atmosphere on radio-wave propagation between satellites and stations on the surface can result in serious errors in geodetic satellite applications. This paper discusses the magnitudes of these errors and the corrective procedures used for the specific case of satellite radio-Doppler observations for geodetic studies.

In order to derive either satellite motion or station position from Doppler measurements one must obtain first the geometric range rate, i.e. the time rate of change of the slant range between satellite and observer. In a homogeneous medium the relation between observed Doppler and range rate can be simply expressed as

$$\Delta f(t) = -\frac{\dot{S}(t)}{u}f \tag{1}$$

* This work was supported by the United States Department of the Navy, Bureau of Naval Weapons, under contract No. W—62—0604—C.

where

$$\Delta f(t) = \text{Doppler shift}$$
$$\dot{S}(t) = \text{geometric range rate}$$
$$u = \text{phase velocity of radio signal}$$
$$f = \text{satellite transmitter frequency},$$

so that if u is known $\dot{S}(t)$ is immediately available. However, in the presence of the atmosphere, the phase velocity will vary along the propagation path. It is then necessary to rewrite equation (1) in the form

$$\Delta f(t) = -\frac{\mathrm{d}}{\mathrm{d}t} \int_{P(t)} \frac{\mathrm{d}s}{u(\xi,t)} f \tag{2}$$

where

$u(\xi,t) = $ phase velocity at point ξ in space and at time t

$P(t) = $ propagation path at time t.

It is not reasonable to attempt to find accurate values of u at each point (ξ,t) of interest, so some alternative form of equation (2) must be found.

As a first step, we note that the integral $\int \mathrm{d}s/u$ can be accurately approximated by introducing the optical path length

$$\Lambda(t) = \int_{P(t)} n(\xi,t)\,\mathrm{d}s$$

so that

$$\Delta f(t) = -\frac{f}{c}\frac{\mathrm{d}}{\mathrm{d}t} \int_{P(t)} n(\xi,t)\,\mathrm{d}s \tag{3}$$

where

$n(\xi,t) = $ the index of refraction at point (ξ,t)

$$n(\xi,t) = \frac{c}{u(\xi,t)}$$

$\Lambda = $ optical path length obtained by application of Fermat's principle.

Secondly we note that for our present purpose, the atmosphere can be considered to consist of two distinct parts, the troposphere and the ionosphere, each having basically different effects on radio-wave propagation. The troposphere has a refractive index greater than unity which depends primarily on air pressure, temperature and humidity and is independent of frequency. The ionosphere has a refractive index less than unity which depends (for the frequency regime of interest here) on the free electron density, the ambient magnetic field and the frequency of the radio signal. The two regions can be treated independently.

Tropospheric Effects

The tropospheric contribution to the Doppler shift can be written

$$\Delta f_{tro}(t) = -\frac{f}{c}\frac{d}{dt}[\Delta S_{tro}(t)] \qquad (4)$$

$\Delta S_{tro}(t)$ is the difference between the portion of the phase path in the troposphere and a corresponding straight line path at the time t.

An initial expression for Δf_{tro} has been derived[1] on the basis of satellite–station geometry and the following simplifying assumptions about the troposphere:

(1) The refractivity N of air [where $N = 10^6(n-1) \approx 300$ at the Earth's surface] is a continuous function of height above the Earth but is independent of horizontal position and of time, within the region and the time interval of a satellite pass. This assumption is violated near a weather front but is otherwise fairly accurate.

(2) The N profile (height variation of N) can be approximated by a theoretical (quadratic) expression, decreasing from its value at the tracking station to zero at a specified height above the geoid, the equivalent height of the troposphere.

(3) Curvature of the signal path is small enough to be negligible, except close to the horizon where data are not used for geodesy.

The resulting expression for Δf_{tro}[1] was incorporated into the APL orbit computing programme, starting in January 1964. A value of Δf_{tro} is computed as a correction for the observed Doppler shift at each data point of each pass, using the geometry of a preliminary orbit and the theoretical refractivity profile described above. Initially a local seasonal mean value of the surface refractivity[2] was used at each station as a starting point for the refractivity profile at the time of a pass; diurnal and weather variations were thus neglected. The equivalent height h_0 of the troposphere was assumed to be 23 km at all stations.

The theoretical tropospheric contribution has the same sign as the Doppler shift itself throughout a satellite pass, but unlike the Doppler shift, its magnitude increases sharply at both ends of the pass, near the horizon. If no tropospheric correction is used, the Doppler residuals for any pass (observed minus theoretical Doppler shift) consistently show the sharp increase in magnitude near the horizon which theoretically characterizes the tropospheric effect.

This is illustrated by Figs. 1 and 2, which show the Doppler residuals for two passes, one at medium elevation and one at a very low elevation.

12

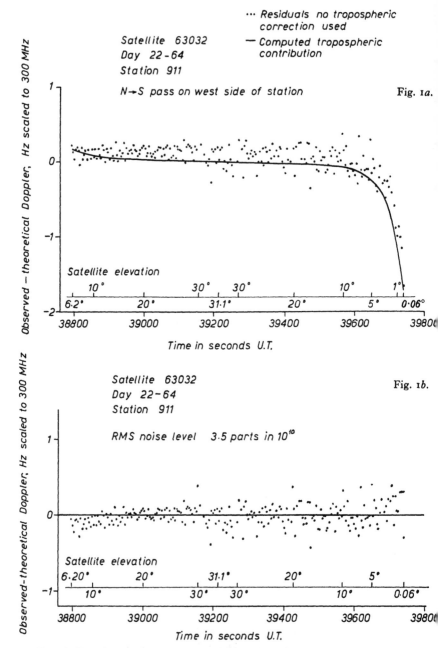

FIG. 1. Doppler residuals for a pass with 31° maximum elevation (observed—theoretical). (a) Without and (b) with the use of tropospheric refraction correction.

Fig. 1*a* and the upper graph in Fig. 2 show the residuals when no tropospheric correction was made and also the theoretical tropospheric correction calculated for that pass. The residuals follow the shape of the theoretical tropospheric error curve (but displaced, since the frequency centring was affected by the large uncorrected refraction). Fig. 1*b* and the lower curve in Fig. 2 show the Doppler residuals which remained when the data were troposphere-corrected. The remaining errors are essentially random, indicating that the systematic errors in the two upper plots were tropospheric in origin.

Two further improvements have been made in computing the tropospheric effect.

First, the local refractivity of air at the tracking station is computed from the equation[3]

$$N = \frac{77 \cdot 6}{T} \left[P + \frac{4810 e_s \times (\text{RH})}{T} \right]$$

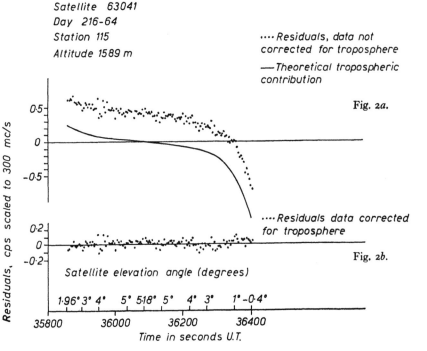

Satellite 63041
Day 216-64
Station 115
Altitude 1589 m

···· Residuals, data not corrected for troposphere

—— Theoretical tropospheric contribution

Fig. 2*a*.

···· Residuals data corrected for troposphere

Fig. 2*b*.

FIG. 2. Doppler residuals for a pass with 5°·18 maximum elevation. (*a*) Without and (*b*) with tropospheric refraction correction.

using weather data sent by the stations along with the Doppler data. In the above equation, T is temperature in degrees Kelvin, P is atmospheric pressure and e_s the saturation pressure of water vapour at the temperature T, both in millibars; RH is the relative humidity. The use of weather data provides the correct starting point for the refractivity profile and should result in a more accurate correction. This is especially important in warm, humid weather. The diurnal variation of surface refractivity in the Washington, D.C., area is likely to be only about 10 N units (3 per cent) peak-to-peak in winter (winter weather effects also being small). In summer, diurnal variations are two or three times as large as this, while a cold front arriving in summer can drop the surface refractivity by 50 N units (15 per cent) within a few hours.

The second improvement stems from a recent study of refractivity profiles obtained from observed upper atmosphere data.[4] It was found that the increase in path length produced by the theoretical (quadratic) refractivity profile is in general not quite equal to the increase produced by an observed profile starting at the same surface refractivity. The ratio between the theoretical and observed effects is, however, a linear function of the surface refractivity for all the 34 profiles which were examined, regardless of geographic location, station altitude, or season. This relation can provide a simple correction factor for improving the quadratic profile results, and is now being introduced into the computation of the tropospheric correction to the Doppler shift.

The tropospheric effect, if uncorrected, produces an 'along-track' error in a geodetic position derived from a Doppler measurement position only if data are not symmetrical about the point of closest approach. It always produces an error in the apparent range from station to satellite at closest approach. Since tropospheric refraction steepens the slope of the observed Doppler shift vs. time curve, the uncorrected troposphere always makes the station appear closer to the orbit than it actually is. The amount of range error for a given state of the troposphere is a function of the maximum satellite elevation angle during the pass, and also depends on how much data near the horizon are included in the computation.

Fig. 3 shows this theoretical range error as a function of pass elevation, for data cut-off angles of $5°$, $10°$ and $15°$ respectively, computed from the corrected quadratic profile as described above, and using a surface refractivity of 320, which is approximately an average value. The tropospheric range error for a $45°$ pass is 12 m if all data are deleted below $15°$ elevation at both ends of the pass; but is more than twice as large (26 m) if data are retained down to $5°$ elevation.

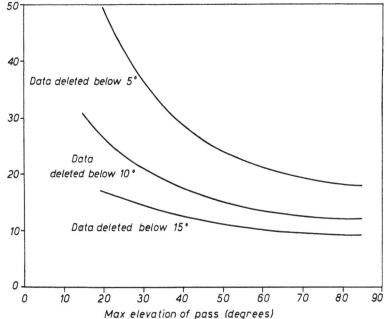

FIG. 3. Range error from uncorrected tropospheric refraction for different data cut-off angles. (500-nautical-mile circular orbit.) Quadratic refractivity profile, corrected:

$$\frac{\int N_{\text{quad}}ds}{a+mN_T}$$

where $N_T = 320$, $h_0 = 23$ km, $a = 31/60$, $m = 1/600$.

The errors are considerably larger than this for lower elevation passes.

Values of surface refractivity which are encountered seldom differ by more than 20 or 25 per cent from the nominal value of 320 used here. A change of 20 per cent in the surface refractivity produces a theoretical change of not quite 10 per cent in the resulting range error. Thus most of the surface conditions which are actually encountered would result in theoretical errors within ± 10 per cent of those in Fig. 3. Fig. 4 shows the effect of temperature and humidity on refractivity, at an atmospheric pressure of 1000 millibars.

It is estimated that the latest form of the tropospheric correction can remove at least 85 or 90 per cent of the tropospheric effect, leaving range errors not more than 10 or 15 per cent of those shown in Fig. 3. If data are deleted below 10° satellite elevation, the maximum error is

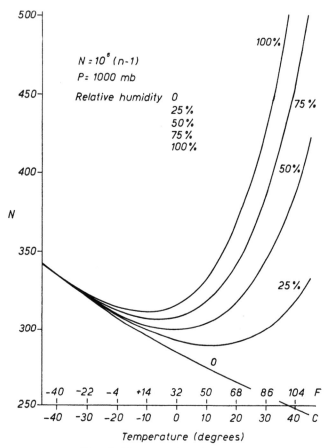

FIG. 4. Tropospheric refractivity.

thus less than 5 m, and the RMS error is probably around 3 m when taken over all passes.

Residual uncorrected errors have several possible sources. The curvature of the signal path has been neglected, but this is extremely small at angles above 5°. Probably more important is the fact that the theoretical and the observed mean profiles are not quite the same shape. As the elevation of the signal path is lowered from 90° toward the horizon, the lower layers of each profile are stretched more than the upper layers, and this differential stretching must have slightly different effects on the theoretical and the actual profiles.

No account has been taken of the difference of an actual instantaneous refractivity profile from the mean profile at that location (e.g. in the presence of a weather front), or of the changes which may occur during a pass. These factors will be sources of noise in the data, but should not have a biasing effect when a large number of satellite passes is being considered.

Ionospheric Effects

The index of refraction for the ionosphere can be approximated by the Appleton–Hartree formula,

$$n_{o,e}(\xi,t) = \left[1 - \frac{f_N^2(\xi,t)}{f^2} \cdot \frac{1}{\alpha_{+,-}} \right]^{1/2} \tag{5}$$

where

$f_N(\xi,t)$ = electron plasma resonance frequency at position ξ
and time t

$$= \left[\frac{N(\xi,t)e^2}{\pi m} \right]^{1/2} \text{ cgs units}$$

$N(\xi,t)$ = electron density at position ξ and time t

e = electron charge

m = electron mass

$$\alpha_{+,-} = 1 - \frac{(f_T/f)^2}{2(1-f_N^2/f^2)} \pm \left[\frac{f_T^4/f^4}{4(1-f_N^2/f^2)} + \frac{f_L^2}{f^2} \right]^{1/2}$$

$f_L = f_B(\xi) \cos \theta$

$f_T = f_B(\xi) \sin \theta$

$f_B(\xi)$ = electron gyro frequency = Be/cm

$B(\xi)$ = Earth's magnetic field at ξ

θ = angle between propagation vector and
magnetic field direction at ξ

and where the subscripts o, e indicate the ordinary and extraordinary indices for the two senses of circular polarization of the propagated wave.

Substitution of the above expression into equation (3) suggests a further simplification into a power series expansion in two inverse powers of the radio signal frequency,*

* W. H. Guier[6] has derived this expansion through terms in a_3 from a rigorous treatment of Maxwell's equations with suitable boundary conditions.

$$\Delta f(t) = \frac{\dot{S}(t)}{c}f + \frac{a_1}{f} + \frac{a_2^{(\pm)}}{f^2} + \frac{a_3}{f^3} + \cdots \tag{6}$$

where a_1 = first order term, proportional to the time derivative of $N(\xi,t)$ integrated along geometric slant range.

$a_2^{(\pm)}$ = second order (Faraday rotation) term which depends on polarization, and on time derivative of $N(\xi,t)$ and θ.

a_3 = third order term which depends on various powers of $N(\xi,t)$ and its spatial gradients.

Equation (6) clearly shows an important characteristic of ionospheric effects, viz., the first term on the right side of the equation which contains the geometric range rate $\dot{S}(t)$ needed for geodetic analyses is directly proportional to the transmitter frequency f, while all other terms are inversely proportional to f. In principle, then, one need only choose a high enough frequency to reduce the undesired terms to a small enough value to be neglected. Unfortunately, practical considerations dictate that one must use lower frequencies to get optimum efficiency with present equipment. Thus it is important to know the values of a_1 which will actually be encountered in geodetic operations.

A considerable body of experimental data is available for a_1 and a_2. The former corresponds to the 'differential Doppler' and the latter is related to Faraday rotation. Both kinds of measurements are widely used in studies of the ionosphere *per se*. However, our knowledge of a_3 is rather fragmentary.

The data indicate that the only terms in equation (6) that generate significant geodetic errors are a_1/f and a_3/f^3, the former being the dominant term. The table below lists ranges of values which have been observed during the recent sunspot minimum, the average range bracketing about $\frac{2}{3}$ of the observations, and the maximum being the largest actually seen. The values tabulated for the sunspot maximum are estimated for the next peak in the cycle (*ca.* 1967–70), and are intended to correspond to a peak similar to that experienced during the 1958 maximum. The predicted values of a_3 are particularly suspect.

	$a_1{}^*$ in $(c/s)^2$		$a_3{}^*$ in $(c/s)^4$	
	avg. range	max.	avg. range	max.
Sunspot minimum	0·1 to 5× 10⁸	10⁹	0–2 × 10²²	2× 10²³
Sunspot maximum	0·1 to 15× 10⁸	3× 10⁹	1–20× 10²²	20× 10²³

* Maximum values of these quantities within a satellite pass.

The following general comments are applicable to the table:

At night only the lower values of a_1 are observed.

The larger values of a_1 are seen only during the day (primarily afternoon) and occur only infrequently at latitudes more than $40°$ away from the geomagnetic equator.

Although the evidence is incomplete, it appears that significant values of a_3 occur only in conjunction with large a_1 and at locations within $40°$ of the geomagnetic equator.

Using the values of a_1 tabulated above, one can calculate the lowest carrier frequency that must be used to ensure that the error from a_1 is less than some particular level. An appropriate level is 5 m maximum error in any one pass, which will reduce to about 3 m when averaged over a number of passes. The contribution from a_3 will be negligible.

a_1 in $(c/s)^2$	Lowest frequency for 5-m position error*
5×10^8	5 Gc/s
10×10^8	7 Gc/s
15×10^8	9 Gc/s
30×10^8	13 Gc/s

* The particular error encountered in a given pass will vary somewhat with the pass geometry and the detailed variation of $a_1(t)$ within the pass.

It would not be necessary to design for the extreme values of a_1, since they occur only very infrequently. However, a single frequency geodetic system intended to meet the stated criterion should not operate below the region of 5 to 7 Gc/s.

An alternative approach is to use two harmonically related frequencies. Simultaneous Doppler measurements on two frequencies permit one to obtain two equations (6) which can be solved simultaneously to eliminate the first order ionosphere term in a_1, so that it is only necessary to cope with a_3. If a 5-m position error per pass is allowed as before, it is possible to compute the lowest frequency for the various a_3 values tabulated above. A convenient frequency ratio for the pair is 2 : 1, this combination resulting in the following:

a_3 in $(c/s)^4$	Lowest frequency pair for 5-m position error
2×10^{22}	140 Mc/s–280 Mc/s
20×10^{22}	250 Mc/s–500 Mc/s
200×10^{22}	450 Mc/s–900 Mc/s

In addition to permitting a substantial reduction in frequency, the two-frequency technique provides a measurement of $a_1(t)$ which is a direct indication of any unusual ionospheric activity that may be experienced.

All of the geodetic studies conducted at the Applied Physics Laboratory have used a two-frequency radio Doppler system. In this system, the satellites radiate two or more continuous frequencies all derived from the same satellite-borne ultra stable oscillator. The first-order (a_1) correction is performed electronically in each ground station in real time. When more than two coherent frequencies are used it is possible to obtain values for $\dot{S}(t)$, $a_1(t)$ and $a_3(t)$ by simultaneous solution of the three simultaneous equations obtained (from equation 6). This in fact is the source of the a_3 tabulated above.

Most of our current geodetic data are being obtained from the use of a 162 Mc/s–324 Mc/s frequency pair. We are also exploring a 324 Mc/s–972 Mc/s combination (to be launched soon on board the NASA GEOS-A satellite*) for possible use during the coming increase in solar activity. The geodetic position errors expected for these combinations for different values of a_3 are listed below:

a_3	Position error in metres	
	162/324 Mc/s	324/972 Mc/s
2×10^{22}	3	0·07
20×10^{22}	30	0·7
200×10^{22}	300	7

Summary

The Doppler shift of radio transmissions from a geodetic satellite as seen from a fixed ground station is assumed in this paper to be adequately approximated by the equation

$$\Delta f(t) = -\left[\frac{\dot{S}(t)}{c}\right]f - \frac{f}{c}\frac{d}{dt}\left[\Delta S_{tro}(t)\right] + \frac{a_1(t)}{f} + \frac{a_3(t)}{f^3} \qquad (7)$$

The term on the left side of the equation is the measured Doppler shift. The first term on the right is the geometric range rate (scaled by f/c) required for computation of either the satellite trajectory or geodetic position. The second term on the right is the error resulting from tropospheric effects, and the remaining terms are the first- and third-order ionospheric terms.

* Launched on 6 November 1965 as *Explorer 29* [Ed.].

On the basis of theoretical analyses and a large body of experimental data, it is known that the errors arising from tropospheric and ionospheric influences on propagation (phase) velocity are grossly in excess of geodetic requirements unless corrected or removed.

Tropospheric errors can be reduced by deleting all data for elevation angles below 10° and by calculating a correction deduced from local weather conditions at each ground station at the time of the Doppler observations. As a further precaution, data are not used if a serious weather disturbance (e.g., cold front) is known to have occurred when the observations were made. The residual tropospheric errors are estimated to be about 3 m averaged over a number of passes.

Ionospheric errors can be reduced to an acceptable level by use of a frequency in the vicinity of 5 Gc/s or higher. However, we have chosen a system based on two coherent frequencies, 162 Mc/s and 324 Mc/s, since these frequencies have proved easier to instrument in a space-craft than the single higher frequency. Simultaneous observation of the frequency pair permits elimination of the first-order ionosphere term in $a_1(t)$. The residual error from the third-order term is estimated not to exceed 3 m when averaged over a number of passes. As a further precaution, $a_1(t)$ can be observed from the dual frequency measurements and used to delete data of those passes for which the ionosphere is seen to be disturbed.

The uncorrected third-order ionosphere error is expected to increase substantially when solar sunspot activity rises *ca.* 1967–70. A 324 Mc/s–972 Mc/s frequency pair is to be orbited on the GEOS-A satellite* for possible use at that time.

Thus the total effect of atmospheric propagation errors in our present geodetic data is estimated to be $(3^2+3^2)^{1/2} \sim 5$ metres.

References

[1] Hopfield, H. S., 1963. The effect of tropospheric refraction on the Doppler shift of a satellite signal. *J. Geophys. Res.*, **68**, 5157–68.

[2] Bean, B. R. & Horn, J. D., 1958. On the climatology of the surface values of radio refractivity of the Earth's atmosphere. National Bureau of Standards, Report 5559 (Boulder Laboratories).

[3] Smith, E. K. & Weintraub, S., 1953. The constants in the equation for atmospheric refractive index at radio frequencies, *Proc. Inst. Radio Engrs, N.Y.*, **41**, 1035–7.

[4] Hopfield, H. S., 1965. Improvement of the tropospheric correction for Doppler data, based on a study of upper air meteorological data. Applied Physics Lab. Report TG-646.

[5] Stratton, J. A., 1941. *Electromagnetic Theory*. (McGraw-Hill Book Company, New York.)

[6] Guier, W. H., 1963. Ionospheric contributions to the Doppler shift at VHF from near-Earth satellites. Applied Physics Lab. Report CM-1040.

* Launched on 6 November 1965 as *Explorer 29* [Ed.].

[7] Pisacane, V. L. & Stuart, J. D., 1965. The effect of collision on the group refractive index of the ionosphere. Applied Physics Lab. Report TG-649.
[8] Defense Research Laboratory Ionospheric Studies, DRL Report 487, J. F. Willman, August 1962; DRL Report 491, J. F. Willman & J. F. Doyle, February 1963.

DISCUSSION

N. J. D. Prescott: The explanation for the difference between the error due to the tropospheric correction given by Dr Weiffenbach and my own estimate is probably that he is concerned with position, while I am concerned with range. Also I have quoted the probable error (0·67 standard deviation).

G. C. Weiffenbach: The figure I have quoted is about the 2σ level.

S. K. Sharma: I am surprised that the error due to uncertainty in the higher spherical harmonics is as great as 75 m. Is this due to the eighth and higher-order harmonics only?

G. C. Weiffenbach: Each orbit allows us to solve for two harmonics. We now have four orbits so we should know the eighth harmonic coefficient, but it is not known how much effect the higher orders have.

F. J. Hewitt: I do not see why the velocity of propagation in the ionosphere affects the Doppler shift at the satellite. Surely it is the velocity of propagation at the satellite which determines the Doppler shift?

G. C. Weiffenbach: This is like all other problems in electromagnetic wave propagation. The electrons in the ionosphere are accelerated, not only in the path, but around it also because of dispersion. These electrons then re-radiate energy and so combine with the main wave, producing a change in the phase velocity. This occurs all along the path and not only at the satellite.

Télémétrie
des Satellites Artificiels
par Laser

R. BIVAS

Saint-Michel l'Observatoire

This paper discusses the principle of obtaining laser echoes from satellites. So far some 150 measures have been made and up to twenty have been on one orbit. For observations over 5 minutes on any one day the m.s.e. of the actual measurement has been of the order of 2 m. It is expected that the system can be used to determine the distance between two points 3000 km apart to 5 m.

De nombreuses expériences de télémétrie optique mettant à profit les propriétés particulières de la lumière 'laser' ont été réalisées ou sont en projet. Le Service d'Aéronomie du C.N.R.S., pour sa part, a mis en place à l'Observatoire de Haute-Provence (Saint-Michel l'Observatoire—Basses-Alpes) le matériel nécessaire pour effectuer la télémétrie du satellite américain S-66. Après de nombreux essais sur cibles terrestres, un premier écho sur satellite a été obtenu le 24 janvier 1965, montrant que l'expérience était possible et que la portée de notre télémètre actuel est de l'ordre de 1500 km.

1. *Principe et Matériel Expérimental*

Rappelons rapidement le principe de la mesure:

Un laser à rubis déclenché par une cellule de Kerr émet, pendant un temps de l'ordre de 30 nanosecondes, une impulsion lumineuse. Le départ de cette impulsion déclenche un compteur. Le faisceau lumineux frappe le satellite équipé de réflecteurs trièdres qui réfléchissent dans la direction d'incidence. Cet écho, collecté par un petit télescope de 36 cm de diamètre et filtré par un filtre interférentiel, est détecté par un photomultiplicateur dont le signal de sortie arrête le compteur. Par ailleurs, on a enregistré l'heure du départ de l'impulsion. On détermine donc ainsi, pour chaque écho, une valeur de la fonction $r(t)$.

L'ensemble du matériel est monté sur une tourelle à commande hydraulique pointée par un opérateur qui dispose d'une lunette binoculaire pour la recherche et la poursuite précise du satellite.

L'émetteur laser fournit une énergie de 0,3 Joule en 30 ns., à une cadence maximum de 15 coups par minute. Le récepteur compte et imprime le temps avec une résolution de 10 ns. L'heure, recalée sur les signaux MSF, est enregistrée à chaque tir avec une précision du millième de seconde.

2. *Résultats Obtenus*

Après le premier écho enregistré en janvier 1965 les expériences ont été reprises de mars à juin et ont fourni environ 150 mesures réparties sur une vingtaine de passages des satellites S-66 et BE-C.

Le nombre des échos obtenus varie beaucoup d'un passage à l'autre (de 20 à 0) sans que ce phénomène puisse être suffisamment expliqué par les conditions de l'expérience. De plus il semble que, dans tous les cas, l'énergie de l'écho soit inférieure à celle prévue par le calcul d'un facteur qui, d'après les mesures effectuées par le Dr Plotkin au G.S.F.C., varie entre 2 et 8.

Je ne pense pas que ceci puisse être expliqué par la position aléatoire du récepteur dans la très complexe figure d'interférence due à la centaine de facettes réfléchissantes du satellite, ni par un phénomène 'd'ombres volantes' dû à la déformation de la surface d'onde par l'atmosphère: en effet, on devrait alors observer non seulement des cas où l'énergie est inférieure à la moyenne calculée, mais aussi des cas où elle est supérieure, ce qui semble ne pas se produire.

Précisons que, outre le calcul, des mesures effectuées sur cible terrestre ont confirmé que la portée limite devrait correspondre à une énergie deux fois plus faible que la portée réellement obtenue qui a été de 1600 km sur S-66 et 2000 km sur BE-C.

La précision des mesures est évaluée à:

(1) ± 20 ns. sur la mesure du temps d'aller et retour de la lumière, soit une erreur relative de 2.10^{-6}.

(2) 1 ms. sur le pointage de l'heure.

(3) 3.10^{-7} sur la connaissance de la célérité de la lumière dans le vide. La cause d'erreur dominante est actuellement la seconde mais elle est quelque peu extérieure au problème de la télémétrie et elle pourra être réduite grâce à un chronographe fournissant le dix-millième de seconde. Il est remarquable de constater que l'erreur principale sera alors due à la constante universelle c et il est donc permis de penser que la comparaison des résultats de la télémétrie laser avec ceux obtenus par une autre méthode fournira une nouvelle détermination de cette constante dont la précision pourrait atteindre 10^{-6}.

Remarquons enfin que, contrairement aux mesures sur ondes radio ou aux mesures optiques angulaires, la présence de l'atmosphère ne perturbe pas la précision de la mesure puisque sa traversée par le faisceau lumineux provoque un éloignement apparent du satellite de l'ordre de 2 m, correction qui n'est pas négligeable mais dont les variations mal connues le sont.

3. *Exploitation Numérique des Résultats*

La très grande précision de la méthode a d'ailleurs été confirmée par les premiers calculs utilisant trois passages observés les 25, 28 et 29 mars 1965 et comportant une trentaine de mesures.

Les mesures ont été traitées selon un programme de corrections différentielles établi par la Division Mathématiques du C.N.E.S. et fondé sur la théorie de Brouwer. On a ainsi déterminé les paramètres de l'orbite osculatrice le 27 mars à oh T.U. à partir des paramètres publiés par la Smithsonian Institution. La comparaison des éléments ainsi déduits de l'observation de trois passages seulement avec les paramètres calculés par la Smithsonian Institution fait ressortir des différences faibles: 3 m sur le demi grand-axe a, 10^{-4} radians sur l'inclinaison i etc.... De plus, la cohérence interne obtenue pour ces 3 passages est de 11 m (écart quadratique moyen). Ceci peut paraître peu satisfaisant.

Cependant la cohérence interne sur un seul passage est de 2 m, ce qui donne une estimation plus exacte de la précision de la mesure physique. Elle n'atteint 11 m que si l'on traite simultanément 3 passages, ce qui paraît traduire une insuffisance de la théorie utilisée pour la précision que l'on recherche. D'ailleurs, d'aussi bonnes cohérences n'ont pu être retrouvées pour les passages de S-66 et BE-C observés en mai et juin, les cohérences obtenues sur un passage restant cependant du même ordre.

Ces résultats n'ont cependant pu être obtenus qu'après élimination de plusieurs mesures, en général situées sur les deux extrémités de la fraction de trajectoire observée, dont on n'avait pourtant *a priori* aucune raison de penser qu'elles étaient moins bonnes que les autres. Il y a là un point non élucidé jusqu'à maintenant.

4. *Projets de Développement*

Outre l'amélioration de notre station de Saint-Michel, au point de vue de la portée et de la fiabilité essentiellement, nous espérons pouvoir mettre en service prochainement une seconde station analogue dans le but d'étudier le problème du rattachement géodésique.

En effet, on voit aisément que, si l'on dispose de quatre stations de télémétrie laser et si l'on suppose connues les positions relatives de trois d'entre elles, on peut déterminer la position de la quatrième par des mesures effectuées sur un même arc d'orbite. Cependant deux stations suffisent à déterminer la *distance* qui les sépare et, également, à contrôler la méthode par comparaison avec les résultats de la géodésie classique.

En l'absence des quatre stations nécessaires à une détermination complète des coordonnées par laser, la détermination précise de la distance de deux stations est déjà une première étape, d'autant qu'elle peut être complétée grâce à des mesures angulaires effectuées par photographie du satellite sur le fond des étoiles. Ces mesures, bien que moins précises peuvent provisoirement apporter à la télémétrie le complément d'information indispensable. La précision de la détermination de la distance de deux stations pourrait être de 5 m sur 3000 km.

DISCUSSION

H. P. Meles: Is it possible to use an atomic clock for timing?

R. Bivas: Yes, but at present we have not one available.

R. C. A. Edge: I congratulate M. Bivas on his work which is a great advance and well ahead of any other work in this particular field. The use of lasers and a corner cube reflector gives accuracy of the order of 2 m and there is no trouble with the ionospheric and atmospheric corrections. This holds great promise, but the difficulty is in getting reflections.

M. Bowman: At the Radio and Space Research Station at Slough we have obtained some reflections using a laser but only a maximum of two on one pass. I also find that the signal strength is less than expected but hope that the launching of GEOS will improve things. The precision of our measurements is only 1 kilometre, but this is a matter of the length of pulse transmitted.

R. C. A. Edge: Perhaps the reason why Mr Bowman has only obtained two reflections on one pass compared with M. Bivas' figure of twenty is due to the difference in observing conditions at Slough and in Haute Provence.

J. W. Wright: Could you say how you determined the accuracy to be 2 m?

R. Bivas: The maximum error determined by terrestrial measurements is ± 3 m. Further evidence has been obtained by calculating orbital parameters to give the best adjustment for fitting 3 trajectories on 4 days. The mean square error over the whole period was 11 m, but over a short period the consistency was 2 m.

J. W. Wright: Is there any possibility of recording the reflected light pulse on a photographic plate?

R. Bivas: Although this is feasible we fear we have not sufficient power.

J. W. Wright: Do you experience any difficulty due to the rotation of the Earth?

R. Bivas: This effect is negligible but there is a substantial shift of the reflected path due to motion of the satellite. If the reflected beam were parallel this would be about 100 m; it is therefore necessary to use a slightly divergent beam of semi angle v/c where v is the velocity of the satellite and c is the velocity of light.

R. C. A. Edge: I notice you use optical tracking which presumably means that you can only observe in the dusk. Is there any danger to the human eye from the reflected beam?

R. Bivas: We do indeed observe in the dusk; the power is much too low to present any hazard.

N. J. D. Prescott: Do you use prediction data?

R. Bivas: Although this is done at Slough we do not find it very useful for our work since the accuracy of pointing is not very high.

Report on Aerodist Surveys carried out by the Australian Division of National Mapping since 1963*

B. P. LAMBERT
Australia

Presented by
J. V. BULEY

The Division's Aerodist equipment was mounted in a helicopter and subjected to preliminary trial during 1963. In the latter half of 1963 and during the 1964 field season it was successfully used on line crossing trilateration surveys.

This year the equipment has been transferred to a fixed-wing aircraft and after trilateration survey is now about to be used to obtain position-controlled photography.

This paper describes the Aerodist work carried out by the Division and concludes with an estimate of the likely resultant accuracy.

Introduction

The Division's Aerodist equipment was mounted in a helicopter and subjected to preliminary trial during 1963. In the latter half of 1963 and during the 1964 field season it was successfully used on line crossing trilateration surveys.

This year the equipment has been transferred to a fixed wing aircraft and after further trilateration survey is now about to be used to obtain position-controlled photography.

This paper describes the Aerodist work carried out by the Division and concludes with an estimate of the likely resultant accuracy.

1963—Equipment Trials

For familiarization purposes the equipment, consisting of a two-channel system, was first mounted in a motor vehicle and tested on the

---

* This report has been prepared by officers of the Division of National Mapping, Department of National Development, Australia.

ground in mid-1963 for line crossing techniques over a length of approximately 36 miles (58 km). The result came within $\pm 1\frac{1}{2}$ metres of the Tellurometer measurement for the whole line.

The equipment was next mounted in a Bell 47J helicopter with the 2 master units and attached aerial control boxes, the pen recorder, triode and junction boxes mounted inside the cabin. The antennae were bolted on a rigid platform fixed externally to the rear skids of each side of the helicopter.

A number of test lines varying from 58 to 203 km were then measured in Victoria. Two of these lines had previously been directly measured with the Tellurometer and the following comparative results were obtained:

Line	Geodetic length (G) (metres)	Aerodist			A–G
		Length (A) (metres)	No. of meas.	Spread of meas.	
Bacchus Marsh } Geelong	58 155·94	58 155·4	18	5·2	−0·5
Porndon } Monmot Hill	75 201·35	75 203·2	4	5·6	+1·9

1963—Operational Surveys

Production work was then undertaken in Central Queensland with the objective of measuring, by line crossing technique, the sides and diagonals of a series of $1° \times 1°$ quadrilaterals which were to be fitted into a surrounding first-order survey.

Two pairs of ground stations were used and were moved by road vehicles; each vehicle being equipped with a 25-watt h.f. Transceiver for both ground-to-ground and ground-to-aircraft intercommunication.

Some difficulties were experienced with the helicopter which, at times, had to operate at altitudes of 2100 m when lines greater than 150 km were measured but only minor equipment troubles were encountered and these were easily attended to by field personnel.

Over a 6-week period thirteen quadrilaterals were measured including four comparisons with first-order Tellurometer traverses. The average number of crossings was five.

Most ground stations were fixed for elevation by third-order levelling but in some instances it was necessary to resort to careful barometric work.

In all operational work the practice has been to compare continuously with the surrounding first-order survey in order to determine any index error or any performance variability in the Aerodist equipment and the following comparisons were obtained during the 1963 survey.

Line	Geodetic length (G) (metres)	Aerodist			A–G
		Length (A) (metres)	No. of meas.	Spread of meas.	
Noakes Lookout} Bluff }	125 935·02	125 935·3	3	0·7	+0·3
Staircase} Planet }	91 522·87	91 519·9	7	3·6	−3·0
Staircase} Dawson }	145 816·62	145 817·3	4	4·7	+0·7
Dawson} Holly }	127 947·21	127 945·8	5	3·7	−1·4

At this time the 'atmospherics' at the airborne station were obtained by holding an Assman–Lambrecht psychrometer in the slipstream and then making rapid readings inside the cabin.

1964 Surveys

In 1964 a third master channel was obtained and the three masters were mounted in the helicopter; this third channel did not function satisfactorily and was not used operationally.

Crossover co-axial switches were fitted to enable the master stations to operate through either of the antennae and thereby enable line crossings to be made in both directions.

Use was made of a prototype psychrometer that had been developed by the Applied Physics Division of the Commonwealth Scientific and Industrial Research Organization. It embodied a vented and baffled venturi to reduce the velocity of air flow and prevent dynamic heating effects. The depression of the wet bulb was determined by means of thermocouples and the output was amplified to a suitable display by means of a d.c. amplifier. These data were referenced to an insulated mercury-in-glass thermometer inside the aircraft and readout was by means of a switched meter.

The first operations in 1964 were unsuccessful and the Aerodist equipment had to be returned for workshop servicing but after complete re-alignment the original two masters worked satisfactorily with only minor interruptions. However, operations with the third master were again unsuccessful.

Over a period of nine weeks eighteen new stations were established in a 1° quadrilateral pattern and 110 lines measured with an average of seven crossings to each line. On sixteen lines comparisons were made with the first order Tellurometer traverse. These comparisons are tabulated on page 360.

1965 Surveys

This year the equipment has been transferred to an Aero Commander 680 E fixed wing aircraft in which the antennae are fitted at the end of rotatable axes and are about 2 ft clear of the fuselage. A simple lever moves them to a vertical position in flight and to a horizontal position for landing and take off.

A commercially manufactured version of the CSIRO psychrometer has been fitted into one window spacing of the aircraft and the three masters, the pen recorder and ancillary equipment have been mounted against the rear bulkhead of the cabin.

Some weeks were spent re-aligning and testing the equipment before it was taken into the field for operational work. (It is now standard procedure to do this and actually check the equipment over a known line before proceeding to the field for operational work.)

On return to the field, once again, the original two masters operated satisfactorily with minor maintenance but the third master was unsatisfactory.

In a period of eight weeks seventy-seven lines were measured with seven to ten crossings each and comparisons were made with eleven Tellurometer traverse lines.

On completion of this operation the aircraft returned to headquarters for installation of positioning camera equipment consisting of a Wild Horizon Camera geared to a Vinten vertical camera and the whole linked with ground sighting device.

The aircraft has now just left for the field where it is intended to try and obtain horizontal positional control for mapping operations. Tentatively this is planned to consist of ground trilateration stations at the 30-minute graticule intersections and controlled photography at the remaining 15-minute intersections.

To the present only the following of this year's comparison lines have been computed (see table on page 363):

Line	Geodetic length (G) (metres)	Aerodist			A–G
		Length (A) (metres)	No. of meas.	Spread of meas.	
Bendemeer NMC 51	84 098·28	84 095·8	8	5·1	−2·5
Antares NMC 51	151 715·33	151 709·3	Rejected (poor trace)		−6·0
Antares Bendemeer	113 410·44	113 408·0	7	3·9	−2·4
Antares Wingebar	107 161·75	107 161·6	5	6·6	−0·2
Wingebar Bendemeer	172 321·70	172 319·1	5	3·2	−2·6
Wingebar Woorut	97 991·95	97 991·6	5	2·5	−0·3
Woorut King Jack	107 481·00	107 481·9	6	2·9	+0·9
Woorut Kaputar	153 018·14	153 017·8	3	2·6	−0·3
King Jack Kaputar	86 185·39	86 183·5	7	7·6	−1·9
Kaputar Gragin	100 100·43	100 099·2	8	2·7	−1·2
Gragin Carpet Snake	88 028·32	88 026·8	7	7·5	−1·5
Carpet Snake Gammie	124 679·69	124 678·2	6	4·9	−1·5
Gammie Fair Hill	117 400·96	117 399·4	7	5·0	−1·6
NMC 51 E 355	106 536·52	106 536·4	7	1·9	−0·1
E 355 Bendemeer	156 100·02	156 099·3	7	4·3	−0·7
Mitchell Bassett	96 221·76	96 222·9	5	3·2	+1·1

| Line | Geodetic length (G) (metres) | Aerodist | | | A–G |
		Length (A) (metres)	No. of meas.	Spread of meas.	
Noake's Lookout Bluff	125 935·02	125 935·3	10	7·8	+0·3
Gammie Fair Hill	117 400·96	117 398·5	10	3·6	−2·5
Texas Gammie	95 254·09	95 251·3	11	6·2	−2·8
Fair Hill Bluff	91 006·94	91 005·8	11	3·9	−1·1

Reduction of Data and Computations

Charts from the line crossing measurements are examined, pairs of distances extracted and a graph plotted. Groups of ten measurements are then selected at equal intervals on each side of the minimum point and these twenty-one readings are computed on a standard programme to give, by least squares, the best fitting parabola.

There is a large volume of work in the extraction of data from the graphs, therefore an order has been placed for a chart reader that will readily convert data into digital form.

Every perimeter has been connected by Tellurometer traverse to the surrounding first-order survey and the computation of geodetic co-ordinates will be by a programmed least-square variation of co-ordinates technique in which the perimeter stations will be held fixed.

Operational Functioning

There is no doubt that knowledgeable and skilled servicing is required in order to keep the Aerodist equipment operational. (This has been provided by the Australian agents.) Each year it has taken some weeks, at the start of the survey season, to get the equipment functioning but once it has started to function properly very little further trouble has been encountered in the field when operating two master stations for line-crossing measurements. It has just not been possible to get the third master to operate satisfactorily.

Analysis of Results

It has been accepted that the errors in the first order Tellurometer traverses are negligible in comparison with the errors of the Aerodist measurements.

A general analysis indicates that in the range approximately 100 to 150 km the size of the errors was fairly constant and from all twenty-five comparisons so far made with the geodetic survey an index correction of +0·9 m has been determined.

The residuals of each observation have been brought to unity weight and a statistical analysis of these unity weight observations gives a standard deviation of ±3·5 metres for a single line-crossing measurement.*

The corrected measurements and their equal unity weights are tabulated in the Appendix.

Likely Precision

It has been concluded, that if in the course of 1-degree quadrilateral-type trilateration nine satisfactory crossings are obtained per line over a period of at least two days, the standard deviation of measurement will be ± 1·2 m which is the equivalent of ± 12 and 8 parts in 10^6 respectively on the orthogonals and diagonals.

It is expected that these ratios will be improved by the large number of conditions introduced in the course of fitting the trilateration to a complete perimeter surround but that they will be adversely affected by the strains introduced by the adjustment of the national geodetic survey.

In this survey the average length between junction points is 200 miles and loop mis-closures have averaged 2 parts in 10^6 while the preliminary computation of the main network, based on equal weighting for azimuth and length, has resulted in adjustments between traverse junction points of an average of 0·4 seconds of arc for azimuth with a maximum of 1·4 seconds, and of 1·8 parts in 10^6 for length, with a maximum of 6 parts in 10^6.

It is hoped that the finally adjusted lengths of the Aerodist 1 degree trilateration network will have a standard deviation of ± 10 parts in 10^6 provided these networks are restricted to a block of about 4° of longitude × 4° of latitude throughout the continental area of Australia.

---

* A similar treatment of the comparisons reported in 'Aerodist Test Project' (*The Canadian Surveyor*, September 1964) gave a standard deviation of ± 2·7 m for a single line-crossing measurement.

APPENDIX

Line	Length (metres)	No. of meas. (n)	A–G (corrected for index error = +0·9 m)	Errors converted to unity weight $\sqrt{n}$(A–G) = V	V^2
Bacchus Marsh Geelong	58 156	18	+0·4	1·7	2·89
Porndon Monmot Hill	75 201	4	+2·8	5·6	31·36
Noakes Lookout Bluff	125 935	3	+1·2	2·1	4·41
Staircase Planet	91 523	7	−2·1	5·5	30·25
Staircase Dawson	145 817	4	+1·6	3·2	10·24
Dawson Holly	127 947	5	−0·5	1·1	1·21
Bendemeer NMC 51	84 098	8	−1·6	4·5	20·25
Antares Bendemeer	113 410	7	−1·5	4·0	16·00
Antares Wingebar	107 162	5	+0·7	1·6	2·56
Wingebar Bendemeer	172 322	5	−1·7	3·8	14·44
Wingebar Woorut	97 992	5	+0·6	1·3	1·69
Woorut King Jack	107 481	6	+1·8	4·4	19·36
Woorut Kaputar	153 018	3	+0·6	1·0	1·00
King Jack Kaputar	86 185	7	−1·0	2·6	6·76
Kaputar Gragin	100 100	8	−0·3	0·8	0·64
Gragin Carpet Snake	88 028	7	−0·6	1·6	2·56
Carpet Snake Gammie	124 680	6	−0·6	1·5	2·25
Gammie Fairhill	117 401	7	−0·7	1·8	3·24
NMC 51 E 355	106 537	7	+0·8	2·1	4·41
E 355 Bendemeer	156 100	7	+0·2	0·5	0·25
Mitchell Bassett	96 222	5	+2·0	4·5	20·25

Line	Length (metres)	No. of meas. (n)	A–G (corrected for index error = +0·9 m)	Errors converted to unity weight $\sqrt{n}$(A–G) = V	V^2
Noakes Lookout⎫ Bluff ⎭ Gammie ⎫	125 935	10	+1·2	3·8	14·44
Fair Hill ⎬ Texas ⎫	117 401	10	−1·6	5·1	26·01
Gammie ⎬ Fair Hill ⎫	95 254	11	−1·9	6·3	39·69
Bluff ⎭	91 007	11	−0·2	0·7	0·49

$$\sum V^2 = 276\cdot65$$

$$\sqrt{\frac{\Sigma V^2}{n-1}} = \sqrt{\frac{276\cdot65}{24}} = \pm 3\cdot4 \, \text{m}$$

DISCUSSION

Owing to shortage of time there was no discussion of this paper.

Adjustments of Aerodist
Observations

B. R. EVANS

Canada

Aerodist is a powerful tool for extending survey control
rapidly over wide areas. The method of adjustment outlined in this paper is suitable for medium-sized computers
with magnetic tape or disk facilities, and the paper includes
a general outline of a programme to perform the adjustment.

The method is developed in connection with continuous
trilateration by Aerodist, but can also be used to adjust data
from other airborne measuring systems (HIRAN, SHIRAN,
etc.). The paper shows how data from continuous trilateration
by Aerodist and data from connecting surveys and azimuth
observations can be adjusted simultaneously. The same
programme can be used for fixing the positions of photo
stations when Aerodist is used with fixed ground stations.

1. *Applications*

Aerodist is an electronic ranging system capable of measuring simultaneously and continuously distances to three ground points; these
measurements are accurate to 5 ppm[8] for maximum distances of about
100 miles. This ability to take ranges on *more* than two ground points has
opened up new possibilities in electronic surveying, particularly continuous trilateration, which for establishing horizontal mapping control is
considerably faster and cheaper than ground surveys, and can be carried
out in terrain where ground surveys are impractical.

Essentially, the procedure is quite simple, as illustrated in Fig. 1. By
means of radio waves, an instrument in the aircraft takes continuous
simultaneous ranges to three ground transmitters at prescribed time
intervals (i.e. from successive points along the flight line). Then, if
the positions of two of the ground transmitters are fixed and the height of
the aircraft is known, the position of the unfixed ground transmitters can
be calculated. This is the basic concept of continuous trilateration by
Aerodist; in practice, one can fix a whole net of new ground points. It is
also possible to adjust other types of surveys simultaneously with data
obtained from continuous trilateration, and these data can be incorporated
in the computational procedures outlined in this paper for continuous

367

trilateration. Examples of such additional surveys are line crossing, precise azimuth determinations and photo-fixing. (For this last, see page
382.)

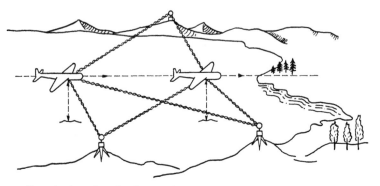

FIG. 1. Aerodist simultaneously ranging on three ground stations
while an airborne profile recorder measures the height of the aircraft
above ground level.

Computer and Simplified Solution Necessary

Continuous trilateration produces a tremendous quantity of data, and
the large number of observations (of the order of 300–700 per figure)
produces redundant observation equations, which make it necessary[5]
to use the method of least squares to obtain the best solution. For these
reasons, an electronic computer is the only practical way of processing the
data.

However, it is the purpose of this paper to develop computational
procedures to substantially reduce the computer time required for the
least-squares solution, by taking advantage of certain characteristics of
the matrix of normal equations. The paper proper will conclude with
some notes on programming this solution.

2. Fixing One New Point on the Ground

Adjusted Length a Function of Latitude and Longitude of End Points

The minimum number of ranges (lines) necessary to calculate the
preliminary co-ordinates of one unfixed ground point is six, arranged as
shown in Fig. 2, top. From the resulting length equations we can calculate
the preliminary co-ordinates (ϕ_3 and λ_3) of the unknown ground station
(number 3).

Then, for any line whose length we want to adjust (from station i to
station j), the adjusted length S_{ij} can be expressed as a function of the

latitude and longitude of its end points:

$$S'_{i,j} + v = f(\phi_i' + \Delta\phi_i, \quad \lambda_i' + \Delta\lambda_i, \quad \phi_j' + \Delta\phi_j, \quad \lambda_j' + \Delta\lambda_j) \quad (1)$$

where v is the correction to be applied to the observed length.

If the line is fixed at one end (i.e. if one of its end points is a known ground station), (1) reduces to

$$S'_{i,j} + v = f(\phi_i' + \Delta\phi_i, \quad \lambda_i' + \Delta\lambda_i) \quad (2)$$

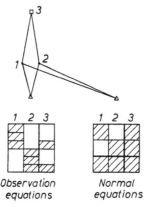

Observation   Normal
equations   equations

FIG. 2. Minimum figure to fix one new ground station with schematic representation of form of matrices produced.

We can linearize the relationship in (1) by means of a Taylor's series expansion in which powers of $d\phi$ and $d\lambda$ higher than the first are ignored, obtaining

$$S'_{i,j} + v = \left(\frac{\partial f}{\partial \phi_i}\right) d\phi_i + \left(\frac{\partial f}{\partial \lambda_i}\right) d\lambda_i + \left(\frac{\partial f}{\partial \phi_j}\right) d\phi_j + \left(\frac{\partial f}{\partial \lambda_j}\right) d\lambda_j$$
$$+ f(\phi_i', \quad \lambda_i', \quad \phi_j', \quad \lambda_j') \quad (3)$$

where $\partial f / \partial \phi_i$ is the numerical value of the differential at point i, and $f(\phi_i', \lambda_i', \phi_j', \lambda_j')$ is the length from i to j, computed from the preliminary positions of i and j by a formula to be given later (page 384).

Translation into Matrix Notation

Before beginning our discussion of the least squares method, we shall rewrite (3) in a form suitable for translation into matrix notation, since this will simplify both the notation and the explanations. Thus

$$\left(\frac{\partial f}{\partial \phi_i}, \frac{\partial f}{\partial \lambda_i}\right)\left(\frac{d\phi_i}{d\lambda_i}\right) + \left(\frac{\partial f}{\partial \phi_j}, \frac{\partial f}{\partial \lambda_j}\right)\left(\frac{d\phi_j}{d\lambda_j}\right) = [S'_{i,j} - f(\phi_i', \lambda_i', \phi_j', \lambda_j')] + v$$
$$(4)$$

or

$$A_{ki}X_i + A_{kj}X_j = b_k + v_k \tag{5}$$

Equation (5) will be the standard form of an observation equation used in this paper and represents the kth row of A in the matrix expression for the entire system of observation equations

$$AX = B + V \tag{6}$$

For a figure containing the minimum number of lines to make the figure rigid, we get a unique solution X for this system of equations. When there are redundant lines (equations), the method of least squares will find the best[5] solution. An outline of this method follows.

Re-arranging (6) we get

$$V = AX - B \tag{7}$$

where V is a column vector of the residuals, or corrections to the observed quantities, X is a column vector $(d\phi_1, d\lambda_1, d\phi_2, d\lambda_2, \ldots d\phi_m, d\lambda_m)^T$ and B is the absolute term.

In the above, it was assumed that all the measurements had the same relative accuracy, and consequently equal weight. This can be ensured if the observation equations (4) for length are written as

$$\frac{1}{S'_{i,j}}\left(\frac{\partial f}{\partial \phi_i}, \frac{\partial f}{\partial \lambda_i}\right)\left(\frac{d\phi_i}{d\lambda_i}\right) + \frac{1}{S'_{i,j}}\left(\frac{\partial f}{\partial \phi_j}, \frac{\partial f}{\partial \lambda_j}\right)\left(\frac{d\phi_j}{d\lambda_j}\right)$$

$$= 1 - \left\{\frac{1}{S'_{i,j}}[f(\phi_i', \lambda_i', \phi_j', \lambda_j')]\right\} + \frac{v}{S'_{i,j}} \tag{4a}$$

Assuming that all length equations for lines measured with the same equipment will have the same weight, equation (4a) is satisfactory for continuous trilateration with Aerodist. However, in practice one may wish in adjusting more complicated figures by this method to adjust simultaneously results of other surveys, including azimuth observations between selected ground stations. In order to incorporate these other surveys into our calculations, it is necessary to assign each observation a weight, w_k.

We are now ready to form the normal equations, i.e. the equations that make $V^T V$ a minimum[5]. The formula for normal equations formed from weighted observation equations is

$$NX = H \tag{8}$$

where $N = (A^T W A)$ and $H = (A^T W B)$, when W is the matrix of weights.

From (8)

$$X = N^{-1}H \qquad (9)$$

The residuals (V) can then be computed by substituting for X in (7):

$$V = AN^{-1}H - B \qquad (10)$$

3. *Continuous Fixing of New Ground Points*

The classical straightforward application of the method of least squares outlined in Part 2 above is not practical for continuous trilateration computations since it requires the inversion of the whole normal-equations matrix N, (equation 9) which will be far too big to be inverted on any but the largest computers. It is possible, however, to employ the least-squares method on a medium-sized computer if certain special characteristics of N are taken advantage of.

We shall therefore discuss several figures more complicated than the minimum figure (Fig. 2), and the matrices they produce, in order to explain these special characteristics of N. Then we shall deal with the problem of fixing a whole net of new ground points, and give an outline of the actual solution.

More Complicated Figures

Fig. 3 schematically represents the situation in which simultaneous ranges are taken on three ground stations (two of which are fixed) from *more* than two air stations (compare Fig. 2). Each small shaded rectangle in the box representing the observation-equation matrix A represents one part of an observation equation in our standard form (equation 5): the rectangles in the last column on the right represent A_{kj}, or the coefficients of the corrections to the unknown ground station; the rectangles in the other columns, descending stepwise on a 'diagonal' from the upper left corner, represent A_{ki}, or the coefficients of the corrections to successive air stations.

In accordance with equation (8), we form the normal-equation matrix (N) by multiplying A on the left by A^T. It is important to know what happens in this multiplication, in order to understand the special characteristics of N mentioned above. Let us consider, therefore, the kth row of A, which consists of A_{ki} in the body of A and A_{kj} in the last column. The multiplication on the left of this row by its transpose is as follows:

$$\begin{bmatrix} A_{ki}^T \\ 0 \\ A_{kj}^T \end{bmatrix} \begin{bmatrix} A_{ki} & 0 & A_{kj} \end{bmatrix} = \begin{bmatrix} A_{ki}^T A_{ki} & 0 & A_{ki}^T A_{kj} \\ 0 & 0 & 0 \\ A_{kj}^T A_{ki} & 0 & A_{kj}^T A_{kj} \end{bmatrix} = N_{ij}^k$$

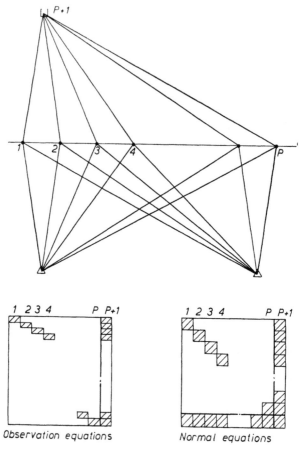

Fig. 3. Fixing one new ground station by continuous trilateration, with form of matrices produced. For simplicity, observations to fixed ground stations are not shown.

N_{ij}^k is shown schematically in Fig. 4. The single row of A has produced four submatrices in N_{ij}, which forms a square on the diagonal of N;

$$N_{ij} = \sum_{k=1}^{m} N_{ij}^k$$

where m is the number of rows from A corresponding to observations linking stations i and j. This square will always lie with $A_{kj}^T A_{kj}$ in the lower right corner of N, and $A_{ki}^T A_{ki}$ on the diagonal, and with $A_{kj}^T A_{kj}$ and $A_{ki}^T A_{ki}$ in the last column and in the last row, respectively. As i

increases, $A_{ki}^T A_{ki}$ will move down the diagonal and N_{ij} will 'shrink' towards the bottom right corner of N.

The sum of a series of such multiplications for each row of A will produce N, which will have the general shape of N in Fig. 3. We can form N in this way, multiplying each row of A on the left by its transpose, and adding the resulting submatrices to N_{ij} separately, because the expression for N_{ij}^k involves only the kth row of A.[3]

Thus, N in Fig. 3 has on its diagonal a series of independent square submatrices, each of which corresponds to one separate air station. Each of the square submatrices in the last column, along with the corresponding one in the last row, corresponds to one range to the unknown ground station. The bottom right square submatrix, however, is a sum receiving one contribution for every air station.

In terms of a computer, this means that each element in N can be formed and an output obtained without being stored, except for $A_{kj}^T A_{kj}$, which must be summed for all k.

Special Characteristics of N

We are now ready to discuss the special characteristics of N which will make practical a least-squares solution even when it would be impractical to invert N itself. We shall achieve this by partitioning N according to the scheme shown in Fig. 5, where N_{22} is

$$\sum_{k=1}^{m} A_{kj}^T A_{kj}$$

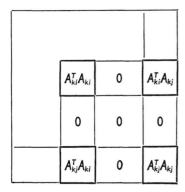

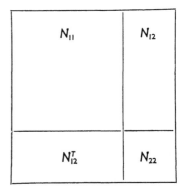

Fig. 4. Contribution to N of one observation equation $A_{ki}X_i + A_{kj}X_j = b_k + v_k$.

Fig. 5. Partitioned matrix of normal equations.

13

N_{12} is the right-hand column, N_{12}^T is the bottom row, both excluding N_{22}, and N_{11} is the rest of N.

We partition N in this way because a solution can now (equation 17) be obtained which requires inversion only of N_{11} and C. Since in Fig. 3 N_{11} is a quasi-diagonal matrix with independent 2 by 2 submatrices on its diagonal, its inversion will require only a series of inversions of these submatrices and will produce another quasi-diagonal matrix with 2 by 2 matrices on the diagonal. Also for Fig. 3, C will be a 2 by 2 matrix. Since the inversion of a 2 by 2 matrix is a trifling calculation, the great advantages of this approach over one involving the inversion of N itself are clear.

We shall now show that the matrices resulting from more complicated figures, including a whole net of new ground points, can also be partitioned in this same way to permit a simplified method of solution, and then we shall outline the method of solution itself. But first, we must consider one key point in the setting up of the figure itself.

In Fig. 3, the reader will notice that the air stations are numbered from 1 to p, and the unknown ground station is numbered $p+1$, one *higher* than the last air station. In terms of the matrix A, this means that all the coefficients of corrections to the ground station (all A_{kj}s) will be in the last or $(p+1)$th column, and therefore that $A^T A$ will have the general shape of N in Fig. 3. Of course, N must be of this general shape if partitioning is to yield a simplified method of solution. Therefore, the basic point in this method of solution is that all unfixed ground stations must be numbered *after* all the air stations.

Fig. 6 represents schematically the situation in which we are simultaneously fixing two new ground stations using a set with four measuring channels. In A, each column representing an air station now contains two contributions, A_{ki} and $A_{(k+1)i}$, each corresponding to a range to one of the unknown ground stations. There is another column, $p+2$, on the right (compare Fig. 3) for the extra unknown ground station. N differs from N in Fig. 3 only in having two columns and two rows in N_{12} and N_{12}^T respectively, and in having two 2 by 2 submatrices in N_{22}, on the diagonal. N_{11} is still quasi-diagonal with 2 by 2 submatrices on the diagonal. Therefore, this N can be partitioned and solved by the same simplified method as can N in Fig. 3.

Fig. 7 represents a situation identical to that in Fig. 6, except that the distance between the two unfixed ground stations is known from some other survey. The two parts of this length equation are represented by the two rectangles at the bottom right corner of A. In this situation, N differs from N in Fig. 6 only in having contributions due to these additional observations in N_{22}, which consequently becomes a full matrix with four 2 by 2 submatrices. However, a matrix this small

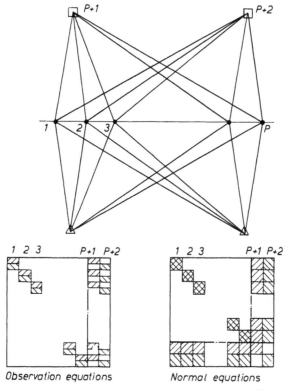

FIG. 6. Fixing of two new ground stations by continuous trilateration, with form of matrices produced. Observations to fixed ground stations not shown.

is no problem at all to invert, and N_{11} is still quasi-diagonal. Therefore, this matrix also can be partitioned to give the same kind of simplified method of solution as those in Figs. 3 and 6.

In Fig. 8, we come to the situation which one would probably be concerned with in practice, that of establishing a net of new fixed ground points by continuous trilateration. Fig. 8(a) shows one possible way of doing this. Here G1 to G6 represent the unfixed ground stations and G7 and G8 represent the fixed ground stations. As the aircraft flies from air stations 1 to P, it records simultaneous ranges to G1, G2 and G7. After the aircraft has passed P, the transmitter at G7 is turned off and those at G3 and G4 are turned on (this assumes four channels in the receiver), and ranges are recorded to G1, G2, G3 and G4. After the aircraft has passed 2P, the transmitters at G1 and G2 are turned off

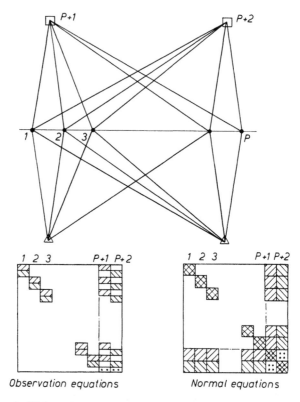

Observation equations   *Normal equations*

Fig. 7. Fixing of two new ground stations with distance (P+1) to (P+2) measured by another survey, and form of matrices produced. Observations to fixed ground stations not shown.

and those at G5 and G6 turned on, and ranges are recorded to G3, G4, G5 and G6 until after the aircraft passes 3P, when G3 and G4 are turned off and G8 is turned on. Ranges are then recorded to G5, G6 and G8 until the aircraft passes 4P.

This scheme of taking observations results in an observation-equations matrix A of the sort schematically represented by Fig. 8(b), where we assume for simplicity that only five ranges are taken in each of the parts of the flight outlined above. Also, in this example, it is assumed that the distances G2 to G4 and G3 to G6 are known from some other survey and are to be included in the figural adjustment. This is represented by the rectangles in the bottom right corner of A in columns (4P+2, 4P+4) and (4P+3, 4P+6) respectively. The normal-equation matrix N is schematically shown in Fig. 8(c). It is

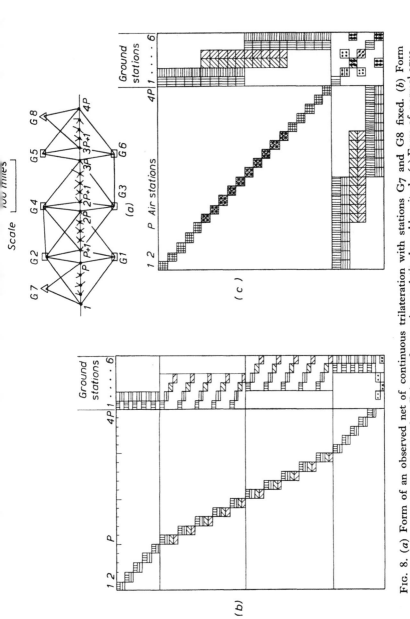

FIG. 8. (a) Form of an observed net of continuous trilateration with stations G_7 and G_8 fixed. (b) Form of observation equations. Matrix of coefficients of corrections to latitude and longitude. (c) Form of normal equations. Matrix of coefficients of corrections to latitude and longitude.

immediately apparent that N_{11} is still quasi-diagonal and that N_{22} is a 12 by 12 matrix consisting of thirty-six 2 by 2 submatrices, which is still small enough to be easily inverted on a medium-sized computer. In general, N_{22} will be $2g$ by $2g$, where g is the number of new ground stations. Since with Aerodist the ground stations can be 100 miles or more apart, and the lines between satellite triangulation stations are of the order of 600–1000 miles, one could, for example, use the above scheme to establish secondary survey control points along one of these lines and keep N_{22} to the order of 40 by 40. This is still quite manageable on (for example) an IBM 1620 (40 K digit storage) which can store and invert matrices of orders up to about 70 by 70. It should also be noted that the additional data for precise azimuth or distance measurements between unfixed ground stations will all be contributed to N_{22} (and will not affect N_{11} or N_{12}, which will thus still not have to be permanently stored), and yet will not increase the dimensions of N_{22}, which we must store anyway.

The Solution

After N has been partitioned as above, we obtain the following matrix equations:

$$N_{11}X_1 + N_{12}X_2 = H_1 \qquad (11)$$

$$N_{12}^T X_1 + N_{22}X_2 = H_2 \qquad (12)$$

From (11) we get

$$N_{11}X_1 = H_1 - N_{12}X_2 \qquad (13)$$

whence

$$X_1 = N_{11}^{-1}(H_1 - N_{12}X_2) \qquad (14)$$

Substituting this expression for X_1 into (12), we get

$$N_{12}^T N_{11}^{-1}(H_1 - N_{12}X_2) + N_{22}X_2 = H_2 \qquad (15)$$

which reduces to

$$(N_{22} - N_{12}^T N_{11}^{-1} N_{12})X_2 = (H_2 - N_{12}^T N_{11}^{-1} H_1) \qquad (16)$$

or

$$CX_2 = K_3 \qquad (17)$$

where

$$C = (N_{22} - N_{12}^T N_{11}^{-1} N_{12})$$

and

$$K_3 = (H_2 - N_{12}^T N_{11}^{-1} H_1)$$

C is a $2g$ by $2g$ matrix, and K_3 is a $2g$ by 1 vector.

From (17) we get

$$X_2 = C^{-1}K_3 \tag{18}$$

where X_2 denotes the corrections to the preliminary positions of the ground stations.

Substituting this expression for X_2 into (14), we get

$$X_1 = N_{11}^{-1}H_1 - N_{11}^{-1}N_{12}(C^{-1}K_3) \tag{19}$$

where X_1 denotes the corrections to the preliminary positions of the air stations.

Now the residuals V can be calculated from equation (7). The whole process is then iterated until the corrections are less than the desired tolerance.

It is worthwhile noting that the inverse C^{-1} from equation (18) can be used to estimate the variance for the adjusted co-ordinates of the ground stations, if an estimate of the variance factor (variance due to weight alone) is available. The diagonal elements of the variance covariance[4] matrix S^2C^{-1} are the estimates of the variance of the ground stations, where

$$S^2 = \frac{V^T V}{(\text{no. of surplus observations})} \tag{20}$$

A Word of Caution

The method of solution outlined above simplifies matters *only* if N_{11} is quasi-diagonal, that is, if all the sub-matrices on the diagonal are independent of each other. This has been true in all the figures discussed above, but here the air bases (distances between air stations) are also measured, and thus the contributions to N_{11} are not independent, and it will be as impractical to invert N_{11} as to invert N itself. In such a case, some other conventional procedure, such as the square root method,[3] must be used to find a solution.

4. *Notes on Programming*

This part of the paper contains an outline of steps involved in determining the adjusted coordinates of the new ground stations, as well as remarks on the inclusion in the adjustment of data from other surveys (see page 367). It is not, however, a discussion of an actual programme, since the exact format of the programme would depend upon the individual preferences and the computing facilities of the user.

Size of the Job

At present, data from the Aerodist are in the form of a continuous

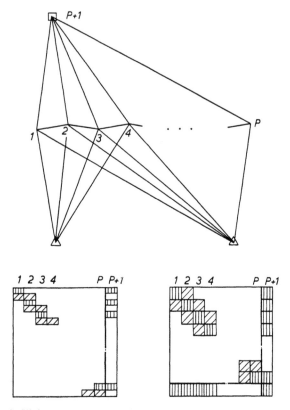

FIG. 9. Fixing one new ground point, when air bases are measured, and form of matrices produced. Observations to fixed ground stations are not shown.

'wiggly trace' which must be interpreted and converted to distances by a technician. Supposing that readings of the trace were made corresponding to every tenth second in the air, for a 100-mile flight line and a speed of 100 m.p.h., we have 360 sets of ranges and 720 unknowns. Attempts are already being made, however, to digitize the output so that it can be stored on magnetic tape. This would enable us to get a range for every second in the air, but it also raises the question of the optimum number of ranges to give both adequate positional accuracy and economy of computation. To settle this question, statistical tests with real data will have to be made. But in any case, the reader will recall that the practicality of the simplified solution outlined above is not affected by the number of air stations, since each observation equation can be contributed separately to the solution and then passed out on tape until needed for computation of the corrections (see page 373).

Computations

The computations will require the following steps.

1. Reduce all the observed ranges to sea level, correcting for meteorological readings,[7] using the formulae on pages 383–4.

2. Compute preliminary co-ordinates for the air stations as follows:

(a) For each range taken from the first air station, compute a corresponding preliminary distance, using the preliminary co-ordinates of the ground stations (which in this step we are holding fixed) and estimated co-ordinates (ϕ_i, λ_i) of the air station, in the formula for the inverse on pages 384–5. Care must be taken that the estimated position of the first air station is on the same side of the line joining the two ground stations as its actual position. If it is not, the iterated position will be a 'mirror image' of the actual position.

(b) Compute numerical values of the differentials in equation (4), which in this case, because we are holding the ground stations fixed, reduces to

$$\left(\frac{\partial f}{\partial \phi_i}, \frac{\partial f}{\partial \lambda_i}\right)\left(\frac{d\phi_i}{d\lambda_i}\right) = [S'_{i,j} - f(\phi_i', \lambda_i')] + v \qquad (21)$$

(c) Form two normal equations in the two unknowns ($d\phi_i$ and $d\lambda_i$) and solve.

(d) Apply the corrections $d\phi_i$, $d\lambda_i$ to ϕ_i and λ_i and calculate the residuals.

(e) If the corrections are still too large, repeat (a) to (d), using the adjusted values of ϕ_i and λ_i. In the event of a failure to converge, programme for a 'suspected blunder in data'.

(f) When the corrections are less than the desired tolerance, compute the position of the next air station by repeating (a) to (e). Unless a new flight line has been started, it will be possible to use the adjusted position of the previous air station as an initial estimate of the position of the next.

3. Correct the distances so that they will all relate to some common point in the aircraft. This corrects for the different positions of the antennae of the various channels, and the correction depends on the position of the aircraft relative to the ground stations to which ranges were recorded from that air station.

13*

4. Form observation equations for every observation that has been found free of blunders in step 2, and for data from any other surveys which are to be adjusted at the same time. We now allow the ground stations to 'move' in the adjustment.

5. Form and solve the normal equations.

6. Print the inverse of C (variance covariance factors).

7. Compute and print the residuals and statistical analysis.

8. Correct the preliminary co-ordinates of all stations (ground and air) in the adjustment.

9. If the adjustment is not yet satisfactory, repeat steps 4 to 8 using the adjusted coordinates until convergence is reached. Again watch for blunders in data in the form of very large residuals, and reject bad data.

10. Terminate computations.

The above outline is intended for a medium-sized computer equipped with magnetic tapes or disks. On a card machine reorganization of steps and further presolution might be desirable.

Airborne Photo Control

Since Aerodist has been widely employed in Canada for photo fixing, this paper would be incomplete without some brief mention of this particular application and its connection with continuous trilateration.[6]

For mapping purposes, a framework of perimeter control points for blocks of aerotriangulation is required. These points can be established from the air with the help of Aerodist. Vertical photographs of identifiable detail are taken along the perimeters of the blocks from 7000 to 10 000 ft with a super-wide angle lens, and a record of each exposure is kept on the Aerodist data output. It is then possible to compute the positions of the photo stations and thus to fix the plumb point of each photo. With this information, the desired control points can later be positioned by photogrammetry.

The coordinates of the photo stations in the above procedure can be computed by the continuous trilateration programme simply by supplying the programme with the adjusted positions of the ground stations in step 2 and coming off the computer after step 3. Any photographs taken during actual continuous trilateration can be fixed by means of X_1, the corrections to the air stations computed during the solution of N (equation 19).

Programme to be Written

As this paper goes to press, a start has been made on writing this programme in FORTRAN for the CDC 3100 computer (16K word memory). This machine will have five tapes, but the programme will use only two, and for the present input will be on cards. The programme will eventually adjust all types of data mentioned in this paper, but the exact organization and recompartmenting of the steps will be determined by tests with data from earlier surveys with the Aerodist.

5. *Formulae*

Formulae Used in Preliminary Computations

The distances measured electronically will be slope distances and will have to be reduced to sea level before they can be inserted in any of the correction equations or geodetic position formulae described below.

The formulae for the reduction to sea level are taken from Geodetic Survey of Canada Publication No. 78 'Geodetic Application of SHORAN', page 59.

The radius of curvature (r) of the electronic beam is greater than that for the Earth (R). As a first approximation it is assumed that the relation $r = 3 \cdot 91R$ is sufficiently close, and a further correction to allow for variation from this relation is obtained from meteorological data.

The geometrical corrections which may be applied at this stage are, then, (*a*) reduction to sea level, (*b*) reduction for slope and (*c*) rectification of the curvature of the Earth and of the SHORAN ray. These are given in this order in the following expression for M.

$$M = S - \frac{S(H+K)}{2R_\alpha} - \frac{(H-K)^2}{2S} + \frac{14 \cdot 3}{15 \cdot 3 R_\alpha{}^2} \cdot \frac{S^3}{24} \quad (22)$$

where the unit is the mile, the last term is a simplification of

$$\frac{S^3}{24}\left(\frac{1}{R^2} - \frac{1}{r^2}\right),$$

and

$M =$ the corrected distance

$S =$ the measured distance, the velocity of propagation having been corrected for meteorological conditions[7]

$H =$ elevation of the airborne station

$K =$ elevation of the ground station (antenna)

$R_\alpha =$ radius of curvature in azimuth α

$\quad = NR/(N \cos^2\alpha - R \sin^2\alpha)$

In the geodetic formulae which follow, a and b denote the major and minor axes of the spheroid, and the e denotes the eccentricity of the spheroid where

$$e^2 = \frac{a^2 - b^2}{a^2}$$

Formulae used for the Direct Solution

Formulae used in the direct solution are from [1] and [2]. They are used for computing the position (ϕ, λ) of a point on the spheroid at a given distance and azimuth from a point of known position.

$$R_2 = -\left(\frac{e^2}{1-e^2}\right) \cos^2\phi_1 \cos^2\alpha_{1,2} \tag{23}$$

$$R_3 = 3\left(\frac{e^2}{1-e^2}\right)(1-R_2) \cos\phi_1 \sin\phi_1 \cos\alpha_{1,2} \tag{24}$$

$$R_4 = -R_2(4-9R_2) - 3\left(\frac{e^2}{1-e^2}\right) \sin^2\phi_1(1-R_2)(1-5R_2) \tag{25}$$

$$\theta = \frac{S}{N_1} - \frac{R_2(1+R_2)}{3!}\left(\frac{S}{N_1}\right)^3 - \frac{R_3(1+3R_2)}{4!}\left(\frac{S}{N_1}\right)^4$$
$$- \frac{R_4 + 3R_3{}^2 + 4R_2 \cdot R_4 - 3R_2{}^4}{5!}\left(\frac{S}{N_1}\right)^5 \tag{26}$$

$$\frac{N_1}{R} = 1 - \frac{R_2}{2!}\cdot\theta^2 - \frac{R_3}{3!}\cdot\theta^3 - \frac{R_4}{4!}\cdot\theta^4 \tag{27}$$

$$\sin\psi = \sin\phi_1 \cos\theta + \cos\phi_1 \cos\alpha_{1\,2} \sin\theta \tag{28}$$

$$\sin\Delta\lambda = \frac{\sin\alpha_{1,2}\sin\theta}{(1-\sin^2\psi)^{1/2}} \qquad \cos\Delta\lambda = (1-\sin^2\Delta\lambda)^{1/2} \tag{29}$$

$$\tan\phi_2 = \left(1+\frac{e^2}{1-e^2}\right)\left[\sin\phi_1 - e^2\left(\frac{N_1}{R}\right)\sin\phi_1\right](1-\sin^2\psi)^{1/2} \tag{30}$$

$$\Delta\lambda = \lambda_1 - \lambda_2 \tag{31}$$

Formulae Used for the Inverse Solution

Formulae used for the inverse computation are from [2], and are used to compute the distances and azimuths in both directions between two points of known position on the spheroid.

$$A = \cos\phi_2 \sin\Delta\lambda; \qquad \Delta\lambda = \lambda_1 - \lambda_2 \tag{32}$$

$$B = \frac{b^2}{a^2}\cos\phi_1 \sin\phi_2 - \cos\phi_2 \sin\phi_1 \cos\Delta\lambda + \frac{N_1}{N_2}\cdot e^2 \cos\phi_1 \sin\phi_1 \tag{33}$$

$$\sin\alpha_{1,2} = \frac{A}{(A^2+B^2)^{1/2}} \qquad \cos\alpha_{1,2} = \frac{B}{(A^2+B^2)^{1/2}} \tag{34}$$

$$N_1 = a(1-e^2\sin^2\phi_1)^{-1/2} \qquad N_2 = a(1-e^2\sin^2\phi_2)^{-1/2} \tag{35}$$

$$\frac{K}{N_1} = \left[\left(\frac{N_2}{N_1} \cos \phi_2 \cos \Delta\lambda - \cos \phi_1 \right)^2 + \left(\frac{N_2}{N_1} \cos \phi_2 \sin \Delta\lambda \right)^2 \right.$$

$$\left. + \left(\frac{N_2}{N_1} \frac{b^2}{a^2} \sin \phi_2 - \frac{b^2}{a^2} \sin \phi_1 \right)^2 \right]^{1/2} \tag{36}$$

$$S_{1,2} = K \left[1 + \frac{1}{24} \left(\frac{K}{R} \right)^2 - \frac{1}{8} F \left(\frac{K}{R} \right)^3 + \left(\frac{3}{640} + \frac{3}{80} H + \frac{1}{4} F^2 \right) \left(\frac{K}{R} \right)^4 \right.$$

$$\left. - \left(\frac{3}{16} F \cdot H + \frac{5}{12} F^3 \right) \left(\frac{K}{R} \right)^5 \right] \tag{37}$$

$$\frac{K}{R} = \frac{K}{N_1} \left[1 + \left(\frac{e^2}{1 - e^2} \right) \cos^2 \phi_1 \cos^2 \alpha_{1,2} \right] \tag{38}$$

$$F = \frac{f \cdot h}{1 + h^2} \qquad H = \frac{f^2 - h^2}{1 + h^2} \tag{39}$$

$$f = \frac{e}{(1 - e^2)^{1/2}} \sin \phi_1 \qquad h = \frac{e}{(1 - e^2)^{1/2}} \cos \phi_1 \cos \alpha_{1,2} \tag{40}$$

Observation Equations

In the following section, let

V	be the correction to an observed quantity,
i and j	the numbers of the two end points of a line,
$\alpha_{i,j}$	the azimuth of a line from point i to point j,
$S_{i,j}$	the length of a line joining point i and point j,
R_i	the radius of curvature along the meridian at point i,
N_i	the radius of curvature in the normal section at point i,
ϕ_i	the latitude of point i,
λ_i	the longitude of point i,

and let

$\alpha'_{i,j}, \alpha'_{j,i}, S_{i,j}$ be the preliminary values obtained by the inverse computations from the preliminary co-ordinates for points i and j.

If the individual observation equations in the matrix equation

$$AX - B = V \tag{41}$$

are written in the form

$$A_{ki} X_i + A_{kj} X_j = b_k + v_k \tag{42}$$

then V_k may be considered to be the correction to be added to the observed quantity if the identities given in the next two sections exist.

Length Correction Equation

If a length correction equation is written in the form of equation (41), then

$$A_{ki} = \frac{1}{S'_{i,j}}(-R_i \cos \alpha'_{i,j},\ N_i \cos \phi_i \sin \alpha'_{i,j}) \qquad (43)$$

$$A_{kj} = \frac{1}{S'_{i,j}}(-R_j \cos \alpha'_{j,i},\ N_j \cos \phi_j \sin \alpha'_{j,i}) \qquad (44)$$

$$b_k = \frac{1}{S'_{i,j}}(\text{Observed } S_{i,j} - \text{Preliminary } S_{i,j}) \qquad (45)$$

when

$$V_k = \frac{V_{i,j}}{S'_{i,j}} \qquad (46)$$

Azimuth Correction Equation

If an azimuth correction equation is written in the form of equation (41), then

$$A_{ki} = \frac{1}{S'_{i,j}}(R_i \sin \alpha'_{i,j},\ N_i \cos \phi_i \cos \alpha'_{i,j}) \qquad (47)$$

$$A_{kj} = \frac{1}{S'_{i,j}}(R_j \sin \alpha'_{j,i},\ N_j \cos \phi_j \cos \alpha'_{j,i}) \qquad (48)$$

$$b_k = -\{\alpha'_{i,j} - [\text{obs. } \alpha_{i,j} - (\lambda' - \text{astro. } \lambda) \sin \phi']\} \qquad (49)$$

when

$$V_k = V_{i,j} \qquad (50)$$

Sign Convention

In all equations in this paper, the following sign convention has been observed:

Latitude   +ive increasing to the pole
Longitude +ive increasing to the west
Azimuth   +ive increasing clockwise from the north point.

References

[1] Bomford, G., 1952. *Geodesy*. Oxford.
[2] Clarke, A. R., 1880. *Geodesy*. Oxford.
[3] Evans, B. R., 1965. 'Least squares programming for band matrices'. Surveys and Mapping Branch Provisional Report No. 65–5, Ottawa, Department of Mines and Technical Surveys.

[4] Gale, L. A., 1965. 'Theory of adjustment by least squares'. *The Canadian Surveyor*, Vol. XIX, No. 1.
[5] Thompson, E. H., 1962. 'Theory of the method of least squares'. *The Photogrammetric Record*, Vol. IV, No. 19.
[6] Tuttle, A. C. 'Aerodist, The Flying Chainman'. A paper presented at the 1965 joint convention of the American Society of Photogrammetry and the American Congress on Surveying and Mapping. Publication pending.
[7] U.S. Department of Commerce, Weather Bureau. W.B. # 235, *Psychometric Tables*, U.S. Government, Washington, 1941.
[8] Yaskowich, S. A., 1964. 'Aerodist test project'. *The Canadian Surveyor*, Vol. XVIII, No. 4.

Laser Applications

Progress in the Application of Lasers to Distance Measurement

K. D. HARRIS

G. & E. Bradley Ltd

Over the last four years, several proposals have been made for using Lasers in systems to measure distance.

There are basically three ways in which a Laser may be used as the source of energy in distance measurement. In all the methods, the light is transmitted to a reflecting or diffusing surface and back to a receiver near the transmitter. The differences occur in the way the light is used in the space between. These three ways are:

(*a*) By using the high coherence of the Laser output to produce an interference pattern over the distance to be measured, it is possible to express the distance in terms of a number of wavelengths of light.

(*b*) By modulating the Laser with some high frequency radio signal, it is possible to compare the phase of the modulation of the transmitted wave with that of the received wave and so calculate the distance.

(*c*) A pulse-modulated signal may be timed over double the distance, so that the distance may be computed, as in conventional radar.

Each of these methods has been tried with varying amounts of success. Generally, the third type (*c*) has proved most practical to date, with devices now being manufactured. Experimental results will be reported for each type in the paper.

Introduction

Since the first days of Lasers five years ago, their potentialities in the fields of communications and radar have been appreciated. However, practical Lasers available in the first three years fell far short of realizing these possibilities. This led to disappointment and a turning away from Lasers in these fields—too much was expected of the art too soon. Now engineering achievements in Laser device technology are step-by-step approaching many of the target requirements.

The first field of useful application of the Laser has been to distance measurement.

There are basically three ways in which a Laser may be used as the source of energy in distance measurement. In all these methods, the light is transmitted to a reflecting or diffusing surface and back to a receiver near the transmitter. The differences occur in the way the light is used in the space between. These three ways are:

(a) By using the high coherence of the Laser output to produce an interference pattern over the distance to be measured, it is possible to express the distance in terms of a number of wavelengths of light.

(b) By modulating the Laser with some high frequency radio signal, it is possible to compare the phase of the modulation of the transmitted wave with that of the received wave and so calculate the distance.

(c) A pulse-modulated signal may be timed over the double distance, so that the distance may be computed, as in conventional radar.

As this paper is in the nature of a general review of the field, it may be useful to go into greater detail of these methods, before discussing particular devices.

Distance Measurement by Optical Interference

Optical interference has been used for many years in the measurement of length. Examples range from the determination of the metre in terms of wavelengths of light to the use of Moiré fringes in machine tool control. These systems consist usually of two surfaces which are illuminated with incoherent light to set up a three-dimensional standing wave pattern between the surfaces. Viewed from outside, light and dark fringes are seen, which indicate positions where the path differences between the light reading adjacent fringes are half a wavelength.

A simple example is the Fabry–Perot interferometer, which consists basically of two partially-silvered parallel mirrors. Parallel monochromatic light entering the space between the mirrors through one mirror is multiply reflected back and forth between them. Interference occurs between these multiple rays so that viewed through a telescope, by transmission or reflection, a concentric system of ring fringes is seen. The diameter of the rings is dependent on the wavelength of the light and the distance between the two mirrors. On changing the separation, the fringe pattern expands or contracts. Movement of half a wavelength shifts the pattern by one fringe; changes of $\frac{1}{10}$ fringe or 3×10^{-6} cm are readily detectable so that we see the Fabry–Perot cell is a very sensitive indicator of change of length.

In order to measure distance we need the number of complete half-wavelengths between the two mirrors and the fraction over. The fraction may be calculated from measurements of the pattern. The integer may be found by repeating the experiment with other wavelengths of light. This is the so-called method of exact fractions.

The usefulness of this method depends on the visibility of the fringes, or how clear and sharp they are. With increasing separation between the two mirrors more and more light is required to be beamed into the cell to maintain good visibility. With conventional sources to get more light the lamp has to be run hotter or has to be bigger, or both. The former makes focusing of the light into the interferometer more difficult, while the latter broadens the emission line, which degrades the fringe pattern. Thus, this system was limited practically to distances of a few tens of centimetres.

The replacement of the lamp by a Laser produces a marked improvement in the distance over which good quality fringes may be obtained. A simple gas Laser giving a few milliwatts output power gives good fringes over distances of tens of metres. The combination of coherence, narrow linewidth and high directivity gives the Laser this advantage. The limit to the distance which can be measured is set by atmospheric disturbances which affect the fringe visibility.

Apart from the use of a Laser simply as an improved light source, further improvements result if automatic readout systems are required. The realization of precise distance measuring equipment of this type is being sought in several different directions. First, in the machine tool field there is still a great need for micrometers which can measure the dimensions being machined continuously, and so allow a correction for tool wear to be fed to the tool. Here it is not primarily the micro-inch accuracy which is attractive, but the ease with which the Laser may be fitted into an automatic system. Second, in the fields of marine, power and aeronautical engineering there is a need for equipment for the accurate mapping of the contours of large parts such as propellers, turbine blades, fuselage and wing sections, etc. These measurements may require accuracies of the order of a thousandth of an inch in distances of order 10 ft, with lateral scanning to plot continuously the contour under test. Third, in civil engineering the increasing need for precise construction of large structures is putting a considerable strain on existing methods of surveying. Examples of such structures are: platforms for large generating plant and atomic reactors; dams, bridges and tunnels; and radio telescopes and particle accelerators.

These projects are already demanding measurement to less than a millimetre over distances of the order of 100 m, often as changes in length.

All these applications can use a single wavelength Laser and fringe counting methods. The development of practical devices in these fields is being carried out in many laboratories. The main problems to be solved concern the production of compact and reliable Lasers which have the required stability of output, and low running costs.

Distance Measurement by the Modulated Beam Technique

Existing equipment uses either a coherent microwave signal derived from a valve source such as a klystron, or an incoherent light source such as a filament lamp or arc tube. The former type is easy to modulate at frequencies up to 100 Mc/s and can be used over very long distances with simple directive aerials and a sensitive, selective superheterodyne receiver. It is not affected drastically by weather conditions. On the other hand, the beam spreads sufficiently to give spurious reflections on occasions, which make the distance readings ambiguous. Modulation at higher frequencies to increase the accuracy becomes increasingly difficult and expensive and is, of course, limited by the frequency of the generator itself. At a modulation frequency of 100 Mc/s with phase comparison accurate to 1°, distance can be measured to about 1 cm.

The existing optical systems, being incoherent, suffer generally from poor signal-to-noise ratio, particularly in daylight, and from the difficulty of modulation at increasingly high frequencies.

The Laser offers improvements to the optical system to bring it up to the standard of the microwave system. That is, the high spectral intensity allows receiver sensitivity to be improved by the addition of narrow band interference filters, so discriminating against unwanted light. The nature of Laser light makes modulation up to at least 10^{13} c/s feasible in time. The beamspread from a Laser transmitter is considerably less than that from a conventional light source, or from a microwave antenna of comparable size. However, to obtain full advantage from this high angular discrimination, careful alignment of the system is required.

Its advantages should make the Laser a most useful component in distance measuring equipment for surveying and geodesy. The limitation due to atmospheric absorption will probably be overcome by using wavelengths (in the far infra-red for example) where the loss due to water vapour can be less than with visible light.

The development of modulation systems based on Lasers is, as with interferometers, dependent on the availability of suitably rugged, long-life devices. The requirement for stability in the former case is not as stringent as for the latter, at least for modulation frequencies up to 10^9 c/s, which are likely to be used in the beginning. Development effort has been directed towards solving outstanding problems in the modulators themselves. Because of the difficulty of modulating a Laser internally, particularly at frequencies above 1 Mc/s, most experimental systems have used Kerr or Pockel cells, in various configurations, which modulate the output of the Laser by changing the angle of polarization. This may be converted to amplitude modulation by a Nicol prism or other polarization analyser. The disadvantages of these systems are: firstly, that the losses in these cells and prisms are high; secondly, the efficiency of the cells is low; thirdly, the cells often work at unpleasantly high voltages (20 kV), which should be avoided in portable equipment if possible.

Recently, other more promising modulators have been investigated using piezo-electric effects and modulation effects within the atoms of the Laser material itself; more work is urgently needed before they become completely satisfactory.

Distance Measurement by the Radar Method

By sending out pulses of light and collecting the diffused echoes with a telescope feeding a photomultiplier, it is possible to extend the techniques of radar into the optical region.

The first requirement from a Laser for such a system is high peak power to get useful range information off diffuse targets. Pulsed ruby Lasers meet this requirement if monopulse or at least low pulse repetition rate operation is acceptable.

Within a few months of operation of the first ruby Laser, a simple range-finding set-up was tested. This system had to be improved considerably before widespread use could be made of it. The main modification was to the form of the output pulse of the Laser. In an unmodified ruby Laser, the output occurs as a random succession of spikes lasting for a time of the order of a millisecond. The height and position of the spikes is quite different from shot to shot. This makes, first, the recognition of the echo pulse difficult if it is weak and, second, precise measurement of the elapsed time very difficult—particularly if done electronically with a high frequency clock.

Schemes to produce single pulses from ruby Lasers have therefore been investigated intensively. The technique of Q-switching has proved very promising in this connection. This uses some means of frustrating the normal operation of the Laser until the populations of the atomic states have been inverted, and then releases all the stored energy suddenly, thereby producing a very high intensity output in a very short time.

The means of achieving Q-switching may be commonly mechanical, electro-optical or chemical. An example of a mechanical Q-switch is a rotating prism used as one of the end reflectors of the Laser optical resonator. Careful timing of the excitation discharge to the rotation of the prism holds up the emission of Laser radiation until the prism is square with the end of the ruby rod, when emission occurs rapidly. Typical results are an increased peak power of more than 100 times in a pulse duration of about 30 mμs, for a rotation speed of 20 000 rev/min.

The most common electro-optical Q-switch uses a Kerr cell with suitable polarizers. With the polarizers initially crossed to heavily attenuate the transmission of light through the ruby, the ruby is excited by the usual discharge. The Kerr cell is then pulsed to rotate the plane of polarization so that the transmission attenuation disappears. Sudden Laser output occurs, with typically up to 1000 times the peak power in a duration of perhaps 10 mμs.

The chemical Q-switch, the most recent addition to the devices at our disposal, consists of a simple cell of liquid which is a solution of a bleachable dye. Popular examples of such dyes are pthalo- and kryptocyanine. These have the characteristic that they normally absorb heavily in the red part of the spectrum, but under high intensity become suddenly transparent. Insertion of such a cell in the optical resonator of a Laser has the same effect as a Kerr cell which operates automatically under the influence of the Laser light itself. The beauty of this device is its simplicity—no power supplies or timing devices being necessary. The disadvantage is that one has little control of when the pulse comes within the time when the Laser material is excited. However, this does not matter for static distance measurement.

The accuracy to which measurement of distance can be made using the radar method depends principally on the rise time of the transmitted pulse, and secondly on the amount of sophisticated electronics which can be accommodated in the equipment. As present Q-switched Lasers have pulse rise times of about 10 mμs, the limiting accuracy of distances measured is then about 3 m. In time, this can probably be improved by a factor of 10, but only at the price of considerable complication in both Laser and electronics. Thus, this method is not in the first class for surveying on the ground. Its main use comes from the fact that measurements can be made single-ended, using the backscattered light off buildings, trees and almost any object which may be discerned by looking through a telescope. Thus the first main use has been military, but it seems probable that surveyors will also find use for this type of instrument in remote areas where travel is difficult.

For distance measurement, the radar method is proving very useful. At distances from 50 to 5000 m, cloud height and structure can be determined with sufficient accuracy for meteorological purposes. Atmospheric phenomena such as inversion layers, clear air turbulence, etc., have also been detected with Laser systems. Satellite tracking experiments are being carried out by several groups. In this application, the distance is so large that the accuracy to a few metres is relatively very high. These experiments have great relevance to the future of geodesy, of course, for measuring distances beyond the horizon.

Conclusions

Developments in Lasers are now making them suitable for certain areas of distance measurement. Most advanced are the systems based on the radar method, with several military field versions available and larger experimental systems being used for satellite tracking and meteorology. Next in state of development are interferometers using Lasers for machine tool gauging, with at least two large systems being commercially available. Least developed is the modulation method, where a few experimental systems have been demonstrated, but saleable equipment requires a better scheme for modulating the Laser. It is unfortunate that this last method is potentially the one most suited to geodetic and survey measurements.

DISCUSSION

R. H. Bradsell: Why is the ADC modulator only suitable for laboratory use, and is there not considerable danger in the use of Lasers?

K. D. Harris: Laboratory devices such as these are not yet ready for incorporation in instruments for use in the field. The Laser is like a gun and in using it one must set up the necessary safety precautions.

J. W. Wright: The range finder to which you referred with an accuracy of only 5–10 m would not seem to be of very great value to the surveyor since detailed topographical survey is now normally done from air photographs. For cadastral survey, identification of points is required which entails a personal visit.

K. D. Harris: While the accuracy is not very high, I would have thought that it might be of value for working in remote areas. It would be possible to use the Laser rangefinder to measure the distance to the tops of neighbouring mountains and thus locate an observer's position.

R. C. A. Edge: I would agree that the rangefinder is not a geodetic instrument though it may perhaps have topographic applications. I believe that the ruby pulse Laser technique has a future role in satellite geodesy, while the CW frequency modulated gas Laser will prove to be an effective source of light for the Geodimeter-type instruments.

K. D. Harris: The greatest accuracy can only be achieved with CW, but a ruby or gas-filled Laser can each be used in a pulse or CW role.

A. Marussi: Instruments capable of measuring variations in distance very accurately would be most valuable for measuring the Earth's crustal movements.

K. D. Harris: This is an important point and Lasers may also be useful for checking movement of machinery by Doppler. We have found that there is no need for the machine to have a highly reflective surface.

J. Kelsey: I think there is a future in using a pulse Laser to determine the height of an aircraft undertaking aerial photography.

Experiments with Lasers in the Measurement of Precise Distances

S. E. SMATHERS, A. C. POLING, R. TOMLINSON
and H. S. BOYNE

U.S. Coast and Geodetic Survey
Environmental Science Services Administration
Rockville, Maryland 20852

Comparable Geodimeter measurements have been made with a mercury vapour lamp and a 0·5 mW Laser over taped distances up to 1750 m and over a previously measured Geodimeter distance of 8 miles. Differences in these measurements are of the order of 1 cm indicating no degradation in accuracy of the Laser measurements. The long-range capability of a continuous 50 mW Laser (6328Å and 6118 Å) was also investigated. Distances of 4 and 10 miles were successfully measured with unusual sensitivity in a moderate haze which would not permit measurement with the standard mercury vapour lamp. Path differences between the two wavelengths are quite small and consistent with recent results from the Central Radio Propagation Laboratory. Improvements in precision are theoretically possible, and are indicated by limited laboratory data. However, as would be expected, the variation in atmospheric refractivity dominates on the long horizontal paths.

1. *Introduction*

Recent improvements in Laser[1] light modulating techniques, photomultipliers and electronic instrumentation[2] make it possible to investigate space geodesy with improved tools and techniques. It appears possible to solve the atmospheric refractive index problem[3] as well as to increase the horizontal distances along the Earth's surface and to satellites.

In August 1963, the U.S. Coast and Geodetic Survey initiated a research and development programme under the technical direction of Dr Hellmut Schmid to improve the accuracy and range of horizontal measurements along the Earth's surface. Guide lines for this effort were provided by Study Group No. 19 of the International Association of Geodesy[4] and by others.[5] Thus, due in large part to this international group, a concerted research effort on the atmospheric refractive index problem was initiated at the National Bureau of Standards, Boulder, Colorado. As part of this effort, a co-operative in-house project at the

Coast and Geodetic Survey and the National Bureau of Standards of Washington D.C. was also initiated to study methods of improving the range of geodetic measurements and to define the limitations of the improvements possible.

The purpose here is to describe the techniques used and results of two groups of Laser experiments designed to determine the feasibility of using present gas Lasers to improve range without a loss of accuracy.

2. *Comparable Mercury Vapour and Laser Geodimeter Measurements*

The purpose of these measurements was to make a simple comparison of the conventional mercury vapour light source which is standard in the model 4D Geodimeter and a gas Laser. A taped base line known with an accuracy of better than 1 part per million (ppm) was taken as the standard for this comparison.

The Laser used in the experiment was a model 130 gas Laser made by Spectra-Physics Inc.[6] The compact $(13\frac{1}{2} \times 4\frac{1}{2} \times 3\frac{1}{2}$ in., 11 lb) device has a power output of 0·1 to 0·5 mW CW at 6328 Å. The output beam is linearly polarized by passage through Brewster's angle windows. The diameter of the emergent beam is 2·5 mm at the exit, and the divergence was 0·1 milliradians.

An oversimplified description of the step-by-step atomic process involved in this Laser is given below. An optical resonator consisting of a 30-cm glass tube with two highly reflective multi-layered dielectric reflectors at either end is filled with helium and neon gas. The beam exit mirror is partially reflecting. Helium atoms are raised to the $1s$ metastable state of 20·61 eV above the ground state by d.c. excitation. The neon atoms are excited to the $3s_2$ state of 20·66 eV by energy exchange collisions with the excited helium atoms which return to the ground state. Thus there is a preferential population of the neon atoms at the $3s_2$ level relative to the $2p_4$ level, and there is a potential gain at the 6328 Å wavelength. Neon atoms at the $3s_2$ level are stimulated by 6328 Å photons to emit in-phase photons of the same wavelength and return to the $2p_4$ state and the ground state where they can be excited again by collision with d.c. excited helium atoms, thereby making the coherent light output a continuous one from the partially reflecting mirror.

The Laser was mounted on the right side of the Model 4D Geodimeter. Horizontal and vertical adjustments were provided for pointing and bore sighting the Laser with the Geodimeter telescope and receiver optics.

A complete electro-optical shutter, consisting of a focusing lens, Kerr cell, objective lens and polarizer, was mounted in front of the

Laser (Figs. 1(*a*) and 1(*b*)). Two 45° mirrors were mounted on an aluminium bar for internal calibration and for directing the Laser light to the distant mirror from a point near the optical axis of the receiver (Fig. 2). The transmitter's 30 Mc/s voltage was used to drive the Laser optical shutter. However, good tuning was not achieved owing to the increase in capacitance of the transmission line.

This experimental unit was thus equipped to provide alternately mercury vapour and Laser light measurements, thus permitting a comparative study of the two light sources over predetermined distances. Measurements were made over a distance of 50 m of the precisely taped base line at Beltsville, Md, to determine the zero constants for the mercury vapour and Laser light sources. Measurements were also made to check the alignment and sensitivity of the photomultiplier to the red line.

Laser data taken with a precisely centred beam in the Kerr cell, qualitatively indicated that improvement in precision of the Laser Geodimeter is fairly certain as indicated by less change in delay line readings due to phase changes; however, this experimental system is not stable enough to permit a complete investigation of this effect.

Three sets of measurements were made over taped distances of 1650 and 1750 m and over a previously measured 8-mile Geodimeter line using both light sources (Table 1). The first set of measurements was

TABLE 1. *Comparative mercury vapour and Laser measurements*

	Mean distance (m)	Mercury vapour–Laser difference (m)	Taped difference (m)
Model I Laser 11 runs	1650·035	−0·006	+0·010
Mercury vapour 14 runs	1650·029	—	+0·016
Model I Laser 1 run	12896·369 (8 miles)	−0·005	+0·013
Mercury vapour 4 runs	12896·364 (8 miles)	—	+0·018
Model 2 Laser 3 runs	1750·055	−0·008	+0·005*
Mercury vapour 3 runs	1750·047	—	−0·013

* Internally calibrated.

Fig. 1(*a*). Electro-optical shutter mounted in front of the Laser.

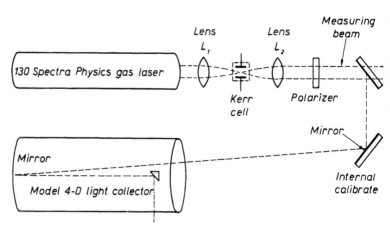

Fig. 1(*b*). Experimental Laser light-source.

FIG. 2. Gas Laser.

made with no internal calibration of the Laser light source; the second and last sets were made with internal calibration. End point measurements of temperature, pressure and humidity were made for correction of the velocity of light. Barrell and Sears' formula for unmodulated light and the group velocity correction of resolution No. 1[4] were used to reduce the data.

Comparison of the first and last sets of measurements shows a continuing need for internal calibration. The 5 mm difference between Laser measurements and the taped distance indicates a small improvement in accuracy; however, the sample was small. The 8-mile measurements, while above the mean of previous Geodimeter measurements by approximately 1 cm, show only a 5-mm difference between the mercury vapour and Laser light sources. The one centimetre most probably can be attributed to a 20 kt wind which prevailed during the measurements. Thus, these data show no significant difference in the accuracy of mercury vapour and Laser measurements. Further, no overwhelming difficulties were experienced in using a Kerr cell with a Laser light source in this fashion.

3. *Precise Distance Measurements with a 50 mW Gas Laser*

The objective of these measurements was to investigate the long-range capability of a more powerful Laser and to study the efficiency of ordinary optics and current photomultipliers being used for first-order traverse measurements. Such measurements have been suggested by many investigators,[7,8] but the state of the Laser art would not permit such measurements.

A Spectra-Physics Model 125 gas Laser[9] was loaned to the Coast and Geodetic Survey for 4 weeks for these experiments. The output was measured at 50 mW at 6328 Å and 6 mW at 6118 Å. The beam diameter was 2 mm and vertically polarized. The wavelength could be changed by altering the Brewster's angle window.

A Kerr cell optical shutter similar to the previous experiments was constructed and mounted in front of the Laser (Fig. 3).

The zero constant was determined by measuring 62 m with the Geodimeter mercury vapour lamp and both lines of the Laser. The Laser was mounted on a table on the top of the Washington Science Centre from which 4 and 10 miles were measured successfully in a moderate fog (Table 2). The visibility was 5 miles. Meteorological data were taken at both ends of the lines for colour corrections and determination of the velocity of light correction.

These data indicate that precise distances can be measured with Kerr cell shutters and excessive coherent light up to 50 mW without degrading the accuracy. Further, the poor visibility which existed on the 10-mile line indicated that 1 mW at 6118 Å is about the lower limit for these conditions.

TABLE 2. *50 mW Laser experimental data*

	Mercury vapour (m)	Laser 6328Å (red) (m)	Laser 6118Å (orange) (m)
Calibration	62·376	62·372	62·378
4 miles	7675·958	7676·001	7676·011
10 miles	—	16779·047	16779·021
Constant	0·450		

Fᴵɢ. 3. Laser and associated equipment.

Scatter in both the atmosphere and in the conventional optics used was very pronounced. The back scatter from the beam could be observed 10 miles or more, such that bore sighting of the telescope was accomplished by simply aligning the cross hairs with the far end of the back scattered beam. The light loss in the optical shutter was 90 per cent for both wavelengths, indicating the need for coated aspheric lenses. The light loss was measured at about 30 per cent in the Kerr cell which should be replaced with a KDP cell.

The beam diameter at 10 miles was approximately 20 ft. and can be significantly improved with a diffraction limited collimator (0·01 milliradians).

Future plans include permanent modification of a model 4D Geodimeter, and construction of a 50 mW light source for operational measurement of distances up to 50 miles.

References

[1] Bridges, W. B. & Chester, A. N., 1965. Visible and u.v. Laser oscillations at 118 wavelengths in ionized neon, argon, krypton, oxygen and other gases. *Applied Optics*, **4**, No. 5.

[2] Jensen, H. & Ruddeck, K. A., 1965. Applications of Laser profiles to photogrammetric problems. Paper presented at the American Society of Photogrammetry, Washington, D.C.

[3] Bender, P. L. & Owens, J. C., 1965. Corrections of optical measurements for the fluctuating atmospheric index of refraction. *Journal of Geophysical Research*, **70**, No. 10.

[4] Edge, R. C. A., 1963. Study Group No. 19 of the International Association of Geodesy. *Bulletin Géodésique* (December), Resolution No. 4.

[5] National Academy of Science—National Research Council; Report of the Ad Hoc Panel on Basic measurements, Washington, D.C., 8 December, 1961.

[6] Spectra–Physics; Model 130 Gas Laser Operations and Maintenance Manual, 1255 Terra Bella Ave., Mountain View, California.

[7] Sybren, H. de Jong, Private Communication; The University of British Columbia, Vancouver, 8 Canada, 8 January, 1965.

[8] Simmons, Lansing, Private Communication; Coast and Geodetic Survey, Washington Science Centre, Washington, D.C.

[9] Spectra–Physics; Model 125 Gas Laser Specifications; 1255 Terra Bella Ave., Mountain View, California.

DISCUSSION

J. C. Owens: What is the advantage of the KDP modulator over the Kerr cell?

S. E. Smathers: With the KDP modulator, less light is lost by scattering.

K. D. Harris: How much power is emitted in the 6118Å orange line?

S. E. Smathers: Initially 6 mW and the total output after going through the shutter is 1 mW.

J. W. Wright: Were the ranges quoted for observing at night and not in the daytime?

S. E. Smathers: Yes.

W. R. C. Rowley: What caused the loss of brightness and scatter in the lens?

S. E. Smathers: The lack of brightness and scatter is due to reflections from the lens surfaces. We plan to fit a better quality lens which will improve the performance.

R. C. A. Edge: I notice that the crystal modulator used in the Carl Zeiss Jena instrument has very good light transmission. Did you consider using it?

S. E. Smathers: Yes, I have thought about it since.

Phase-Modulated-Laser Alignment Interferometer

D. BLUMFIELD

Presented by

A. F. P. SCHRYVER

Elliott Bros., London

A recently developed short gas Laser is capable of being frequency-modulated over a range of up to 800 Mc/s centred on the optical frequency of $4 \cdot 7 \times 10^{14}$ c/s. In the proposed alignment interferometer, an interference pattern is produced by two parallel beams of light, which are derived from the Laser and arranged to have a large relative optical path-length difference. As the optical frequency is swept, the fringe pattern is scanned in space and only a single photo-detector is required at each remote station, to determine the time of maximum illumination in relation to the phase of the Laser modulation, and hence its position with respect to the fringe system. Ambiguities can be eliminated by the use of two alternative beam separations. The proposed system should be capable of angular resolutions of the order of microradians.

The possible applications of continuous Lasers for distance and angular measurements have been pursued vigorously ever since the helium–neon gas Laser was developed. Basically the systems developed are modifications of existing microwave techniques but employing a much higher carrier frequency of some 5×10^{14} c/s. This paper describes an alignment interferometer which takes advantage of this high carrier frequency to obtain very precise alignment over distances ranging from tens of centimetres to 1 km or so. It has many applications in civil engineering such as the alignment of linear particle accelerators and rocket sledge tracks, and in mechanical engineering in the alignment of large machinery or for the control of machine tools. For example, a machine tool can be made to follow a light beam and thus machine more accurately than its slide allows.

The system described was developed to enable a rocket sledge track 1 km long to be aligned straight to within 0·25 mm. The application calls for no fixed line to which the track is aligned, but a simple addition to the system allows alignment to any desired datum.

The system consists of a modified double-slit interferometer in which interference patterns are produced in the same way as in the original double-slit experiment described by Young. Fig. 1 is a diagram of the arrangement of this experiment and shows how the interference patterns are produced. It can be seen that alternate bright and dark lines parallel to the slits corresponding to constructive and destructive interference are obtained. The spacing of the fringes is a function of the

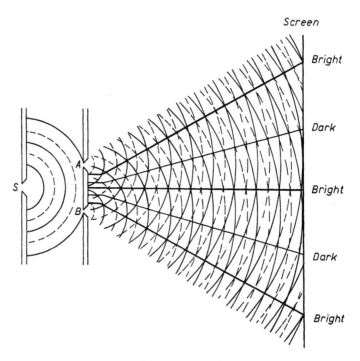

FIG. 1. Young's experiment.

wavelength of the light, the distance between the slits and the distance from the slits at which the fringes are observed. As the locus of these fringes is used as a line from which measurements can be made, it is necessary to verify that it is a straight line. Fig. 2 shows the derivation of the equation of the locus. It will be noticed that the locus of the point P can be expressed in terms of $(l_2 - l_1)$ which for the case of the first destructive interference fringe must be equal to $\lambda/2$ and for the nth dark fringe $n\lambda/2$ providing both slits emit radiation that is exactly in phase.

Hence the locus of the nth dark fringe can be expressed by the equation

$$x = \frac{2\,yd}{n\,\lambda}\left[1 - \frac{1}{2x^2}(d^2+y^2)\right]$$

For normal values found in practice, this equation represents a straight line with a small error term which can be shown to be negligible for most applications. The locus of a fringe can, therefore, be taken as the measuring datum in the system.

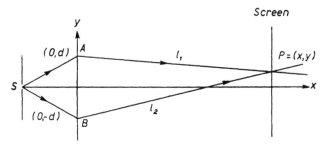

FIG. 2. Derivation of the equation of the locus.

$$l_1{}^2 = x^2+(d-y)^2$$
$$l_2{}^2 = x^2+(d+y)^2$$

$$l_2-l_1 = x\left[1+\left(\frac{d+y}{x}\right)^2\right]^{\frac{1}{2}} - x\left[1+\left(\frac{d-y}{x}\right)^2\right]^{\frac{1}{2}}$$

$$= \frac{2yd}{x} - \frac{yd}{x^3}(d^2+y^2)+(0)\frac{1}{x^5}$$

$$\simeq \frac{2yd}{x}\left[1 - \frac{1}{2x^2}(d^2+y^2)\right]$$

Thus to align, for example, the rocket sledge track, the following arrangement can be used. Two parallel, coherent light beams corresponding to light emerging from the two slits are sent down the track. By the time the beams have reached the first measuring station, they have overlapped owing to diffraction and have consequently produced an interference pattern similar to that in Fig. 1. There would be, say, ten measuring stations at intervals of 100 m down the track and, at each, a detector would be used to measure the distance from the reference face of the track to the bright fringe that is taken as the datum. As it is necessary to enclose the beams and the detectors in a pipe to avoid atmospheric effects, it would obviously be preferable to make the measurement by moving the fringes rather than the detectors. This can be done by varying the relative phases of the two light sources—that is, phase modulating one of the beams with respect to the other. If an electrical signal can be used to control the modulation, the phase difference between this signal and that produced by the detector gives a measure of the distance of the detector from the datum bright fringe. So all that is required at each measuring station is a detector that can be swung into the light path when a measurement is required. This method has the advantage that the phase difference is independent of the signal amplitude. We can detect easily one twentieth of a complete fringe, that is a phase difference of 18° between the modulating signal and the detector signal; thus to obtain alignment to 0·25 mm at 1 km, the fringe spacing must be 20 × 0·25 mm, i.e. 5 mm at 1 km. Substituting this into the equation for the locus of the fringes, we find that the two beams of light (or the slits) must be roughly 10 cm apart. Further, the error term in the equation of the fringe locus represents in the worst case, that is at the first measuring station, only 0·025 mm, which is negligible in this case.

We may not be sure that the detectors are aligned to within half a fringe initially, so we may have to use a coarse set of fringes first. These are easily obtained by reducing the distance between the two beams of light, but there are nevertheless some practical difficulties in doing this, a possible solution to which is mentioned later.

So far we have made no mention of Lasers, and why, then, has this sort of system not been used before Lasers existed? The reason is simple —a monochromatic source of sufficient intensity and of sufficient coherence time has not been available for long distance interferometry such as this. However, if we use a helium–neon gas Laser the coherence time is enormous and adequate light is emitted. Also, as the frequency of the radiation from the Laser can be varied slightly by altering the length of the optical cavity, this suggests an easy way of obtaining phase modulation of one beam with respect to the other. The sort of arrangement required is shown in Fig. 3.

The output of the Laser is a narrow parallel beam of light and it is not necessary to use actual physical slits. In this arrangement, the relative phases at A and B depend on the number of wavelengths in the path AB. By varying the frequency of the light, the relative phase of A and B is controllable. The 'parallel' beams spread owing to diffraction and, at the first measuring station, the beams overlap. This arrangement would be difficult to align and does not give the desired independent control of the separation of beams and path length difference so it can be replaced by that shown in Fig. 4. The system provides for two alternative beam separations with substantially the same path difference, and is constructed from a number of solid glass or quartz blocks mounted in contact.

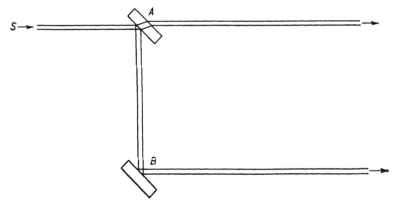

Fig. 3. Arrangement required for phase modulation of beams; A, half mirror; B, full mirror.

Now as the Laser frequency can be changed by about one part in a million, one million wavelengths in the path difference will be required to obtain a phase change of 360°. Hence, the distance 2AF must be equal to one million wavelengths—that is about 65 cm in air and about 15 cm in glass.

A frequency controlled helium–neon gas Laser has been developed by Elliott Bros. for use in systems such as this, and Fig. 5 shows a photograph of this Laser. Three invar rods $\frac{1}{2}$ in. in diameter with lengths of 40 turns per inch thread at each end are held at the corners of an equilateral triangle by the two spacers. The mirrors forming the ends of the cavity are each mounted on circular plates supported on the three invar rods by means of special nuts that allow the mirrors to be adjusted for alignment. At the upper end of the cavity, a cylindrical piezo-electric transducer is interposed between the mounting plate and the mirror. A similar element is mounted on the opposite side of the plate and carries a dynamic balancing weight equal to the weight of the mirror. Above this weight a photocell is mounted. The Laser is supported axially by two asbestos braces. At the bottom of the cavity there is a prism that turns the Laser beam through 90° so that it emerges horizontally.

Frequency control of the Laser radiation is effected over the width of the Doppler-broadened atomic line by altering the cavity length by applying a suitable voltage to the piezo-electric mirror mounting. The photo-cell monitors the change in amplitude due to the frequency variation and provides information to enable the centre frequency to be kept constant.

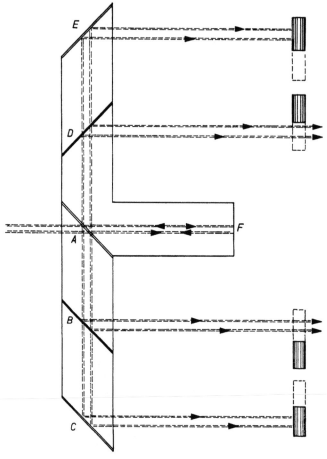

FIG. 4. Arrangement required to give independent control of beams and path-length difference. Surfaces *A*, *B*, and *D* are half-silvered; Surfaces *E*, *F*, and *C* are fully silvered.

Calculation on the power level expected at the detectors for the rocket sledge track application shows that with sensitive areas measuring from 0.25×2.5 mm to 2.5×25 mm, a sufficiently large signal would be available from a Laser output of 0.5 mW, which is the approximate power output of the Laser described above.

An improvement in general convenience is obtained using the Wall-mark effect position-sensitive photocells which utilize the lateral photo-electric effect. In this case, zero output from the photocell indicates

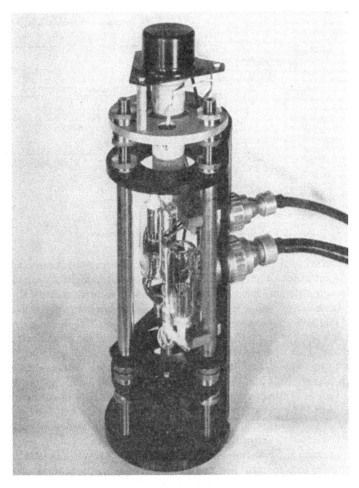

FIG. 5. Frequency-controlled helium–neon gas Laser.

that the bright fringe is centrally placed in the aperture; if otherwise, the polarity of the electrical output indicates the sign of the error.

This type of detector can be used with an electronic servo system to set the Laser frequency so that a bright fringe falls centrally on the detector as is shown in Fig. 6. The angle of the bright fringe is then proportional to the voltage applied to the piezo-electric mirror mount. If we swing each detector into view in succession and record the angles, it is easy to compute the deviation from linearity at any point. The

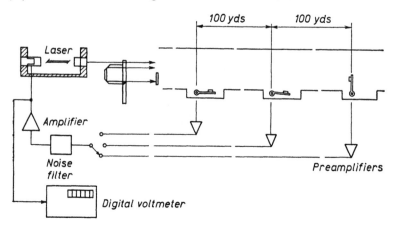

Fig. 6. Use of position-sensitive photocells.

sequencing of the measurement and the processing of the results can be made automatic. The system as described measures angles in one plane only. To accommodate the other plane, it would be convenient to use the same Laser with a second set of prisms oriented at right angles to this one and a second set of detectors also oriented at right angles to the first set, but employing the same mountings and the same electronic systems. This system could be applied to the control of machine tools but, in this application, it is desirable to have simultaneous readout in both planes. This can be done by using polarized light and the two-axis type of Wallmark detector. As plane polarized light does not inter-fere with light polarized in the orthogonal plane, we can set up two pairs of parallel beams, each pair being polarized in orthogonal planes and the resulting diffraction pattern is therefore the sum of the patterns due to each pair separately. As gas Lasers produce plane polarized light, we could conveniently use two gas Lasers each producing a pair of parallel beams and would then have a pencil of straight lines set up with independent control of their elevation and azimuth. The interference pattern would look like a square array of dots. We could mount the single detector on the machine tool slide and use two servo systems and take a record of the Laser control voltage as the slide is traversed. This would give a record in two planes of the linearity of the motion, or pro-vide a reference line from which the machine can be controlled that is independent of the slide linearity.

The control of machine tools in this way is perhaps the most attrac-tive use of the interferometer, as it requires no elaborate setting-up procedures to ensure that measurement is made from the same bright fringe at each station, the datum fringe being tracked as the carriage moves down the slide. In fact it could be found necessary to carry out initial alignment of the system using discrete measuring stations, by tracking a fringe down the pipe to ensure that the detectors at each station are measuring to the same datum fringe.

In conclusion, it must be emphasized that we have examined only a very limited field of application using interference patterns in measuring techniques—for example, alignment to a straight line is not essential; if the characteristics of any required curve are known, the location of any structure built to this curve can be verified using the same apparatus.

There must be many other applications of this alignment interferometer and it is felt that it will provide a very powerful tool in distance measurement.

DISCUSSION

R. C. A. Edge: That was a most interesting paper and I would like to congratulate you on it. The alignment of accelerators is a problem which often faces geodetic organizations. Could this technique be used?

A. F. P. Schryver: I think the alignment could be built up with a series of straight lines. Difficulty in knowing which fringe was being observed could result in ambiguities.

J. J. Gervaise: In CERN we have aligned a Laser with a rotating cell as a target. The accuracy obtained with this apparatus was 0·07 mm, for a 50-m path, in the vertical and horizontal planes. For accelerators, we do not want to work in vacuum pipes and we have to find the maximum range we can use in the open air, because for an accelerator of 300 GeV, whose circumference is about 8 km, one needs about 40–50 km of vacuum pipe, depending on the type of measurements one has to make.

P. Berthon Jones: Professor Van Heel used a graticule with one slit covered to produce an asymmetric pattern which allowed one to find to which part of the interference pattern one was pointing. Later the slit could be uncovered for the measurement.

I would point out that the placing of the double slit close to the single slit required high angular stability. Instead one could use a single detector at one terminal, the single slit at the other, and move the double slit to the points where the alignment was required. This method was accepted by Professor Van Heel.

A. F. P. Schryver: Angular stability requirements were $\frac{1}{20}$ second of arc but on a system of this type it was worth spending the money needed for a suitable pillar.

J. C. de Munck: In Holland we used a defraction screen of concentric rings which avoids any ambiguity in the position.

A. F. P. Schryver: That would be a good method.

T. S. Moss: Why not vary the phase delay in the Laser?

A. F. P. Schryver: We did do that. It does mean that a frequency stabilizer is needed.

Measurement
of Short Distances

Use of Lasers for Displacement Measurement

W. R. C. ROWLEY

Standards Division, National Physical Laboratory
Teddington, U.K.

The measurement of short lengths, as in the standardization of survey tapes at the NPL, can now be done by means of Laser techniques in place of the former long and tedious methods.

Distances of the order of 50 m can be measured to a high degree of accuracy by a reflector system travelling at 10 ft. per sec.

The introduction of Lasers greatly speeds up measurements but it does nothing to overcome the problem of the refractive index of air, although over short distances it should be possible to calculate its value to 1 in 10^7.

Laser Interferometry

Although the measurement of length by interferometry using optical radiations is a long-established and well-developed technique, the introduction of the Laser light source has opened the door to a much wider range of applications. The remarkable properties of Laser radiation lead to changes in the established interferometric techniques. Simpler optical systems can often be used because of the uniphase wavefront, and the narrow beam diameter relaxes tolerances of optical alignment. The narrow bandwidth eliminates the necessity to design interferometers having small path differences and the great brightness of the Laser source is of great advantage for photoelectric detection. Laser interferometry lends itself particularly to fringe counting techniques. Whereas previously the feeble intensity of the available monochromatic sources limited the maximum counting rate, often to no more than a few hundred counts per second, counting rates of more than a million per second are possible with a Laser source.

NPL Scale-measuring Interferometer

The special features of Laser radiation have led the National Physical Laboratory to develop a measuring machine which determines automatically to a high order of accuracy the positional errors of graduations on scales up to one metre in length. Measurements are made by a fringe counting technique, the observations being made photoelectrically whilst the scale is moved steadily by a transporting mechanism along the bed of the machine.

A two-beam interferometer, shown in Fig. 1, is used in this apparatus. Movement of the carriage carrying the corner cube or reflector system changes the optical path difference. The interferometer is aligned so that the superposed beams are parallel and their interference causes the whole field to change intensity uniformly. The intensity variations are strictly sinusoidal for uniform movement, one cycle corresponding to a carriage movement of half the wavelength of the Laser radiation. Thus by counting the number of intensity variations, the distance of movement of the carriage can be determined. In the measuring machine a scale is mounted on the same carriage as the corner cube and is viewed by a photoelectric microscope fixed to the semi-reflector block. Signals, from this microscope, are measured by gate counters which operate from the interferometer signals so that the positions of the scale lines are determined relative to the zero scale line. There is also a phase measuring system for determining fractional parts of the interferometer signals.

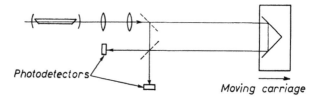

FIG. 1. A two-beam interferometer.

This measuring machine has been carefully designed and engineered on strict metrological principles to enable measurements of the highest accuracy to be made. Its purpose is to measure length scales of various kinds, up to a metre long, with an accuracy of one part in 10^7 and better. It is a semi-automatic machine which can be programmed to measure only the desired graduation lines on the scale and to record the readings on punched paper tape as the carriage moves steadily at constant speed. As a measuring machine it is rather unusual in that it makes its measurements on the scale whilst it is in steady movement.

Application to Longer Distances

With a Laser source there is no difficulty in obtaining interference effects over paths of 50 m, and in principle over paths several orders of magnitude greater than this. If an interferometer of the type described above is set up with a bi-directional counting system (using two photodetectors in phase quadrature) then any movement of the reflector system along the optic axis can be measured and recorded. At the NPL it is intended to use this kind of arrangement to measure Earth movements, both elastic movements and those across known geological fault lines. In the study of this application it is hoped to collaborate with a geophysical group at Cambridge.

It is of course possible to measure distances of the order of 50 m by traversing a reflector system all the way from one end to the other. Moreover, it seems quite a practical proposition; because of the great intensity of Laser radiation, it is possible to get good signal-to-noise ratios in wide band detector and amplifier systems. Thus fast counting rates are possible and probably movements of 10 ft. per second or more could be used. The advantage of using interferometry over such distances is the great accuracy of the technique. Because of this and also the simplicity of the measuring procedure it is quite probable that the NPL will use this Laser interferometer technique for the measurement of surveying tapes. At the moment surveying tapes which are required for projects demanding particular accuracy are measured at the NPL by comparison with reference tapes which have been measured in terms of a 4-metre bar, which is in turn measured against a standard one-metre bar. The interferometric technique will afford a more direct, accurate, and convenient measurement procedure.

Accuracy Limitations

The importance of interferometric measurement is on account of its high accuracy. Indeed the most accurate way of measuring or defining length is through interferometry. It is because of this that the primary length unit (the metre) was redefined in 1960 in terms of the wavelength of the radiation corresponding to a particular energy level transition of the krypton-86 atom. Thus the primary length standard is an optical radiation and it is natural to consider using interferometric techniques whenever very precise length measurement is required.

The accuracy of interferometric measurement using Laser radiation is limited by three main effects.

1. Wavelength Stability

Although the Laser radiation is at any time extremely monochromatic, the actual value of the wavelength is liable to fluctuations unless some system of stabilization is employed. The wavelength is determined not only by the energy level scheme of the active atoms, but also by the precise length of the optical cavity. Thus a variation of perhaps two parts in 10^6 is possible. Such variations can, however, be reduced by electronic feedback controls which stabilize the wavelength to the centre of the energy transition. With the existing controls, a long term wavelength stability of at least one or two parts in 10^8 may be achieved. This is comparable with, though perhaps not quite as good as, the krypton-86 primary standard. It is nevertheless quite adequate for most purposes.

2. *Refractive Index*

It is unfortunate that the properties of Laser radiation do nothing to overcome the problem of the refractive index of the air through which the light passes. The accuracy of a measurement can only be as accurate as the knowledge of the refractive index. Over the length of path envisaged for interferometric measurement it is not difficult to sample the average atmospheric conditions, which should enable the refractive index to be calculated to one part in 10^7. Greater accuracy can be realized under laboratory conditions by the use of a refractometer.

3. *Interferometer Geometry*

Because a gas Laser emits its radiation in a beam of quite small diameter, it is tempting to try to use narrow diameter beams throughout the whole optical system. There is danger, however, in using too small a beam diameter. Diffraction makes it impossible to keep a beam of light truly parallel. The plane wavefront of a parallel Laser beam soon becomes spherical as the beam diverges, the effect being accentuated by smaller initial diameters. As a result, the superposed wavefronts in an interferometer cannot in general have the same curvature at different values of path difference, and this gives rise to an error in the apparent distance and a loss of visibility. This error is calculable and need not be significant if adequate beam diameters are used. It is usually adequate for the beam width to be several times the minimum for a confocal cavity of similar dimensions to the interferometer. For interferometers more than one or two metres in length, the increase of beam diameter by diffraction encourages the use of progressively larger diameter beams so that the diffraction error becomes correspondingly less important.

Conclusion

The use of Laser light sources makes interferometric measurement a much more practical technique for long distances than hitherto. Probably distances of several hundred metres could be measured directly by fringe counting techniques in suitable conditions. The advantage of the method lies in its inherent accuracy.

Acknowledgement

The work described in this paper was undertaken as part of the research programme of the National Physical Laboratory and is published by permission of the Director.

DISCUSSION

K. Poder: Have you considered using this system to measure the value of gravity?

W. R. C. Rowley: Dr Cook of the NPL recently completed a measurement of the acceleration due to gravity using the system in which a glass ball is thrown upwards and its passage is timed between two positions of accurately known vertical separation. Such an experiment could be carried out using a corner cube reflector and a fringe counting system. This arrangement would have the advantage that the whole movement could be followed continuously by recording the interferometer signals against a high-frequency timing signal. The method has much to commend it but although we have considered using this system we do not intend to proceed with it at the present time.

G. C. Weiffenbach: An experiment using a corner cube in free fall was recently described in the *Journal of Geophysical Research.**

M. J. Puttock: A similar experiment to establish 'g' by the free flight of a corner cube with fringe counting is in an advanced stage in Australia. Is the high rate of counting referred to in your paper bidirectional?

W. R. C. Rowley: No.

F. T. Arecchi: Similar experiments are also being carried out in Italy.

J. J. Gervaise: To what extent would this technique be suitable for use outside the laboratory? The reflector system must be moved along a straight path and this may require the establishment of guidance rails.

W. R. C. Rowley: In the interferometer which I have described the corner cube reflector must be moved along a straight path and in all our experiments so far this has been done by rails. By changing the interferometer system into a Michelson arrangement and using a large fixed mirror at normal incidence to make the light return through the corner cube along its outgoing path, we can make an interferometer which is insensitive to lateral movements of the corner cube. In this case one can envisage making a corner cube of large proportions, preferably a hollow one formed of separate mirrors, which could be picked up and carried by hand along the optical path. It might be necessary to arrange a stretched guiding line to indicate the approximate position of the beam.

* Vol. 70, No. 16, August 1965, pp. 4035-8.

Long-Distance Interferometry
with an He–Ne Laser*

F. T. ARECCHI and A. SONA

Laboratori Centro Informazioni Studi Esperienze, Milan, Italy

Interference fringes have been observed up to an optical path difference of 120 m (mirror separation 60 m) without a substantial loss of visibility. Laser output is split by a beam splitter on to a Michelson interferometer through a telescope system which reduces the angular spread.

By changing the path difference continuously, in the first experiments the fringe visibility was subjected to a periodical change, going down to zero every time the path difference was an odd multiple of the cavity length. This is due to interference between different axial modes. To avoid this effect we have made both a transverse and an axial selection. The isolation of a TEM_{00} pattern is accomplished by adjusting two movable diaphragms near the mirror. The selection of a single axial mode is done with an external plane mirror which acts as a second Fabry–Perot. This technique does not reduce the output power, as occurs when a near-threshold operation is used to isolate a single mode. On the contrary, a power increase is observed in the useful mode.

An electronic system for measuring and analysing experimental results is described and limitations on circuit bandwidth, observation time and precision are related to fluctuations within the interferometer system and associated circuitry.

1. *Introduction*

We report a long-distance interferometry experiment where the source is an He–Ne Laser working at the 6328 Å transition in single mode operation.

Interference fringes have been observed up to an optical path difference of 120 m (mirror separation 60 m) without a substantial loss of visibility.[1]

* Work done with the financial support of the Consiglio Nazionale delle Ricerche. Reproduced with the kind permission of the Polytechnic Institute of Brooklyn, publisher of the *Proceedings of the Symposium on Quasioptics*, of which this paper forms a part.

An experiment of this kind had already been performed with an infra-red Laser over a 9-m path difference[2] and a 100-m interference experiment with a visible gas Laser has recently been reported.[3]

Three characteristic times can be associated with the Laser interferometer.

(*a*) The coherence time (in the classical sense) which limits the transit time between the Michelson mirrors. Our experiments show that a transit time of the order of 10^{-6} sec is possible.

(*b*) A time of the order of 10^{-3} to 10^{-1} sec which characterizes the elastic fluctuations of the interferometer frame and the acoustical and thermal fluctuations of the traversed medium, and which contributes a limitation on the observation time T in order to reach a given precision.

(*c*) A time of the order of minutes which characterizes the thermal drifts of the frame and the stability of the laser itself.[4]

Much work has to be done to improve the times (*b*) and (*c*) above. For the present experiments, an electronic observation system has been designed which optimizes the available output despite noise and fluctuation considerations.

2. *Experimental Apparatus and Results*

The experimental set-up is shown in Fig. 1. We have used a confocal cavity of 2 m length. The second surface of the mirrors is corrected in order to have a plane wave output. The gas discharge is fed by a current stabilized d.c. supply in order to avoid any fluctuation or 50 c/s line ripple.

The Laser output is split by a beamsplitter M_1 on to a Michelson interferometer through a telescope system which reduces angular spread. The two Michelson mirrors A and B are made as two cat's-eye systems for easier alignment. This allows one to span a long distance on a conventional optical bench without readjusting the system.

In the first experiments, by continuously changing the path difference, the fringe visibility was subjected to a periodical change, going down to zero every time the path difference was an odd multiple of the cavity length. This is due to interference between axial modes, as has been previously pointed out.[2] To avoid this effect, we have made both a transverse and an axial mode selection. The isolation of a TEM_{00} pattern is accomplished by adjusting two movable diaphragms S_1, S_2 near the mirrors. The selection of a single axial mode is done with an external plane mirror (99 per cent reflectivity), which acts as a second Fabry–Perot.[5,6] The result was checked by observing the beat frequencies on a spectrum analyser. The position and alignment of this external mirror were adjusted so that the strengths of the first four beats (75, 150, 225 and 300 Mc/s, respectively) were reduced by more

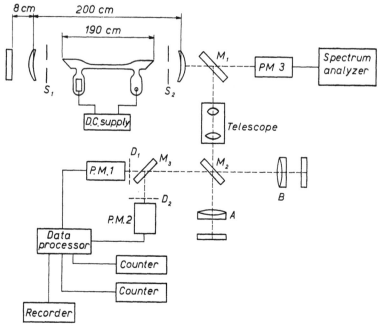

Fig. 1. Experimental set-up.

than 20 dB. This technique does not reduce the output power, as occurs when a near-threshold operation is used to isolate a single mode. On the contrary, a power increase is observed in the useful mode.

Fig. 2(a) shows the fringe pattern observed with this technique for a 6-m path difference. At this point, the visibility would have been very poor if observed without the second Fabry–Perot. Fig. 2(b) shows the fringe pattern for a path difference of 120 m.

Fig. 3 shows the output from a phototube when the separation is 50 m. Line *a* gives the dark value; lines *b* and *c* give the intensities at the two Michelson arms; the fluctuating line *d* gives the shape of the fringes which changes owing to random fluctuations of the optical length. The measured visibility is 86·2 per cent and fits with the value calculated, taking into account the difference in amplitude between the channels and the ratio of the aperture width to the fringe spacing.*

The interference fringes are handled by an observation system designed for two different basic purposes: first, to check the stability

---

* A visibility measurement through a densitometric analysis of a photo-graphic plate is less straightforward, because one should take into account the nonlinear characteristic of the recording emulsion.

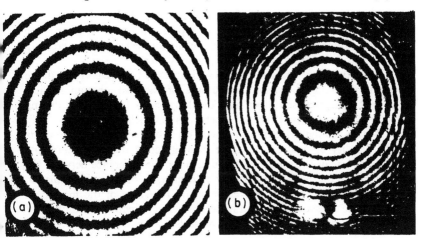

Fig. 2. (*a*) Interference fringes for a 6-metre path difference (Laser with the second Fabry–Perot), exposure time 1/50 sec; (*b*) Interference fringes for a 120-metre path difference, exposure time 1/100 sec.

in time of a fixed length up to 60 m with the resolution of an eighth of a wavelength; second, to allow an absolute measurement of a length between 0 and 60 m in number of eighths of wavelengths. The fringe image is projected on to a pair of photomultipliers P.M. 1 and P.M. 2 through annular slits in diaphragms D_1 and D_2 concentric to the fringe

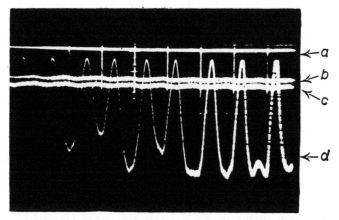

Fig. 3. Output from a phototube viewing the fringes, with a ratio slit-width/fringe-spacing of about 0·4. (*a*) Zero line; (*b*) and (*c*) Intensities of the two separate beams; (*d*) Intensity of the interference pattern moving randomly. Time scale 5×10^{-3} sec/division.

system and narrower than the fringe spacing. The two diaphragms select two regions of the fringe pattern phased about 90° apart in space in order to obtain information both on the number of fringes crossing the apertures and on the sense of the movement.

Since the fringe spacing changes as the path difference is changed, the phase difference between the outputs of the two photomultipliers cannot be kept constant with fixed diaphragms. An improved dephasing system, proposed by Peck and Obetz[7] is now being tested.

A two-channel electronic system, described elsewhere,[8] records the number of quarters of fringe passing across the diaphragm slits in one sense or the other. For fixed mirrors, the difference between the two counters yields the stability information. For moving mirrors, the number recorded in the counter and the sense of the movement yields the measurement of the scanned length with a resolution of $\lambda/8$ and a maximum speed of 10^5 pulses/sec.

3. *Circuit Design and Logic of the Observation System*

The system for observing and analysing the interferometer measurements provides storage and transmittal of experimental results in real time.

An electronic system extracts information on rate, direction and number of fringe pattern changes from the annular apertures in the diaphragms D_1 and D_2 in Fig. 1 which are illuminated by the interferometer output. The apertures are concentrically related rings spaced approximately a quarter of a fringe apart within the fringe pattern area to provide electrical outputs from photomultipliers P.M. 1 and P.M. 2, which are in time quadrature to each other for any change in spacing or position of fringes in the interferometer pattern. The photocell outputs produce a Lissajous pattern on an oscilloscope when connected to its x and y terminals. A complete loop of the pattern corresponds to a full cycle of the fringe pattern and thus a change of path length of λ, or an effective relative shift of the mirrors of $\lambda/2$. The instantaneous position of the spot on the Lissajous circle is a function of the path difference, with period λ. The sense of movement of the spot corresponds to the sense of motion of the moving mirror.

The Lissajous pattern provides a two-dimensional representation of the signals applied to two trigger (discriminator) circuits connected in the same manner as the oscilloscope terminals.

The trigger circuit threshold levels are set to correspond with quadrants of the oscilloscope trace and thus change their state every time the trace crosses the threshold level (respectively, a–a for channel A, and b–b for channel B). Let us call A and $\bar{A}$ the 'on' and 'off' states of the discriminator A; A' and $\bar{A}'$ the transitions $\bar{A} \to A$ and $A \to \bar{A}$; and do the same with B. In Fig. 4, the two threshold lines cross the Lissajous pattern to form a set of eight parameters, four associated with stationary states and four with transitions. When the interferometer is in a stable position, the corresponding point will be in one of four quadrants, and can be located by two stationary parameters. When one mirror is moving with respect to the other, the motion is defined by the four crossings of the *a–a* and *b–b* diameters. These crossings are represented by the combination of the stationary state of one discriminator with the transition of the other. A full clockwise rotation of the spot, starting from the first quadrant, will give rise to the following series of events:

$$BA', \ A\overline{B'}, \ \overline{BA'}, \ \bar{A}B'$$

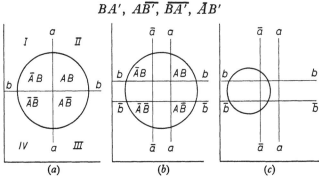

Fig. 4. Lissajous patterns built with the two phototube outputs; $a-a$ and $b-b$, threshold lines of the discriminators.

where each pair of letters denotes an 'and' operation associated with the switching of quadrant, the first letter corresponds to the state of the discriminator which does not change, and the second to the transition of the other discriminator. Similarly, a full counterclockwise notation will give rise to the following events:

$$\overline{AB'}, \ \bar{B}A', \ AB', \ B\overline{A'}$$

The first set of signals is logically summed on to an output X and the second on to an output Y, so a train of pulses is obtained from X and Y, for clockwise and counterclockwise rotations, respectively.

One can easily see that, for any fluctuation in the interferometer around a stable position, the oscillating Lissajous spot yields an equal number of counts at the two outputs X and Y. Therefore, the count difference is unaffected by spurious fluctuations with zero average, but changes by one every time there is an effective path difference of $\lambda/4$, corresponding to a $\lambda/8$ shift in the moving mirror.

15*

Hysteresis splits the switching levels introducing an imprecision smaller than the resolving power $\lambda/8$ [Fig. 4(b)]. Therefore the system is insensitive to hysteresis effects, unless a situation topologically equivalent to Fig. 4(c) occurs. Here one of the two discriminators can no longer switch and both X and Y outputs are zero, independently of the movement of the point on the Lissajous pattern.

These considerations are still valid when the two signals on the phototube are not equal in amplitude or out of quadrature, that is, when the Lissajous is no longer a circle, unless the topological equivalent of Fig. 4(c) occurs.

The block diagram of the circuitry is shown in Fig. 5.

A series of 'and' and 'or' circuits to handle the eight signals from the two discriminators, yields the following outputs:

$$X = A\overline{B'} + \overline{BA'} + \bar{A}B' + BA'$$
$$Y = \overline{AB'} + \bar{B}A' + AB' + B\overline{A'}$$

The X and Y pulses are sent to a reversible decimal counter which yields the count difference as digital output or analog output to a chart recorder. A digital record of a stability and length measurement has been given[1] over an observation time of about 50 sec, and with an integration time of 1 sec (the technique used has been described previously[1]).

4. Characteristic Times of an Interferometric Measurement

A visibility measurement yields information on frequency stability (longitudinal coherence, in the classical sense) for time intervals of the order of the transit time of the light in the device. For example, with a working limit of 300 m, this time is around 10^{-6} sec.

In addition, there is the contribution of the elastic fluctuations of the interferometer frame and the acoustical and thermal fluctuations of the crossed medium. As one can see from Fig. 3, the main contributions of these factors is a random movement of the fringes on a time scale of 10^{-3} to 10^{-1} sec (depending on the working conditions). One can minimize these effects by putting the mirrors and the beam splitters of the interferometer on isolated pillars and sending the Laser beam through evacuated pipes.

Actually, working in normal laboratory conditions on common optical benches, these effects, although present, give rise only to counts with zero average with the circuit logics of Section 3. Despite this zero average, this contributes to reducing the resolving power, as shown in the noise analysis of Section 5.

Finally on a long time scale (> 10 sec), there are drifts due to the

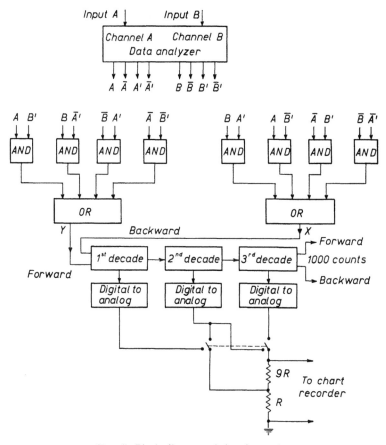

FIG. 5. Block diagram of the electronics.

lack of stability of the Laser itself and to thermal drifts of the interferometer arms. One can reduce the second effect with isolated pillars.
The best data on Laser stability thus far[4] give a reproducibility in
frequency, over long periods of time, of one part in 10^9.

When not using evacuated pipes, a much more severe limitation is
present due to air turbulence. The fluctuations associated with the
second characteristic time are due to the thermal motion of the air
molecules in quiet air, as can be realized with some care in a laboratory.
When the air is stirred up, a long-distance measurement becomes
almost meaningless. Professor Toraldo and his group are making an
extensive analysis of these effects.[9,10]

5. *Noise Considerations and Limits on Counting Speed and Observation Time*

Noise contributes a thickness to the Lissajous circle and two considerations put limits on it:

    (*a*) The thickness of the circle should never be such as to make possible a random pattern within it, which acts as in Fig. 2(*c*).

    (*b*) The thickness corresponding to the R.M.S. of the noise should not rise above the hysteresis level, in order to minimize noise triggered switching.

Otherwise the phototube noise can induce successive switching of the discriminator in a time shorter than the dead time of the following reversible counter. This dead time is $\tau_2 \simeq 10 \ \mu s$ as compared with $\tau_1 \simeq 1 \ \mu s$ of the system logic.

The two limitations can be summarized as follows:

$$i_s - \alpha i_n > i_H > \beta i_n \tag{1}$$

where i_H is the hysteresis current; i_n is the R.M.S. output noise from the photomultiplier tube; i_s is the signal current; and α and β are shape factors defined below. These boundaries limit the region of possible values for i_H.

Some considerations are now given for α and β.

Calculation of β

The electronic circuitry employs the cascading of two different sets of circuits: the logics, with a bandwidth f_1 and a corresponding dead time for counting $\tau_1 = 1/(2f_1)$, and the reversible counter with a dead time $\tau_2 = 1/(2f_2)$. The average rate of noise pulses can be measured on one channel for different values of i_H around I_0, and is given by:

$$f = f_1 \exp - \left(\frac{i_H}{2 i_n \sqrt{2}} \right)^2 \tag{2}$$

The probability of more than one pulse on the same channel within the time is given by:

$$p \simeq f \tau_2 \tag{3}$$

because we shall design the i_H values such that $f \tau_2 \ll 1$.

The number of wrong counts per second due to noise is given by (4) and (5):*

$$n = fp = f^2 \tau_2 \tag{4}$$

    * Furthermore, the average repetition rate should be reduced by a filling factor

$$\eta = 4 i_H / (2 \pi i_s)$$

corresponding to the probability for the Lissajous spot of being in the hysteresis region. But for the sake of simplicity we shall assume $\eta = 1$, which is the least favourable condition.

Furthermore any fluctuation in the optical path with zero average will give an extra number of counts per second F. The total number of wrong counts in the counter is then given by:

$$n = (f+F)^2 \tau_2 \qquad (5)$$

As the wrong counts occur with the same probability in both senses, they give rise to a random walk in the position of the counter. After a time T, the position of the counter will be shifted from the correct one by a quantity which has zero average over a number of equally repeated measurements and a variance $\sqrt{(nT)}$. A consequent error arises in the measurement, because after a time T the position of the counter is affected by a standard deviation $(f+F)\sqrt{(\tau_2 T)}$. The ultimate precision of this measurement is given by $F\sqrt{(\tau_2 T)}$. The extra error $f\sqrt{(\tau_2 T)}$ caused by the electronic circuitry can be minimized by setting the hysteresis at a level such that $f < F$. From (4), this condition corresponds to

$$\frac{i_H}{i_n} > 2\sqrt{(2)} \Big/ \left(\ln \frac{f_1}{F} \right) = \beta \qquad (6)$$

For instance, let us consider two different external conditions, with $F_1 = 10^3/\text{sec}$ and $F_2 = 10^2/\text{sec}$, respectively. In our system, $f_1 = 10^6/\text{sec}$. In Fig. 6 we have shown upper and lower limitations $i_s - \alpha_{i_n}$ (a typical value is $\alpha = 2$ as has been shown[8]) and βi_n vs. i_s for a stability measurement, taking into account that the noise current at the anode of a photomultiplier is given by:

$$i_n = \sqrt{\left(2eGI_0 f_1 \frac{\delta}{\delta - 1} \right)}$$

where f_1 is the bandwidth of the electronics before the discriminators; δ is the secondary multiplication factor of the dynodes; G is the photo-tube gain; I_0 is the d.c. component of the phototube current at the threshold level of the discriminator; and the signal current i_s is related to the d.c. component I_0 through the fringe visibility V:

$$V = \frac{I_{\max} - I_{\min}}{I_{\max} + I_{\min}} = \frac{i_s}{I_0}$$

The set of parameters used is:

$$V = 1, \quad G = 10^6, \quad \delta = 4, \quad \alpha = 2, \quad \beta = 8$$

and two values for f_1: 10^6 and 10^5 c/s.

A limit to the longest observation time T arises from equation (5). If the noise contribution has been minimized, the random fluctuations

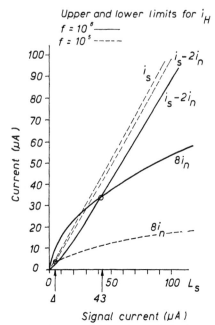

FIG. 6. Noise boundaries to the values of the hysteresis current of the discriminators.

still contribute a variance $F\sqrt{(\tau_2 T)}$, which has to be compared with the number of counts $L = d/(\lambda/8)$ associated with the path difference d, in order to yield a relative precision of the measurement. For instance, if $d = 100$ m, $\tau_2 = 10^{-5}$ sec, $F = 3\cdot 10^2$ sec^{-1} and we want a precision better than 10^{-8}, the maximum observation time is $T = 100$ sec.

Acknowledgement

We wish to thank Professor E. Gatti for helpful discussions during this work.

References

[1] Arecchi, F. T. and Sona, A., 1964. *Nuovo Cimento*, **32**, 1117.
[2] Morokuma, T., Neflen, K. F., Lawrence, T. R. and Klucher, T. M., 1963. *J. Opt. Soc. Am.*, **53**, 394.
[3] Grundzinski, R. and Paillette, M., 1963. *Compt. Rend. Acad. Sci.*, **257**, 3842.
[4] Jaseja, T. S., Javan A. and Townes, C. H., 1963, *Phys. Rev. Lett.*, **10**, 165.
[5] Kleinman, D. A. and Kisliuk, P. P., 1962. *Bell Syst. Tech. J.*, **41**, 453.
[6] Kogelnik, H. and Patel, C. K. N., 1962. *Proc. Inst. Radio Engrs*, *N.Y.* **50**, 2365.

[7] Peck, E. R. and Obetz, S. W., 1953. *J. Opt. Soc. Am.*, **43**, 505.

[8] Arecchi, F. T., Lepre, G. and Sona, A. A new reversible high speed fringe counter for Laser interferometry (to be published in *Alta Frequenza*).

[9] Consortini, A. Ronchi, L., Scheggi, A. M., and Toraldo di Francia, G., 1963. *Alta Frequenza*, **32**, 790.

[10] Toraldo, G. *et al.*, Influence of atmospheric scattering on the line-width of a Laser beam (to be published in *Alta Frequenza*).

DISCUSSION

R. Bivas: Could you give details of line width and frequency stabilization?

F. T. Arecchi: In our single mode Laser the line width and relative short term stability was in the order of 200 c/s whereas the long term stability is better than 1 part in 10^8 when using the Gordon and Labuda stabilization system.

The Application of Gallium Arsenide Light-Emitting Diodes to Electronic Distance Measuring Equipment

H. D. HÖLSCHER

National Institute for Telecommunications Research: C.S.I.R.

The gallium arsenide light-emitting diode is considered as a possible carrier source for short range EDM equipment of high resolution, mainly because of its wide band modulation capability. It is shown that the effects of atmospheric attenuation due to absorption and scattering under 'average' conditions are reasonably small for short range-work.

A system is described which uses such a diode as a carrier source with 75 Mc/s pattern frequency modulation. With suitable phase measuring accuracy at a convenient low comparison frequency, a 1-mm resolution can be obtained.

Some initial short-range results will be presented.

1. *Introduction*

Since the room temperature GaAs light-emitting diode has become available, it has been considered as a possible carrier source for distance measuring equipment mainly because of its wide band modulation capability and the ease with which such modulation could be achieved. Units are now available that have a c.w. light output of some 39 mW at room temperature in the near infra-red region. The wavelength of the emitted light is approximately 0·9 to 0·92 micron, and the spectral width is 0·05 micron; i.e. the light is incoherent.

This paper deals firstly with the problems associated with the use of IR as a carrier in EDM and the expected range that can be achieved with diodes presently available. The second part of this paper describes a laboratory model of an instrument which uses a diode as carrier source and is intended for short range work of high accuracy.

2.1 *Atmospheric Transmission of Near Infra-red*

Optical links are affected by atmospheric conditions owing to molecular absorption by gases in certain bands and to scattering of the radiation by suspended particles.

The GaAs diode radiation falls within window I of the IR spectrum, i.e. between 0·72 and 0·94 micron. There are no strong absorption bands in this region and the overall path attenuation is mainly due to scattering, except in the case of high relative humidity and high temperature where absorption owing to water vapour takes place.

The following two sections dealing with absorption and scattering effects on IR radiation are mainly based on the work of Lange, Schuldt and others.[1]

2.2 *Water Vapour Absorption*

The molecular absorption of IR radiation can conveniently be related to humidity, which is easily measurable.

The transmittance as affected by absorption is mainly dependent upon the partial pressure of water in the path length. The method which is most often used is to specify the total amount of precipitable water in the path in millimetres. If the humidity is known, the number of precipitable millimetres of water (*W*) can be calculated. Fig. 1 (after

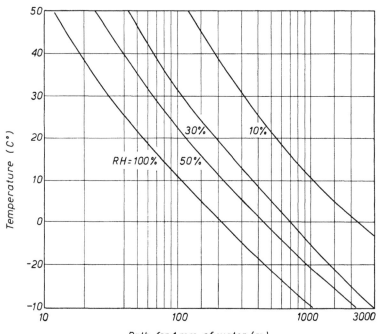

FIG. 1. Path-length for 1 mm of precipitable water versus temperature for various values of relative humidity.

Lange) shows the relationship between path length and precipitable water versus temperature for various values of relative humidity.

Lange has developed empirical expressions which can be used to compute the transmittance Ta for given values of water concentrations. These expressions are:

$$Ta = \exp(-AW^{1/2}) \quad \text{for } W < W_I \tag{1}$$

and

$$Ta = k \left(\frac{Wi}{W}\right)^{\beta} \quad \text{for } W > W_I \tag{2}$$

where W is as defined above, and $W_I = 54$, the value of W for which the absorption in window I changes from 'weak band' to 'strong band' absorption. A, k and β are constants and for window I are given as $A = 0\cdot0305$, $k = 0\cdot8$ and $\beta = 0\cdot112$. Equations (1) and (2) yield the best fits to the measured data for the conditions stated. Ta is plotted versus W in Fig. 2. From Figs. 1 and 2 it is seen that Ta, under reasonably normal conditions, will never be less than $0\cdot8$ for a 1 km path length.

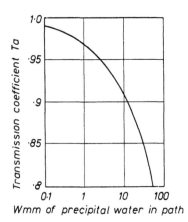

FIG. 2. Transmission coefficients.

2.3 *Scattering of Infra-red*

Assuming Rayleigh scattering (i.e. where the scattering particles are much smaller than λ), the transmitted intensity as affected by scattering can be expressed as

$$T_s = \exp(-A\lambda^q \cdot x) \tag{3}$$

for a path length x. A and q are constants determined essentially by the size and distribution of the scattering particles.

When visibility is exceptionally good $q = 1.6$, while for average visibility $q = 1.3$. When conditions are such that the visual range is less than 6 km, a good value for q is given by

$$q = 0.585V^{1/3}$$

where V is the visual range in km.

The constant A can also be expressed in terms of visual range (evaluated at $\lambda = 0.55\,\mu$) as shown below

$$A = \frac{3.91}{V}(0.55)^q$$

Hence, substituting in equation (3), we obtain

$$T_s = \exp\left[-\frac{3.91}{V}\left(\frac{\lambda}{0.55}\right)^{-q}x\right] \tag{4}$$

It is useful at this stage to introduce the concept of vacuum range, R_v, i.e. the maximum range of the optical link for $T_s = 1$.

If P_{min} is the minimum detectable power and P_0 the transmitted power, then for a vacuum path of $2R_v$

$$\frac{P_{min}}{P_0} \propto \frac{1}{(2R_v)^2}$$

and, for a scattering path length of $2x$

$$\frac{P_{min}}{P_0} \propto T_s \cdot \frac{1}{(2x)^2}$$

where T_s is evaluated for a range of $2x$.

Hence

$$R_v = xT_s^{-1/2} \tag{5}$$

Equation (5) is presented in graphical form in Fig. 3.

Note that $2R_v$ and $2x$ are used above to account for the fact that we are concerned with a system using a reflector at one end of the line.

From the foregoing it can be seen that the effects of scattering will only result in serious attenuation of the IR radiation over short path lengths in extreme conditions. If the design vacuum range is 10 km, then a working range of 1 km will be possible when visibility conditions are such that the visual range is 5 km.

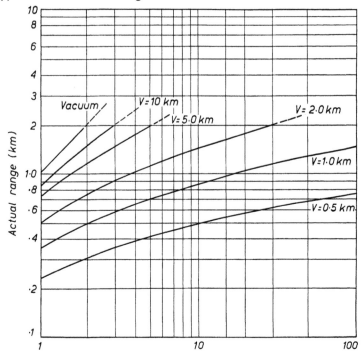

Fig. 3. Actual range versus vacuum range for different values of visual range for a reflected path.

3. Expected Range

The GaAs diode used has a maximum IR output of 30 mW for a forward current of 2 A. The light output is very nearly linearly related to the forward current. With the present modulator, an effective modulation power of approximately 8 mW is achieved. This corresponds to a total modulation component of flux of 4×10^{16} photons/sec.

A photo diode is used as demodulator of the intensity modulated IR signal in the receiver. The Si PIN photodiode with a large reverse bias has a quantum efficiency of $\frac{1}{3}$ to $\frac{1}{2}$ and very wide band operation can be achieved with proper loading.

When operated well below cut-off frequency, the diode signal current for an effective modulation component of incident flux, F photons/sec, is given by

$$i = \eta q F \tag{6}$$

where

$$i = \text{diode signal current}$$
$$\eta = \text{quantum efficiency}$$
$$q = \text{electronic charge}$$

The photodiode noise power is given by[2]

$$P_{pd} = kT_dB + 2qI_0|f(w)|^2B \cdot r_L \qquad (7)$$

where

r_L is the diode load resistance
$f(w)$ is a transit time reduction factor inherently less than one
T_d is the diode temperature
and
B is the effective bandwidth.

The first term in (7) is the thermal noise due to the diode series resistance and the second term is the shot noise of the junction. This term is small compared to kT_dB at VHF and higher frequencies and hence the effective noise temperature of the diode is approximately 300°K if the diode is operated at room temperature.

The effective receiver bandwidth at the input to the phase-sensitive detector is, in this case, the bandwidth of the comparison frequency amplifier because of the hetrodyning process used (see Section 4). The pre-detection bandwidth is kept as great as possible to reduce phase shifts at the pattern frequency.

For a sufficiently accurate measurement of phase it has been established experimentally that the S/N voltage ratio at the input to the phase-sensitive detector should be 10 : 1, i.e. an S/N power ratio of 100 : 1.

If the comparison frequency amplifier bandwidth is B_1 and the receiver noise factor is 5, then the equivalent noise power at the input is

$$P_N = 5 \times 2kTB_1 \qquad (8)$$

For an output S/N ratio of 100, the minimum input signal required is therefore

$$P_{Rmin} = 10^3 kTB_1 \qquad (9)$$

For $B_1 = 1$ kc/s, $T = 300°$K we have

$$P_{Rmin} = 4 \cdot 14 \times 10^{-15} \text{ watts} \qquad (10)$$

This is considerably in excess of the diode equivalent noise power, which will be $2kTB_1$.

Now, from (6), the input power for an incident flux F is

$$P_R = (\eta qF)^2 r_L \qquad (11)$$

where r_L is the diode load resistance.

Equating (10) and (11) gives the minimum required received flux, F_{min}, as

$$F_{min} = \left(\frac{P_{Rmin}}{r_L}\right)^{1/2} \cdot \frac{1}{\eta q} \tag{12}$$

For $r_L = 50$ ohms and $\eta = \frac{1}{3}$,

$$F_{min} = 1 \cdot 7 \times 10^{11} \text{ photons/sec} \tag{13}$$

For the special case of a perfect specular reflector in a vacuum path and neglecting transmission losses in the optical system, the range equation can be written as

$$(2R_V)^2 = \frac{F_O}{F_R} \cdot \left(\frac{d_R}{Q_T}\right)^2 \tag{14}$$

where

R_V is the range to the reflector
F_O is the transmitted flux in photons/sec
F_R is the received flux in photons/sec
d_R is the receiver aperture
Q_T is the transmitter beamwidth

The maximum vacuum range is found by substituting F_{min} in (14) for F_R. Also, for the laboratory model we have $F_O = 4 \times 10^{16}$ photons/sec, $d_R = 0 \cdot 1$ metre and $Q_T = 1/100$ radian. Hence,

$$R_{Vmax} = 2 \cdot 42 \text{ km}$$

Hence, a range of 1 km under good visibility conditions should be possible.

This range is considerably greater than that achieved in practice. This is thought to be owing, firstly, to poor optics in the experimental model and, secondly, to the fact that the small prismatic reflector is not suited to short range work when transmitter and receiver apertures are displaced. In any event, a large improvement is expected with improved transmitter optics, since for the source size of approximately 1 mm a far better beamwidth than 1/100 radian should be possible.

4. *Principles of Operation of the Laboratory Instrument*

The principle of operation of this system employs hetrodyning techniques, so that the phase delay in the modulation (or pattern) frequency can be measured at some convenient low frequency. The transmitter and receiver sections are at the same site, and a retro-reflector is used at one end of the line to be measured.

A block diagram of the system is shown in Fig. 4.

The carrier wave source in this case is the GaAs light-emitting diode, which is amplitude modulated by the pattern frequencies. These pattern frequencies are crystal controlled and stable to better than 1 part in 10^6.

In the laboratory model, a lens system is used for obtaining a narrow beam of IR radiation. The receiver and transmitter apertures are equal and displaced vertically from one another.

The reflected signal received is focused on to a silicon photodiode which is used as the light demodulator. The pattern frequency f_1 is subsequently amplified to improve S/N at the first mixing process,

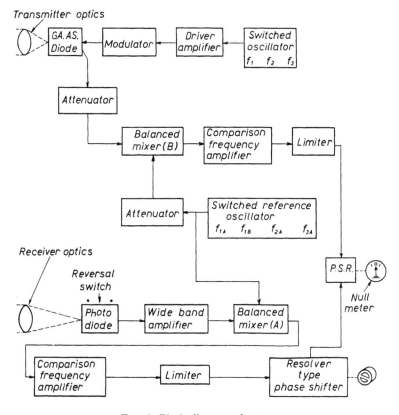

FIG. 4. Block diagram of system.

where a reference oscillator (frequency f_{1A}), spaced from the transmitter modulation frequency f_1, by the comparison frequency f_{c1}, is used as local oscillator drive to the mixer to obtain the difference frequency f_{c1}.

The reference oscillator is also fed via an attenuator to a second balanced mixer, which has as drive signal a sample of the voltage developed directly across the GaAs diode. This drive signal to the second mixer could also be derived via a sampling photodiode placed in the transmitter optics in order to eliminate phase delays between diode voltage and light output modulation.*

The output from this second mixing process is the second comparison frequency f_{c2}.

The two comparison frequencies are then separately amplified and filtered, and the phase differences for various pattern frequencies as described below are read on a resolver and null meter type of phase indicator. To eliminate phase errors due to filtering, the two channels are closely matched for response.

The use of difference patterns only does much to ease the problem of phase shift in the RF section of the instrument.[3]

The spectra for the two mixing processes are as shown in Fig. 5. It should be noted that in the present system the different pattern frequency configurations are applied sequentially. However, with more sophisticated techniques, it is possible to obtain all these patterns simultaneously.

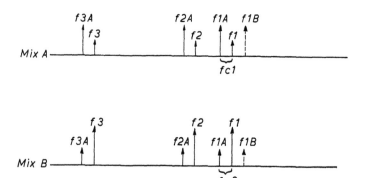

FIG. 5. Spectral diagrams for mixing processes A and B in the four different pattern configurations.

* Since the paper was written it has been established that the method of sampling the GaAs diode voltage or current is unsuitable owing to instabilities in the relationship between the phase of the sampled diode current or voltage and the phase of the IR intensity modulation.

The configuration f_1, f_{1A} provides a reference phase. Then the pattern frequencies resulting in the following configurations, referred to f_1, f_{1A} are:

$$f_1, f_{1B} \rightarrow 2_{f1} \qquad \text{fine pattern}$$

$$f_3, f_{3A} \rightarrow f_1\text{-}f_3 \qquad \text{medium pattern}$$

$$f_2, f_{2A} \rightarrow f_1\text{-}f_2 \qquad \text{coarse pattern}$$

If f_1 is chosen as ≈ 75 Mc/s, $f_1\text{-}f_3$ as $\approx 7\cdot5$ Mc/s and $f_1\text{-}f_2 \approx 0\cdot75$ Mc/s, then these patterns result in a 1, 20 and 200 metre indication for one full rotation of phase. (Note that a double path length change is involved.)

The pattern f_1, f_{1A} could be read directly if the display zero were known to give a 75 Mc/s (or 2 metre) pattern. However, it will be directly affected by the RF phase shift in the receiver and GaAs diode. The difference patterns on the other hand are affected only by differential phase shifts.

One serious source of possible error exists, however, if some transmitter modulation power is fed direct to the receiver via an 'electrical' path—i.e. not via the reflected light path. To obtain the required accuracy of phase indication ($\frac{1}{3}°$ for 1 mm resolution) the level of such a spurious signal must be less than 0·1 per cent of the wanted signal. This is difficult to achieve and maintain over a long period of time. However, if the spurious signal is reasonably small, the effect of such an unwanted coupling can be considerably reduced by the following method.

Let, in Fig. 6(*a*), the vectors M and R represent the modulation signal

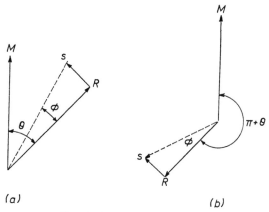

(a)   (b)

FIG. 6. Signal vector diagrams.

and the wanted received signal phases respectively. The angle θ represents the delay due to the path length. Let S represent the unwanted spurious signal in an arbitrary phase, introducing an error ϕ in the indicated phase ψ_1.

Hence

$$\psi_1 = \theta - \phi$$

If now the phase of the received signal is reversed without influencing the spurious signal, the vector diagram is as shown in Fig. 6(*b*). The phase angle indicated will now be

$$\psi_2 = (\pi + \theta) + \phi$$

Hence, adding ψ_1 and ψ_2 and subtracting π (i.e. 50 units on a dial of 100), we obtain

$$\psi = 2\theta$$

This is twice the true phase, free of contamination error. It is of course important that the reversal of the received signal is achieved in the circuits before any contamination has taken place.

5. *Description of Laboratory Model*

The instrument is constructed with the optical system suspended in a yoke, the sides and lower portion of which house the crystal ovens and oscillators and the phase measuring circuits. The final modulator for the GaAs diode is mounted on the transmitter lens housing so as to allow the shortest possible connecting links to the diode. The receiver modulation amplifier and associated mixer is mounted directly behind the receiver photodiode in the receiver lens housing.

The optical design for this model is considered far from ideal and it is expected that a tenfold improvement in beam width can be obtained with an improved optical system alone.

The power requirement for the instrument is 12 V at 1·5 A. A small 45-V battery is also used as bias supply for the silicon photodiode.

The warm-up time for the ovens is approximately 6 minutes. The pattern frequency oscillator and the reference oscillator are switched separately.

A commercial corner-reflector (retro-reflector) was used for all the tests made. However, it is not an ideal type of reflector for a system with T and R displaced.

6. *Future Development*

The discrepancy between predicted and actual range performance of the experimental model will be investigated both from the optical aspect and from that of the receiver photodiode performance.

In order to relax the stringent requirement of phase measurement accuracy to obtain millimetre resolution, it is felt that it would be advantageous to introduce higher pattern frequencies even for the same resolution as at present. Hence, if 750 Mc/s patterns are used say, and if indeed the GaAs diodes can handle these modulation frequencies, then 1-mm resolution would be more practical with a phase-measuring accuracy of one degree. Alternatively, of course, higher resolution could be obtained under these conditions with a complex phase-measuring system.

Acknowledgements

The author wishes to acknowledge the invaluable assistance of Mr I. D. Munro and the workshop team under Mr C. H. Meredith of the National Institute for Telecommunications Research, in the construction of the experimental model; also, the many discussions with his colleagues at the National Institute for Telecommunications Research on the problems encountered.

Finally the author wishes to thank the South African Council for Scientific and Industrial Research for permission to publish this paper.

References

[1] Kruse, McGlauchlin & McQuistan, 1962. *Elements of Infra-red Technology.* John Wiley and Sons.
[2] Sommers, H. S., Jnr., 1963. Demodulation of low-level broadband optical signals with semi-conductors. *Proc. Inst. Electr. Electron. Engrs, Amer.* **51**, 140–6.
[3] Wadley, T. L., 1958. Electronic principles of the Tellurometer. *Trans. S. Afr. Inst. Elect. Engrs,* **49**, Part 5, 143–161.

DISCUSSION

F. T. Arecchi: Do you use a coherent diode?

H. D. Hölscher: No.

T. S. Moss: What is the delay and the gain bandwidth of the amplifier?

H. D. Hölscher: The delay has not been measured but the stability is good. The amplifier gain is 35 dB and the bandwidth 50 Mc/s.

T. S. Moss: Could it be said that the stability is proportional to the total delay?

H. D. Hölscher: Yes.

Final Plenary Session

Chairman: PROFESSOR A. MARUSSI

Resolutions

No. *1 for IAG*

This Symposium, considering the possibilities that have been demonstrated for the furtherance of measurement techniques using Lasers, notably by interferometry, pulsed or modulated beams, recognizing the application to the precise measurement of short distances and variations of distance, measurement of medium distances, as well as measurement of ranges to satellites, expresses the wish that research into this aspect be energetically continued. Close collaboration between geodesists and physicists is urged.

No. *2 for IAG*

This Symposium, considering that the lack of sufficient information about refractive index along the path of electromagnetic distance measurements is the most serious of the remaining sources of error, and recognizing that a very promising approach to determining the influence of refraction by atmospheric dispersion measurements has been made, and that this technique could be applicable also to measurements by means of Lasers, expresses the wish that research along such lines be encouraged.

No. *3 for IAG*

The participants in this Symposium express their appreciation of the work done on the measurement of test nets using various Electromagnetic Distance Measurement Instruments, and urge that this valuable research should be energetically continued.

No. *4 for IAG*

This Symposium expresses the wish that the proceedings of the Symposium shall be published.

No. 1 for Symposium

The participants in the Symposium on Electromagnetic Distance Measurement thank the Royal Society for their invitation to the United Kingdom. They thank the Organizing Committee for the most excellent arrangements made for the Symposium and the organizations which have assisted in making it a success, especially the Ordnance Survey. They thank the University of Oxford and most especially the Department of Engineering Science for their hospitality and for the use of the facilities of the Engineering Laboratory. The lady participants also thank the Ladies' Committee for their entertainment.

No. 2 for Symposium

This Symposium, regretting the retirement of Major General R. C. A. Edge from the Presidency of Special Study Group No. 19 due to the pressure of his other commitments, expresses its admiration for his excellent leadership of the Study Group from its beginning, and especially thanks him for the arrangements he has made for the present Symposium.

Applauded by Symposium

Telegram to Brigadier G. Bomford (*President of IAG*)

The delegates at the Symposium on Electromagnetic Distance Measurement are most sorry that you have been unable to be with them during this week and express their hopes for your speedy recovery.

Telegram to Dr E. Bergstrand

After the presentation of the paper by Dr Bergstrand, the proposal by Professor Asplund that a message of greeting should be sent to Dr Bergstrand (who was unable to be present) was carried by acclamation.

List of Participants

Australia

Berthon Jones, P.
Buley, J. V.
Lambeck, K.
Puttock, M. J.

Austria

Bretterbauer, K.
Rinner, Prof. Dr-Ing. K.

Belgium

Baetslé, Prof. P.-L.

Canada

Saastamoinen, J.
Yaskowich, S. A.

Denmark

Bedsted Andersen, O.
Poder, K.

Eire

Cox, Dr R. C.
Walsh, Capt. M. C.

France

Bivas, R.
Commiot, J.
Deck, M.
Louis, M.

Germany (GFR)

Deichl, Dr K.
Draheim, Prof. Dr-Ing. H.
Grosse, Dr-Ing. H.
Jeske, H.
Kirschmer, Dr G.
Kuntz, Dr-Ing. E.
Lichte, Prof. Dr-Ing. H.
Messerschmidt, Dr E.
Moller, Dr-Ing. D.
Nottarp, C.
Preyss, C. R.
Seeger, Dr-Ing. H.
Walter, H.

Germany (GDR)

Amon, Dipl.-Phys., G.
Deumlich, Dr-Ing. E. R.
Strosche, H.

Ghana

Small, L. G.

India

Sharma, Major S. K.

Italy

Arecchi, Prof. F. T.
Marussi, Prof. A.
Mezzani, L.
Vullo, A.

Netherlands

Bruijn, D. C. de
Burki, W. C. J.
Don, Lt-Cdr C.
Halder, A. Th. van
Meerdink, E. F.
Munck, J. C. de
Palgen, Prof. J.
Verstelle, J. Th.

New Zealand

Lang, R. J.

Norway

Jelstrup, G.

Poland

Krynski, Dr S.
Krzemiński, W.

South Africa

Cabion, P.
Hewitt, Dr F. J.
Hölscher, H. D.
Jones, Dr B. M.
Webley, J. A.

Spain

Nuñez de las Cuevas, Dr-Ing. R.

Sweden

Asplund, Prof. L.
Brook, I.
Schöldström, R.
Sundqvist, S.

Switzerland

Fischer, Dipl.-Ing., W.
Gervaise, J. J.
Hossman, M.
Meles, H. P.
Strasser, Dr G. J.
Ward, A. H.

Thailand

Pachimkul, Col. S.

Tunisia

Kallal, A.

United Kingdom

Armstrong, J. A.
Bassett, T. W.
Bere, C. G. T.
Bill, Cdr R.
Bowman, M.
Bradsell, R. H.
Brown, C. C.
Carmody, Lt-Col. P. J.
Chiat, B.
Collins, W. G.
Connell, D. V.
Convey, T.
Denison, Brig. E. W.
Edge, Major-General R. C. A.
Evans, Dr S.
Froome, Dr K. D.
Gibbons, C. W.
Gowans, J. M. D.
Hall, J. L.
Harris, K. D.
Hawkins, P. O.
Hollwey, J. R.
Jackson, J. E.

United Kingdom (cont.)

Jelly, Lt-Col. J. S. O.
Kelsey, Col. J.
Marples, V. P.
Methley, B. D. F.
Miles, M. J.
Mosley, W. H.
Munsey, D. F.
Niblock, Capt. J.
Olliver, J. G.
Prescott, Major N. J. D.
Rawlence, J. R.
Richards, Major M. R.
Robbins, Dr A. R.
Robinson, Dr G. D.
Rowley, Dr W. R. C.
Schryver, A. F. P.
Scott, L.
Silva, E. H. T.
Smith, A. D. N.
Smith, J. R.

United Kingdom (cont.)

Taylor, P. J.
Trigg, C. F.
Wright, J. W.

United States of America

Culley, F. L.
Humbrecht, Col. G. W.
Krahmer, E. A.
Krogh, S.
McCall, J. S.
Owens, Dr J. C.
Rice, D. A.
Smathers, S. E.
Weiffenbach, Dr G. C.
Wood, L. E.
Woodring, G. R.

West Indies

Wason, H. R.

Lightning Source UK Ltd.
Milton Keynes UK
UKHW020019210722
406167UK00009B/863